THE ALL-AMERICAN TURKEY SHOW

Participants of the All-American Turkey Show gathered with their prize poultry entries at the Grand Forks City Auditorium on North Fifth Street. Photo used with permission of Grand Forks Herald.

THE ALL-AMERICAN TURKEY SHOW

When Grand Forks, North Dakota, Was the Turkey Capital of the World, 1924–1942

by
Gordon L. Iseminger

NDSU NORTH DAKOTA STATE UNIVERSITY PRESS

Fargo, North Dakota

North Dakota State University Press
Dept. 2576, P.O. Box 6050, Fargo, ND, 58108-6050
www.ndsupress.org

The All-American Turkey Show: When Grand Forks, North Dakota, Was the Turkey Capital of the World, 1924–1942
By Gordon L. Iseminger

First Edition
First Printing

LCCN: 2022946210
ISBN: 978-1-946163-55-4 (Hardcover)
ISBN: 978-1-946163-67-7 (Paperback)

Cover design by Jamie Trosen
Interior design by Deb Tanner

The publication of *The All-American Turkey Show: When Grand Forks, North Dakota Was the Turkey Capital of the World, 1924–1942* is made possible by the generous support of donors to the NDSU Press Fund, the NDSU Press Endowed Fund, and other contributors to NDSU Press.

Kimberly Wallin, Director
Suzzanne Kelley, Publisher
Kyle Vanderburg, Assistant Acquisitions Editor
Mike Huynh, Graduate Assistant in Publishing
Emma Borah, Che Flory, and Tatum Hoff, Editorial Interns

Printed in the United States of America

Publisher's Cataloging-in-Publication Data
(Prepared by Cassidy Cataloguing Services)

Names: Iseminger, Gordon L., author.
Title: The All-American Turkey Show : when Grand Forks, North Dakota, was the turkey capital of the world, 1924-1942 / by Gordon L. Iseminger.
Description: First edition. | Fargo, North Dakota : North Dakota State University Press, [2024] | Includes bibliographical references and index.
Identifiers: ISBN: 978-1-946163-55-4
Subjects: LCSH: Agricultural exhibitions--North Dakota--Grand Forks--History--20th century. | Turkeys--Showing--North Dakota--Grand Forks. | Turkeys--United States--History. | Women farmers--North Dakota--Grand Forks--History--20th century. | BISAC: NATURE / Animals / Birds. | HISTORY / United States / State & Local / Midwest (IA, IL, IN, KS, MI, MN, MO, ND, NE, OH, SD, WI) | HISTORY / United States / 20th Century.
Classification: LCC: S555.N92 G6 2024 | DDC: 630.7478416--dc23

To all the Ann Marie Lows
who, with their turkeys,
made Grand Forks, North Dakota,
the Turkey Capital of the World and
the home of the All-American Turkey Show,
even though, as Low acknowledged,
"[T]urkeys are a lot of darned hard work."

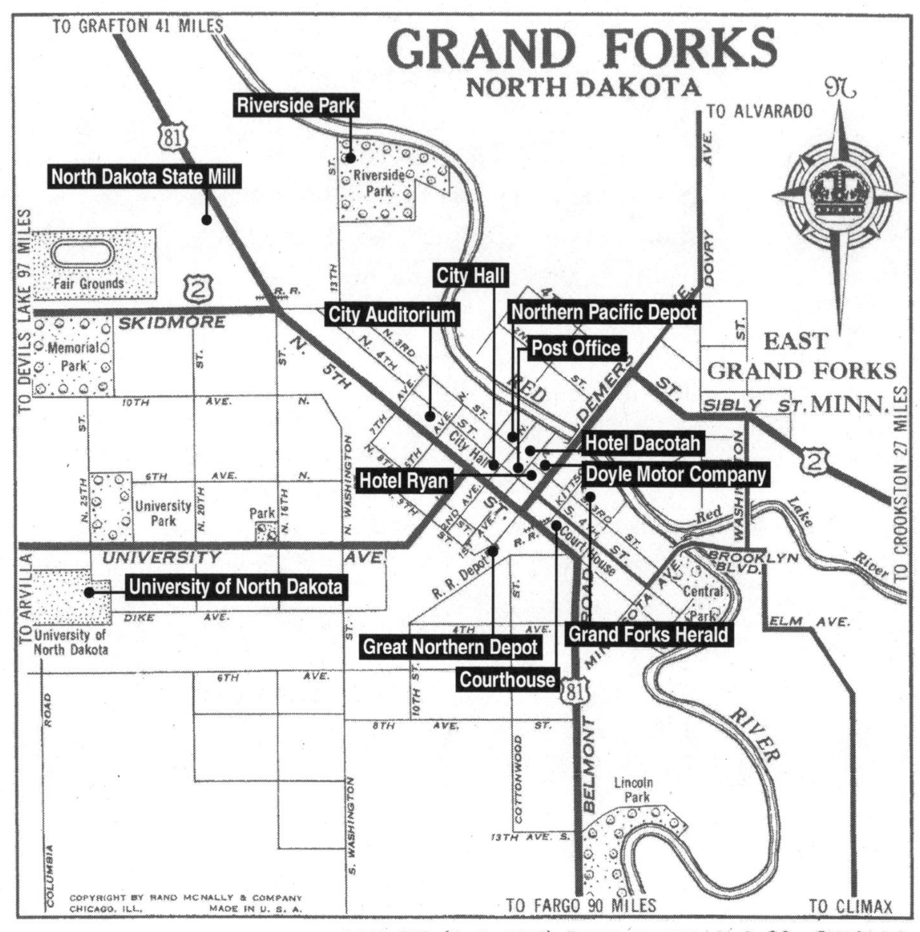

ADV. 759 (A. D. 7923) RAND McNALLY & CO., CHICAGO

TABLE OF CONTENTS

ACKNOWLEDGMENTS

I thank the Grand Forks Herald for introducing me to the All-American Turkey Shows by including a picture of one of the shows in the book titled, *Grand Forks—Proud People, Proud Heritage*. This study is the product of that introduction.

I also acknowledge the assistance and many kindnesses extended to me by the staffs at the following places of research: Chester Fritz Library, University of North Dakota, Grand Forks, North Dakota, especially the departments of Special Collections, Interlibrary Loan, and Periodicals and Microfilm; the North Dakota Institute for Regional Studies, North Dakota State University, Fargo, North Dakota; and the National Agricultural Library, Beltsville, Maryland.

I thank my wife, Trudy, and my son, Carl, for their assistance with the photographs and other materials used to illustrate the manuscript.

Finally, I cannot thank Deneen Marynik enough for deciphering my rough drafts and for typing the manuscript.

PREFACE

"Mighty oaks from little acorns grow."

(with apologies to Geoffrey Chaucer, *Troilus and Criseyde*, 1374)

A comment on National Public Radio, a quartzite monument on North Dakota's boundary with South Dakota, and an image in a pictorial history of Grand Forks, North Dakota, are but three examples of something that caught my attention and aroused my curiosity. In each case, research proved to be the best way to satisfy the curiosity. Although not my original intention, the research not only satisfied my curiosity, it also resulted in public presentations, conference papers, and publications. Allow me to explain.

National Public Radio's Scott Simon is featured on Saturday mornings on NPR's *Weekend Edition*. One Saturday morning, Simon reported that he had visited Minot, North Dakota, where he learned that silk trains had once passed through the city on the Great Northern Railway. I had never heard of silk trains. I wanted to find out what silk trains were, and research proved to be the best way to do it.

My research turned out to be as frustrating as it was fascinating, because I could not find material on silk trains in the sources where I expected it to be. Ralph W. Hidy, et. al, for example, had just published what was purported to be the definitive history of the Great Northern Railway, *The Great Northern Railway: A History*. The authors did not have an entry for silk trains in their index, nor does the word silk appear in the text.

Using Great Northern Railway records, newspapers published in towns on the railroad's route, periodicals, and other materials, I learned that the silk carried on the trains originated in Japan and that the silk, transported across the Pacific Ocean by ship, was landed in Seattle, Washington, and transferred to waiting Great Northern Railway trains.

I followed the Great Northern Railway silk trains on their fast and often hazardous runs from the Pacific coast to markets in New York and other eastern cities. My research did more than satisfy my curiosity. It resulted in conference papers and in my giving a number of public presentations on the topic. I was also invited to speak on the silk trains at several conferences of railroad associations.[1] All those results from hearing Scott Simon's brief comment on National Public Radio.

In the fall of 1979, I visited Ole Breum in Rutland, North Dakota. Breum was the area's self-appointed historian, and he volunteered to take me on a tour of the area's historic sites, among them, a quartzite monument standing on the boundary between the states of North Dakota and South Dakota. I had lived in either South Dakota or North Dakota for almost fifty years, and I had grown up near Sioux Falls, South Dakota, where quartzite was quarried, but I had never heard of these monuments, let alone seen one. To state that my curiosity was aroused would be an understatement. My curiosity was aroused even more when Breum informed me that the entire length of the boundary between North Dakota and South Dakota was marked by these impressive quartzite monuments, one every half mile.

To learn more about these monuments, I turned first to the deans of the states' historians. To my dismay, neither Elwyn B. Robinson in his *History of North Dakota* nor Herbert S. Schell in his *History of South Dakota* had included the boundary monuments in their discussions of the struggles for statehood and the eventual division of Dakota Territory into two states. They were not alone in their omission.

To satisfy my curiosity, I began what turned out to be several years of patient, often frustrating, but wonderfully interesting and rewarding research. I conducted my research in surveyors' notes, government documents, newspapers, and other materials in my office, but it was the monuments and their markings that drew me again and again to the boundary. Before completing my research, I had twice walked almost the entire length of the boundary (the Badlands excepted) between North Dakota and South Dakota (360 miles and 45.35 chains), examining and photographing (yes, and caressing) each monument.

---

[1] My article "Silk Trains on the Great Northern Railway" was published in *Minnesota History* (Spring 1994).

I shared the results of my research in conference papers and in a number of public presentations. I was also invited to speak on the monuments and the survey of the boundary at conferences of geographers, surveyors, and engineers. In 1988, my book, *The Quartzite Border: Surveying and Marking the North Dakota-South Dakota Boundary, 1891–1892*, was published by the Center for Western Studies, Augustana College, Sioux Falls, South Dakota. Continuing to be of interest to readers and still being used as a reference by surveyors, geographers, and engineers, the book is now in its third printing. All these scholarly endeavors from my seeing a solitary quartzite monument standing on a prominent knoll south of Rutland, North Dakota, in 1979.

And now to the image in a pictorial history of Grand Forks, North Dakota.

In 1999, the Grand Forks Herald published *Grand Forks—Proud People, Proud Heritage*. The picture on page 101 of the book (see Frontispiece) attracted my attention and aroused my curiosity. The picture bore the caption, "Participants of the All-American Turkey Show gathered with their prize poultry entries at the Grand Forks City Auditorium on North Fifth Street."

By 1999, I had lived in Grand Forks for almost forty years, and I had never heard of the All-American Turkey Show, nor had I ever seen a domesticated turkey in the vicinity. I was determined to learn something about the All-American Turkey Show.

The picture prompted questions. When was it taken? From how wide an area did the exhibitors come, and were they professional breeders or were they farmers? If farmers, why were they raising turkeys and exhibiting them in a state where wheat was king? Was the All-American Turkey Show held annually and, if so, when were the first and the last shows held? Why were the All-American Turkey Shows held, and why were they held in Grand Forks, North Dakota? Which organizations promoted the shows and why? Did the organizations keep records of the shows and, if so, where were they? These and other impertinent questions begged for answers, but where could I find them?

In my search for publications on turkeys, I found Andrew F. Smith's *The Turkey: An American Story*, published in 2006. Smith is a food historian and, to appropriate a hackneyed expression, his book explains everything one has always wanted to know about turkeys but was afraid to ask. In the section titled "Turkey Breeds," Smith noted that turkeys were not only exhibited in poultry shows and at state and county fairs, they were also exhibited at turkey shows, among them, the All-American Turkey Show held in Grand Forks, North Dakota. Smith's comment, although helpful, raised questions for which I still needed answers. For example, in what years were the All-American Turkey Shows held?

The source to which I turn for information on a topic dealing with North Dakota failed me in this instance. Robinson, in his *History of North Dakota*, did not include an entry for turkeys in his index, nor did he include them in his discussions of how the state's farmers were encouraged to diversify their operations.

Another source to which I turn when seeking information on a topic dealing with North Dakota is *North Dakota: A Guide to the Northern Prairie State*, compiled by workers of the Federal Writers' Project of the Works Progress Administration. The section of the *Guide* describing Grand Forks included as an annual event the All-American Turkey Show, noting that the show was usually held the last week of January. Based on the entry, I could assume that the All-American Turkey Shows had been held long enough and often enough that those compiling the *Guide* felt compelled to include them in the list of the annual events held in the city. I could also now associate a date with the All-American Turkey Shows. The *Guide* was published in 1938.

As always, when in need of information on a topic dealing with Grand Forks, I turned to Curt Hanson, archivist and head of the Elwyn B. Robinson Department of Special Collections in the University of North Dakota's Chester Fritz Library. After searching through the collection's holdings and finding nothing on the All-American Turkey Shows, Hanson warned me that the All-American Turkey Show would not be an easy event to research. I should have heeded his warning. Research proved to be even more difficult than either Hanson or I anticipated, in part, because the official records of the shows have, apparently, been either lost or destroyed.

Hanson had, however, found two references to the shows in Marilyn Hagerty's That Reminds Me columns in the *Grand Forks Herald*. One, titled "All-American Turkey Show marked 1931," noted that "A mammoth

Bronze yearling owned by Mrs. Henry Botz of Cando, N.D., was grand champion of the 1931 All-American Turkey Show here." Some of "the most prominent turkey breeders in the country," the *Grand Forks Herald* had reported, had exhibited their turkeys and "thousands" of people had come "to see the birds in what was called the greatest turkey show in history."[2]

Another of Hagerty's That Reminds Me columns was even more helpful. "There was more than politics around Grand Forks" in 1939, she observed, as "the city was holding its annual turkey show Jan. 16–21. . . . Approximately 400 of the highest-priced turkeys in the United States and Canada were entered in the annual classic of the turkey world."[3] Even more informative was the notation that "Mr. and Mrs. Al Johnson of Wayzata, Minn. strutted off with the All-American Grand Championship. And it was the first year in the 16-year history of the event for participants to win the top prize twice." I could now assume that the first All-American Turkey Show was held in 1924 and that the shows continued at least until 1939.

Curt Hanson also found the first reference to the American Turkey Journal in the Grand Forks City Directory of 1932. The journal, issued monthly, was published by the Page Printing Company, located at 105 South Third Street in Grand Forks.

Only three libraries in the country have complete runs (1932–1942) of the *American Turkey Journal* in their holdings: the New York Public Library, the Oregon State Library, and the National Agricultural Library (Beltsville, Maryland). Only the National Agricultural Library would loan their copies. Wayne Thompson, National Agricultural Library archivist, arranged the loan with Kerry Hackett and Joan Miller of the Interlibrary Loan Department in the University of North Dakota Chester Fritz Library. Knowing that many events had been canceled—some of them discontinued altogether—because of World War II, I could now be reasonably certain that the All-American Turkey Shows were held for nineteen years, from 1924 to 1942.

Having two dates, 1924 and 1942; and two major collections—the *Grand Forks Herald* and the *American Turkey Journal*—I was confident that I could find answers to the questions I had about the All-American Turkey Shows. My reading of Andrew F. Smith's *The Turkey* raised new questions. Again, I was warned, and, again, I should have heeded the warning.

---

[2] *Grand Forks Herald*, January 17, 2006.

[3] Marilyn Hagerty, That Reminds Me, *Grand Forks Herald*, January 21, 2014.

"Many problems," Smith wrote in his acknowledgments, "confront anyone writing about the history of the turkey. Among the mass of information available are tens of thousands of references, descriptions, and depictions in American and European literature, newspapers, diaries, letters, legal documents, paintings, engravings, poems, zoological works, and cookbooks. It is impossible for any one individual to examine each of these."

Realizing that I was not the "one individual" capable of examining every facet of the history of turkeys, I relied on Smith to answer my questions, such as the following: because wild turkeys were native only to the Americas, when had they been domesticated and by whom? When and by whom had domesticated turkeys been taken to Europe? Why had turkeys spread throughout Europe so quickly, and why had they become an important part of the food supply in only a short time? Why, when the Atlantic Seaboard teemed with turkeys, did colonists bring turkeys with them when they established settlements in the New World? How and why did domesticated strains of turkeys get to the Midwest? And why were these turkeys in North Dakota in sufficient numbers to warrant a turkey show in Grand Forks?[4]

The turkey, in the *Standard of Perfection*—the bible of poultry breeders—is considered to be a single breed of poultry, with six (at the time of the All-American Turkey Shows) recognized and accepted Standard strains or varieties: Bronze, Narragansett, White Holland, Bourbon Red, Black, and Slate. These were the only varieties that could be judged and entered into competition for awards at the All-American Turkey Shows. Dressed turkeys of different strains look very much alike and their flesh tastes much the same. The most distinctive and identifiable feature of a turkey strain or variety, therefore, is its unique plumage—the color, shading, and pattern of its feathers. Color was a contentious issue during the years of the All-American Turkey Shows, but it was an absolutely vital consideration when breeding and judging turkeys. When feather color ceased to matter, as was the case when the meat-type Broad Breasted varieties were developed, the All-American Turkey Shows no longer served

[4] Not only did I use Smith's *The Turkey* to answer my questions, I also used the book as a model as I composed the manuscript. My chapters follow Smith's arrangement of his material, and the content of chapters was suggested by some of the titles he assigned to the chapters into which his book is divided. For example: Chapter 1. The prehistoric Turkey; or, How the Turkey Conquered North America. Chapter 2. The Globe-trotting Turkey; or, How the Turkey Conquered Europe. Chapter 3. The English Turkey; or, How the Turkey Cooked the Christmas Goose.

much of a purpose. They would eventually have been discontinued, even had World War II not intervened.

With most of my questions answered, I was prepared to explain why Grand Forks, North Dakota, became the home of the All-American Turkey Show and the Turkey Capital of the World. The resulting middle chapters herein contained are devoted to the explanation and to the history of the All-American Turkey Shows that were held between 1924 and 1942.

Because the All-American Turkey Shows were the first major event of the year in Grand Forks and because the shows drew exhibitors, turkeys, visitors, business, and attention to the city, detailed accounts of the shows were published in the *Grand Forks Herald*, the *American Turkey Journal*, and *Turkey World*. By immersing myself in these accounts, I was able to attend each of the nineteen All-American Turkey Shows held between 1924 and 1942 together with the turkeys, their exhibitors, show officials, and visitors.

I became so familiar with many of the regular exhibitors at the All-American Turkey Shows that I would have recognized them had I met them on the street. I could have discussed Bronze turkeys with Mrs. W. J. Janda of St. Hilaire, Minnesota; Narragansetts with Mrs. William Eddie of Northwood, North Dakota; and Bourbon Reds with Sadie B. Caldwell of Broughton, Kansas. I learned that Alfred Malmberg of Crookston, Minnesota, raised Bronze turkeys and that he held the record of having exhibited at every one of the nineteen shows. John O. and Amanda Allen were interviewed so often for having won so many awards with their Bronze turkeys that I could have found my way without difficulty to their farm near Radium, Minnesota. Ray Andrews also raised Bronze turkeys, and he and his turkeys were in the news so often I could have found my way easily to his place near Petersburg, location of the largest turkey farm in North Dakota.

George E. Lamm of Philip, in South Dakota's west river country, transported his prize Bronze turkeys to the All-American Turkey Shows in the dead of winter in his All-American Express, a Depression-era truck that likely lacked both heater and defroster. Having been raised on a South Dakota farm and having experienced more than eighty Dakota winters, I know what it meant for Lamm to ride in a cold truck cab on the trip of five hundred miles between Philip, South Dakota, and Grand Forks, North Dakota, driving on dirt and gravel roads and taking detours when his way was blocked by snowdrifts.

Judge Walter Burton of Arlington, Texas—among the best storytellers in turkey circles—had stories on every poultry judge in the country, and I laughed along with others when he told them in his characteristic Texas drawl. I listened to Mrs. August Swenson of Gilby, North Dakota, give her inimitable impressions of turkey calls whenever she was called upon. As the quality of turkeys improved with each show, which was among the purposes of the All-American Turkey Shows, I agonized with the judges at the judging tables when they were forced to divide points into quarters when choosing the champion birds in each class. I attended the banquets held on the last evening of the shows, and I enjoyed the toe-tapping music provided by Professor John E. Howard and student musicians from the University of North Dakota Music Department.

Being male, I was not allowed to enter the Hen's Nest, that quiet place in the City Auditorium in which women exhibitors and their female guests could seclude themselves during the week of the All-American Turkey Shows. Nor was I allowed to attend the Hen Club banquets, held on an evening during the week of the show and at which no man was welcome. I did, however, attend each of the Hen Club picnics held between 1931 and 1942.

The picnics, held on a Sunday each July in the Grand Forks tree-shaded Riverside Park, were an opportunity for members of the turkey clan to gather for an "afternoon of turkey gobbling and talking." I sat and ate with turkey folks at the long tables that "groaned," laden with good food. The cold watermelons and the ice cream served in the late afternoon were as much a treat for me as they were for the farm kids who may rarely have seen or tasted either.

Although not my intention, my research on the All-American Turkey Shows—begun only to satisfy my curiosity—resulted in public presentations, conference papers, and this book.

CHAPTER 1

The Turkey Capital of the World

(with apologies to no one)

IN THE YEARS BEFORE World War II, Grand Forks, North Dakota, with good reason, claimed the title, The Turkey Capital of the World. Turkeys were a big business in the city. The *Grand Forks Herald* headline on November 18, 1934, announced "25 [railroad] cars of turkeys shipped from the city." This report was on only one day and in 1934, a year when turkey numbers had been reduced 30–40 percent by drought. Each of the refrigerator cars contained twenty thousand pounds of dressed turkeys, a total of a half-million pounds of meat.

The birds had all been raised in the Red River Valley and surrounding area and were on their way to markets and Thanksgiving dinners on the East Coast. The turkeys would command premium prices because of their reputation for quality. Many more thousands of pounds of dressed turkeys were shipped by freight and by trucks to Minneapolis and St. Paul in Minnesota and to other large cities within a few hundred miles of Grand Forks.

The Thanksgiving movement of turkeys, for years an annual event in Grand Forks, had begun on November 9. At the rate of three thousand per day, live turkeys arrived in the city in large trucks from an area with a radius of 150 miles from Grand Forks, from Pembina near the Canadian border on the north to Steele and Traill Counties on the south. The city's produce plants added employees and additional shifts to dress and prepare the turkeys for market. About 95 percent of the turkeys produced in the area were dressed in Grand Forks because the city's produce firms were well equipped and because flock owners had no means to cool the dressed birds and keep them refrigerated.

No edition of the *Grand Forks Herald* during the month of November failed to include mention of turkeys, if nothing else than to quote the prices paid locally and on markets in cities such as New York and Chicago. Prices were four to five cents per pound higher in November 1934 than they had been at the same time in 1933, partly because fewer turkeys had been produced, but also because the production of chickens, geese, and ducks was also down. With dressed young toms and hens bringing twenty to twenty-two cents per pound—up to twenty-six cents in New York—turkeys were worth stealing.

The city's police chief, Harry Gregg, reported several thefts of turkeys in the area east of the Red River, and he promised that turkey thieves, when apprehended, would face vigorous prosecution. Gregg asked turkey producers to report thefts immediately and to report any suspicious activity they observed around their premises. He also advised flock owners to devise ways of marking their turkeys, so they could be identified if they were stolen.

In addition to being a hub for turkey distribution, Grand Forks was also the home of the All-American Turkey Show, the first, always the largest, and—for a time—the *only* poultry show in the nation devoted exclusively to turkeys. Described as the First and Foremost of All Turkey Shows, The Turkey Classic of America, and The Final Court of Appeal of All Turkeydom, the shows were held for two to six days, usually in January, from 1924–1942.

Truly "All-American," the shows drew exhibitors from the east, west, and gulf coasts and states in between and from several Canadian provinces. Premier turkey producers braved distance, blizzards, snow-blocked roads, and temperatures of -35°F to -45°F to attend the shows; renew acquaintances; learn how to improve their strains and increase production and profits; exhibit their birds; and compete for more prize awards and trophies than were offered by any other livestock show in the nation.

The shows in 1940 and 1942 had entries from eighteen states and three Canadian provinces, evidence of the show's prestige and reputation. From 1932 to 1942, no show failed to include entries from every turkey-producing state in the country. Between 1924 and 1942, the All-American Turkey Show scored publicity coups that no other poultry or livestock show in the country could match.

The *American Turkey Journal*, published in Grand Forks by the Page Printing Company from February 1932 to February 1942, was the All-American Turkey Show's official publication. A monthly publication of thirty-six to forty pages with an international list of subscribers and devoted exclusively to turkeys, the *American Turkey Journal* covered every facet of the turkey industry, including lengthy editorials; helpful articles on turkey breeding, production, and marketing; announcements of events of interest to subscribers; turkey club news; ads for hatching eggs and breeding birds; and advertisements for turkey feeds, medicines, and equipment.

George W. Hackett, of Wayzata, Minnesota, among the nation's foremost experts on turkeys and recognized by many as the country's premier turkey-show judge, edited the *American Turkey Journal* and wrote the helpful monthly editorials during its ten years of publication. For many years the driving force behind the All-American Turkey Show, Hackett managed the show until 1940, when he reluctantly relinquished the position for health reasons.

In the first issue of the *American Turkey Journal* in February 1932, Hackett announced the journal's purpose. "This," he wrote, "is the formal introduction of THE AMERICAN TURKEY JOURNAL, dedicated to AMERICA'S GREATEST BIRD, and devoted to the SINGLE PURPOSE of serving a great industry and the promotion of the same." He professed to having "an unlimited faith" in the future of the turkey industry, and he pledged the journal's "uncompromising support" for Standard turkeys, for judging standards, and for the improved breeding programs that had transformed "the original diminutive wild turkey" into the present varieties of "marvelous turkeys."

Capitalizing on their location in Grand Forks and enjoying the enthusiastic support of the city's residents and businessmen, the All-American Turkey Show and the *American Turkey Journal* exemplify developments in the turkey industry in the first half of the twentieth century.

Intended at their inception to promote the improvement of turkey varieties and to maintain their purity, to demonstrate to area farmers that raising turkeys could be profitable and a way to diversify their operations, to publicize how much the turkey industry contributed to the nation's farm income, and to encourage consumers to eat more turkey meat and not only on holidays and festive occasions, both were successful to the point of being done in by their success.

By 1942, both had served their purpose and had outlived their usefulness. Producers had improved turkeys to the point that judges at the All-American Turkey Show were reduced to dividing points into fourths when awarding prizes for the best birds in each class when judged according to the exacting and unforgiving specifications demanded by the *American Standard of Perfection*.

By the 1940s, the All-American Turkey Shows were being held for little reason other than there seemed to be no good reason not to hold them. Even the traditional Friday evening banquet was now more of a homecoming banquet than the concluding event of a significant and exclusive turkey show. World War II brought both the All-American Turkey Show and the *American Turkey Journal* to unexpected, but merciful, ends.

The All-American Turkey Show owed part of its success to its being located in Grand Forks, a city that, with its many amenities, was as well suited as it was well situated to serve as a world turkey capital. Grand Forks, with a population approaching twenty thousand in 1940, lay in the heart of the Red River Valley, often described as the American Valley of the Nile because of the depth and fertility of its soil.[1]

The valley was famous for its potatoes, and approximately twenty thousand carloads were shipped from the city each year. A large packing plant provided a market for cattle, hogs, and sheep, and two large creameries processed cream from a wide area adjacent to the city. Because of its latitude, the area also produced clover with a high nectar content, and the

---

[1] Because of its geographical location, Grand Forks, according to the *City Directory*, was "The Farmers' Market Center of North Dakota and Northern Minnesota." Two flour mills provided markets for hard spring wheat, the state's principal crop. Barley also yielded well, and hogs finished on the grain made fine-tasting premium bacon, with firm and well-distributed fat. The area's soil and latitude were particularly well suited for high yields of sugar beets of unusually high sugar content, and a local plant processed the beets from several thousands of acres in the Red River Valley.

fine flavor of the honey made from it found a ready market. World records of honey production had been set by local bee colonies.

Even before the first All-American Turkey Show in 1924, the area had become a "Leader in Turkey-Raising," and the *City Directory* noted that "this new farming enterprise" was "growing faster" in North Dakota "than anywhere else in the country." Words well spoken. In fact, the commentary continued, North Dakota had become "one of the leading turkey-raising states in the Union" and the All-American Turkey Show became "the best-known exclusive turkey show in the world."

In the late nineteenth century, most turkeys were grown in New England and in the mid-Atlantic states. By 1890, the center of turkey production had moved to the Middle West and the Southwest, where clean ranges helped to reduce disease and where ample space and feed were available. In 1890, Illinois led the states in turkey production, with Iowa and Missouri close behind. By 1900, Texas was in first place and Missouri in second place. Texas remained in first place until 1940, when California replaced it. In 1920, North Dakota ranked eighth among the states in turkey production, but in 1930 it was second to Texas, with Minnesota in third place.

It is significant to note that the rapid advance of Minnesota and North Dakota in the rankings of turkey-producing states coincided with the beginning of the All-American Turkey Show in Grand Forks. That the purpose of the All-American Turkey Show was to promote turkey raising speaks well of the show's influence. Hackett, highly respected and well regarded by those involved in the production of turkeys, spoke to this in a July 1929 article published in *Turkey World*. "The whole Northwest," he observed, "depends upon the turkey industry from an economic standpoint," and "the importance to which the [All-American Turkey Show] has grown in the business, civic, and social life of Grand Forks and the surrounding country, precludes the remotest probability that this annual event will be less than it is at the present time."

Not until 1935 did California displace North Dakota from second place in turkey production. Minnesota retained its position in third place. Turkey production was again moving westward, to states on the West Coast.

Because of its geographical location, Grand Forks was the distributing center for northern North Dakota, northern Minnesota, and eastern Montana. The largest fireproof storage warehouse in North Dakota, South Dakota, and Montana, containing two acres of floor space and equipment for handling large quantities of freight and produce, provided distribution and storage facilities. Several large wholesale houses were located in Grand Forks, and the *City Directory* listed thirty manufacturing businesses, with a total of 1,800 employees.[2]

The *City Directory* also noted that "a very active and efficient Chamber of Commerce" was "always working for the best interests of the city" to make it "a hospitable convention and exposition center." The city's large convention halls and adequate hotel facilities handily accommodated "the thousands of visitors who come to attend these gatherings."

Grand Forks owes its name to its location at the confluence of two rivers, the Red Lake River flowing from northern Minnesota and the Red River—among the few rivers that flow north—with its source near North Dakota's border with South Dakota. Frequented by Native Americans and contested by the Chippewa and the Sioux, the confluence was marked on maps and named Les Grandes Fourches by French explorers and trappers who selected it as a rendezvous point. In 1800, Alexander Henry, Jr., the North West Fur Company agent at Pembina, established a camp at the point where the two rivers meet and began conducting a brisk trade in furs. By 1820, both the Hudson's Bay Company and the American Fur Company were also conducting trade in the area.

By the mid-nineteenth century, Les Grandes Fourches had become a stopping-off point for the trains of Red River oxcarts that carried trade goods between St. Paul, Minnesota, and Fort Garry (present-day Winnipeg), Manitoba. In 1868, Les Grandes Fourches was a point on the mail

---

[2] The city's trade area had a radius of over 75 miles, with a population of 150,000. The *City Directory* also listed seven hotels, with a total of 550 rooms; two banks and one trust company; twenty-five churches representing the major denominations; four parks; one daily newspaper, both morning and evening editions, with approximately 22,500 subscribers; three hospitals, with two hundred beds; three movie theaters, with a total seating capacity of 2,200 people; 6,057 telephones in service; and a municipal auditorium with 1,800 seats. The University of North Dakota was also located in Grand Forks, with medical and law schools. The public library contained nearly 20,000 volumes, and the city could boast of 68 miles of streets, 39.68 miles paved.

route to Fort Garry, and, on June 15, 1870, the United States government established a post office at Les Grandes Fourches and changed the name to Grand Forks.

Although then officially a post office, in 1870 Grand Forks was barely deserving of a name. That same year, however, the Red River and a flatboat race between competing freight companies breathed the breath of life into the name, and it became a townsite with potential and a promising future.

James J. Hill, the railroad builder, and Captain Alexander Griggs, a Mississippi River steamboat captain, began a flatboat business on the Red River and entered the competition of freighting goods to Fort Garry. Powered by the river's current, flatboats were dismantled and sold for lumber when they had reached their destination. On a freighting trip to Fort Garry in late 1870 and in a race with a competitor, Griggs was forced to tie up his flatboats for the night when he had reached the forks of the Red Lake and Red Rivers. By morning, the river had frozen over, and Griggs and his men were forced to spend the winter of 1870–1871 at the confluence.

They unloaded the freight goods and built a shed to protect the goods from the elements, using lumber from the dismantled boats. Frozen in, and perhaps realizing that the site offered possibilities as a future town, Griggs built a log cabin and took possession of a quarter section of land. In 1875, Griggs platted the townsite, thereby becoming the "Father of Grand Forks." In acknowledgment of his paternity, Griggs's statue stands on the courthouse lawn in Grand Forks. Griggs's hand is on the wheel of his steamboat, and his eyes are fixed on the river, ever watchful for snags, rocks, the changing current, and, perhaps, the boats of competitors.

Grand Forks was at first a river town, with its streets (still confusing to many people) laid out parallel with the river instead of with the cardinal points of the compass. The town became an important docking point for the steamboats plying the Red River between Moorhead, Minnesota, the usual head of navigation, and Winnipeg, Manitoba. Formed in 1872 by Hill and his associates, the Red River Transportation Line monopolized the steamboat business on the river. Red River steamboats did a brisk business during their day, transporting thousands of settlers to the

Statue of Captain Alexander Griggs, the "Father of Grand Forks."
Image courtesy of the Grand Forks Historic Preservation Commission.

Red River Valley and thousands of tons of merchandise, manufactured goods, and agricultural produce.

The steamboat era was coming to an end in 1878 when railroad lines connected Winnipeg and St. Paul, but the *Grandin* was built the same year, primarily to haul grain, by the owners of a huge bonanza farm, and the *Grand Forks* was built in 1895. In the spring of 1912, the *Fram*, the last steamboat on the river, broke from its moorings at Grand Forks and sank. The era of the steamboat had ended. Not only had steamboats provided tales and a touch of romance, they had also aided in settling the Red River Valley. They also created an interest in the area for Hill, the key figure in North Dakota's railroads.

Two railroads served Grand Forks: Hill's Great Northern and the Northern Pacific. In 1934, twenty-two passenger trains per day arrived in the city, offering service and providing links to points all over Canada and the United States. Grand Forks was a freight division point, as well as a railway junction center, and both railroads handled heavy volumes of freight.

Some of the passengers arriving in the city were among the country's best-known turkey breeders, there to attend the annual All-American Turkey Show. Their prize turkeys constituted some of the freight in the baggage cars. The turkeys were shipped in cages designed to protect the birds' plumage and tail feathers. It was critical that neither be damaged, because both figured heavily in the judging and the awarding of prizes and trophies. Both railroads accorded special treatment to turkeys, if not to their owners.[3]

The Great Northern's car repair shop, among the largest in the country, was 360 feet long and 84 feet wide. A crew of eighty men repaired thirty cars per day, 90 percent of them boxcars. The Great Northern's roundhouse, with one of the largest turning tables in the world (to accommodate the largest of the transcontinental locomotives) had a capacity of thirty-eight locomotives. Grand Forks was also the headquarters of

[3] Both the Northern Pacific and the Great Northern had impressive depots and freight houses (both are now on the National Register of Historic Places) in Grand Forks and extensive marshalling yards. That of the Great Northern consisted of twenty-five sets of tracks, each a mile long and each long enough to hold more than one hundred boxcars. Seven yard crews of five men each worked to make up the trains.

the Great Northern's Dakota Division, the largest railroad division in the world, with over 1,800 miles of main-line trackage under the supervision of the Grand Forks office.

Breeders transporting their turkeys to the All-American Turkey Show in trailers pulled by their automobiles, rather than by rail (the trailer quite likely fashioned from the chassis of a Ford Model T or Model A), had the option of using one of two highways, both legendary. Graveled or hard surfaced, the highways were designated on highway maps as all-weather roads, meaning they were passable during wet weather and they were plowed in the winter.

One of the highways entering Grand Forks was the Theodore Roosevelt International Highway. When former President Theodore Roosevelt died on January 6, 1919, a number of trail associations were inspired to designate routes in his honor, among them one organized in Duluth, Minnesota, on February 7, 1919. The Theodore Roosevelt International Highway grew out of the association's efforts.[4]

On this "Rooseveltian highway," according to a 1921 tour guide, a trip was "A Pilgrimage for Every Patriotic American, Every Lover of Nature, Every Admirer of Theodore Roosevelt."[5] The Meridian Highway, later converted to U.S. No. 81, was a truly international highway, the only primary north-south highway girding the nation's heartland. The highway, passing through North America's breadbasket, traced the 97th meridian from Winnipeg, Manitoba, to Laredo, Texas, and it was eventually extended to Mexico City.[6]

---

[4] Spanning parts of the United States and Canada from Portland, Maine, to Portland, Oregon, the transcontinental highway had a length of 4,060 miles. An article published in the *Roosevelt Highway Bulletin* on October 27, 1921, titled "America's Leading Highway," explained that the name was "a suitable memorial" to one of the nation's "greatest builders and statesmen." The designation was also intended to promote tourist travel to parts of the United States that had had few tourists in the past.

[5] Eventually, U.S. Highway No. 2, for most of its length, replaced the Theodore Roosevelt International Highway. Designated Tour 6 in the American Guide Series, *North Dakota: A Guide to the Northern Prairie State*, from Duluth, Minnesota, to the Montana border, a distance of 367 miles, the highway, entirely paved or hard surfaced, entered Grand Forks through East Grand Forks, Minnesota, passing over the Red River on the Sorlie Memorial bridge.

[6] Organized in 1911 in Salina, Kansas, the International Meridian Road Association had the modest goal of building a road on which "a full wagon-box load or a car at high gear" could meet a vehicle going in the opposite direction. As with the Theodore Roosevelt International Association, the Meridian advocates wanted to promote tourism and to encourage northerners to travel to the sunny south in the winter and southerners to travel to the lakes and the cooler climate in the north in the summer.

Designated as Tour 1 in the American Guide Series on North Dakota, the highway entered North Dakota near Pembina and paralleled the Red River to Wahpeton and the Bois de Sioux River to the South Dakota border. According to the wording of the tour, the highway ran "through unbelievably flat green fields, broken here and there by an occasional farmstead." Graveled, or with a bituminous surface, the highway entered Grand Forks after passing through a number of towns, all offering "accommodations of all types."

Steeped in legend, these two international highways, carrying turkey breeders and their prize birds, as well as tourists, intersected at DeMers Avenue and Fifth Street in what was the center of Grand Forks when the city ruled the turkey roost. Within four or five blocks, easy walking distance, from this intersection were located all the city's retail outlets, hardware and furniture stores, automobile dealers and auto supply stores, implement dealers and repair shops, lumber yards, restaurants, hotels, barber shops and beauty salons, doctor and dentist offices, hospitals, banks and financial institutions, movie theaters and recreation parlors, schools and churches, county and federal office buildings, and railway and bus depots. The City Auditorium, site of the All-American Turkey Show, was also within easy walking distance of this important highway junction and the city's business center.

Built in 1909 at a cost of thirty thousand dollars and holding 1,800 people, the auditorium was among the busiest and most popular convention and entertainment centers in the area. Conventions included those of the Women's Christian Temperance Movement, Grand Army of the Republic, Elks, druggists, bankers, and implement dealers. A popular location for auto shows, balls, recitals, concerts, and dramatic presentations, the auditorium also drew some of the nation's best known and most controversial speakers, including politicians William Jennings Bryan, Eugene V. Debbs, Theodore Roosevelt, and Robert M. LaFollette.

Intersecting at DeMers Avenue and Fifth Street, the former Roosevelt and Meridian highways were joined until Skidmore Avenue, passing by the City Auditorium located at North Fifth and Fifth Avenue North, at little more than an arm's length. The former Theodore Roosevelt Highway proceeded west on Skidmore, and the former Meridian Highway proceeded north on Mill Road, passing by the North Dakota State Mill and Elevator, established in 1922 as part of the Nonpartisan League's industrial program. In addition to Dakota Maid Flour, the mill produced breakfast cereals and a full line of poultry feeds.

To test its line of turkey feeds and to demonstrate to flock owners that it paid to use its products in their turkey-breeding programs, the mill maintained a large flock of turkeys. During the week of the All-American Turkey Show, mill officials invited those exhibiting their birds to the mill to learn about the mill's turkey feeds and feeding programs, to view the mill's turkeys, and to be the mill's luncheon or dinner guests, at which the mill's prize turkeys were served.

It is not a stretch to assert that the All-American Turkey Show, capitalizing on its location in Grand Forks with its many amenities, contributed to changing American diets and eating habits. Andrew Smith, writer on matters culinary, noted in his book *The Turkey: An American Story* those individuals who were involved in the genesis of the modern turkey industry. Many of them had exhibited their prize turkeys at the All-American Turkey Show, among them Wallace Jerome of Barron, Wisconsin.

Wallace Jerome was born in Spooner, Wisconsin, and he graduated from Barron High School in 1928. Jerome started raising turkeys in 1922, with fourteen turkey eggs and two brood hens. Only one tom turkey was hatched. Yet a young man, Jerome started to work for the Wisconsin Department of Agriculture as a poultry and egg inspector, and he eventually received a degree in Poultry Husbandry from the University of Wisconsin—Madison.

In 1933, Jerome showed his Bronze turkeys for the first time at the All-American Turkey Show, where he made a strong impression when one of his Bronze yearling tom turkeys was named Grand Champion. Jerome returned to this Turkey Classic of America in 1934 with a much larger entry of his fine Bronze turkeys and made an even stronger impression, winning the much-coveted Master Breeder Award with its gold medal of exclusive design. Jerome's turkeys received a number of other awards at the All-American Turkey Show in 1934, including the sweepstake's prize for the Breeder's Display in the Bronze class.[7]

Discerning consumers' changing tastes and preferences, Wallace Jerome started Jerome Foods, Incorporated, of Barron, Wisconsin. Jerome Foods became the Turkey Store Company, branded as The Turkey Store. The Turkey Store Company was sold to Hormel Foods, which was merged with Jennie-O-Foods, to form Jennie-O-Turkey Store and Products, among the nation's largest producers of turkeys and turkey products.

[7] Jerome had a long history of being recognized in the agriculture industry. In 1929, he was named the Outstanding 4-H Poultry Club Member for Wisconsin, and, in 1973, he was honored with an Honorary State Farmer Award from the Wisconsin Vocational Agriculture FFA Program. The poultry industry also recognized his many accomplishments. Jerome received the Distinguished Service Award from the College of Agriculture and Life Sciences at the University of Wisconsin—Madison, and he also earned a Distinguished Agriculturist Award from University of Wisconsin—River Falls in 1976.

CHAPTER 2

America's Wild Turkeys

(with apologies to ***Meleagris gallopavo silvestris***)

THE TURKEYS VYING FOR attention and competing for awards at the All-American Turkey Show were all descended from America's wild turkeys. The turkeys, now domesticated and classified as strains or varieties, reached Grand Forks, North Dakota, by a circuitous route spanning two continents and an ocean, a journey that took four hundred years to complete.

Turkeys are indigenous only to America, and large flocks of them once ranged from what is now the Canadian province of Ontario to Veracruz, Mexico, and from the Atlantic coast to the American southwest. Only in what are now the Pacific states of California, Oregon, and Washington, and in Idaho, Montana, Utah, Nevada, and Wyoming were wild turkeys not native.

On their epic journey up the Missouri River and to the Pacific in 1804–1806, Lewis and Clark did not encounter wild turkeys north of what is now central South Dakota. Nor did they mention wild turkeys in the account of their winter stay at Fort Mandan, about forty-five miles north of the present-day city of Bismarck. Because the winter was long and cold and because wild turkeys make excellent fare, doubtless the men of the Corps of Discovery would have been grateful to have found wild turkeys in the area.

By the time Europeans arrived in Central and South America in the early sixteenth century, Indigenous peoples had domesticated wild turkeys. Domesticated turkeys were introduced to Spain in 1519, and from there they spread across Europe, reaching England in 1526. Birds from the domesticated English and European stocks were later brought to North

America by early colonists and served as foundation strains, along with wild stocks of the northeastern sub-species of wild turkeys. The six turkey varieties listed in the *American Standard of Perfection*, those competing for awards at the All-American Turkey Show, were all produced in what is now the United States, with the possible exception of the White Holland turkey. Differing primarily in plumage, the six varieties were Bronze, Narragansett, White Holland, Bourbon Red, Black, and Slate.

Wild turkeys are among the largest and most distinctive members of the *Galliformes*, a group of ground-nesting game birds that includes grouse, partridges, and pheasants. Andrew Smith, in his book on turkeys, noted that the American wild turkey and the Asian pheasant are close enough genetically that they can be mated through artificial insemination and produce offspring. When Europeans, familiar with the guinea fowl and peafowl, arrived in the New World they found a bird similar to, but not exactly like, what they were accustomed to seeing. What these Europeans saw and were attempting to describe were American wild turkeys.

In its bulletins, the National Wild Turkey Federation, founded in 1973 and dedicated to the conservation of the wild turkey, identifies six major subspecies of wild turkeys. They differ primarily in the color of their plumage and range, and, somewhat, in their size.

The eastern wild turkey, *M. g. silvestris-silvestris* for "forest" turkey, is the most numerous. It ranges widely from the Atlantic Coast to the Mississippi River, and this was the turkey species first encountered in the wild by the Puritans and the founders of Jamestown. The Florida wild turkey, *M. g. osceola*, is found in the southern half of the state and was named for the Seminole chief, Osceola. Merriam's wild turkey, *M. g. merriami*, was named for C. Hart Merriam, the first chief of the United States Biological Survey. Its range is the mountain regions of the West, from Colorado to Mexico. *M. g. intermedia*, the Rio Grande wild turkey, differs from the eastern and Merriam's turkey by being "intermediate," hence the name. Its range is the south-central plains states from South Dakota to Texas and northeastern Mexico. The fifth recognized subspecies, the Gould's *M. g.*

Mexicana is found in northwestern Mexico and parts of southern Arizona and New Mexico.

The southern Mexican wild turkey, *M. g. gallopavo*, is the only wild turkey not native to what is now the United States or Canada. Considered to be the nominate subspecies, and now probably extinct, it is the accepted forerunner to the domesticated turkey taken to Spain by Spanish explorers and conquerors in the mid-sixteenth century. From Spain, the turkey spread to France and later to England as a farmyard bird and as the centerpiece of feasts for the wealthy. By 1620, the turkey was common enough that the Pilgrims could bring it with them from England, unaware that the bird's close relatives were already occupying the forests of Massachusetts.

Wild turkeys possess the five senses of smell, taste, touch, sight, and hearing. The senses are not equally developed, however, nor do turkeys make equal use of them. Wild turkeys have a poorly developed sense of smell, the least important and useful of their senses. As can humans, turkeys discern tastes such as sour, sweet, bitter, and salty, but, having few taste buds, their sense of taste is poor. The sense of touch is used primarily when feeding. When scratching among leaves, for example, they can feel an acorn or nut under their toes.

Wild turkeys have an almost uncanny ability to locate the source of a sound. Their immediate response is to look in the direction from which the sound came, allowing them to react quickly to predators, and they can hear lower-frequency and more-distant sounds than humans can. Wild turkeys have excellent daylight vision, and they use sight to locate and identify food and predators, especially when hearing is impaired by wind and rain. With eyes on the sides of their heads, they have monocular, periscopic vision, and they can judge distances quickly, even when running or flying. Turkeys also have better peripheral vision than humans do, and, by rotating their heads, they have 360-degree fields of vision.

Wild turkeys are handsome, impressive birds, with a regal, distinctive carriage. Adult males (toms) have dark, fan-shaped tails with feathers all of the same length. Tips of the tail feathers are usually chestnut, brown, or white. Juvenile males (jakes) have fan-shaped tails as well, but with longer feathers in the middle. Males have glossy plumage, mostly greenish bronze, with a complex coppery-gold sheen that is beautifully iridescent in the sunlight.

Feathers on the neck, breast, and body are tipped with a band of velvety black, accentuating the sheen of the rest of the plumage. Wing and tail feathers have alternating dark and light bands. Females (hens) have feathers that are duller overall, in shades of brown and gray. Parasites can dull coloration, and, in males, coloration may be a sign of health, important during mating season when hens are selecting mates.

Adult males have large, featherless, reddish, purplish, or blue heads with red throats and red wattles on their throats and necks. Heads and upper necks have fleshy, reddish, pliable growths called caruncles. The long, fleshy object over a male's beak is called a snood. The length of the snood may be a sign of the tom's health and virility, and hens may favor mates with the longest snoods.

When males are excited—especially during mating season in March and April—heads, neck, wattles, and snoods become engorged with blood and turn bright red. Adult males typically have beards—tufts of long, coarse, hair-like appendages (modified feathers)—growing from the centers of their chests. The beards of old males may trail on the ground. Juvenile males have shorter beards than adults, and only 10–20 percent of hens have beards.

Each turkey foot has three toes in front, with a shorter, rear-facing toe in back. Males have short, heavy spurs, one behind each of their lower legs. Unlike chickens, however, they fight mostly with their beaks. In self-defense or when fighting off predators, they may kick with their legs and use the spurs as weapons. They also ram with their bodies and are able to deter predators as large as mid-sized mammals. Adult males, but not hens or juvenile males, have an interesting appendage that ornithologist John James Audubon, writing in 1831, called the "breast sponge." During mating season, when the male eats little, the appendage is filled with the fat that sustains him. By the end of the mating season, the fat is used up, and the male has sustained a significant loss of weight.

Wings, with spans of four to six feet, are small, relative to the size of the birds. Nevertheless, wild turkeys are agile fliers, capable of accelerating to speeds of fifty-five miles per hour for flights of up to one mile. They fly close to the ground and typically for shorter distances, usually only to escape from predators. They also fly into the branches of trees to roost at night. Rather than flying, wild turkeys prefer to use their powerful legs for walking or running, moving with a long, straddling gait. They can trot at speeds of twelve miles per hour, but, when frightened, they can run twenty-five miles per hour. Able to float, turkeys can swim streams and large bodies of water.

Standing three to four feet tall, adult wild turkeys normally weigh from eleven to twenty-four pounds and measure thirty-nine to forty-nine inches in length. Typically much smaller than males, adult wild turkey hens normally weigh five to twelve pounds and measure thirty to thirty-seven inches in length. Records of tom turkeys weighing over thirty pounds are uncommon, but not rare.

Wild turkeys make a number of sounds, described as gobbles, clucks, putts, yelps, cutts, whines, cackles, and kee-kees. In the spring, adult males gobble to announce their presence to hens and competing males, a sound that can carry up to a mile. Males also emit a low-pitched drumming sound, produced by the movement of air in the air sack in their chests, similar to the booming of prairie chickens. They also produce a sound called the "spit," a sharp expulsion of air from the air sack. Hens yelp to let males know where they are.

Omnivorous, wild turkeys forage on the ground and by climbing shrubs and small trees. They prefer nuts from trees, such as oak and hazel, and when chestnuts ripen, wild turkeys congregate around the trees and gorge themselves. They also eat seeds, rose hips, berries such as juniper and bearberry, grapes, plant tops, spiders, insects, grasshoppers, tadpoles, snails, slugs, worms, beetles, ants, caterpillars, and small reptiles such as lizards and snakes. When food is scarce during the winter, they eat almost anything they can find. Wild turkeys can seemingly adapt to virtually any surroundings, but they prefer open, mature forests with a variety of tree species that offer food and cover. They have difficulty with deep snow when it covers their food sources.

Wild turkeys have life spans of five to fifteen years, but adults and poults have a number of predators, including coyotes, wolves, cougars, bobcats, hawks, and golden eagles. Predators of eggs and newly hatched poults include raccoons, skunks, opossums, foxes, other rodents, and snakes. Poult mortality is greatest the first fourteen days of life, especially of poults roosting on the ground. Predation of poults decreases significantly after six months, when they have grown nearly to the size of adults.

Male wild turkeys are polygamous, attempting to mate with as many hens as possible. They display for hens by puffing out their feathers, spreading their tails, and dragging their wings, a behavior known as strutting. Their heads and necks become brilliant red, blue, and white, colors changing with the turkeys' mood, solid white heads and necks indicating when the toms are the most excited. A hen may signal her readiness to mate by separating herself from the other hens and doing a little dance and flattening herself on the ground.

Courtship begins in March and April, when turkeys are still flocked together in wintering areas. Some males have been observed courting in groups, often with the dominant males gobbling, spreading their tail feathers, and drumming and spitting. Dominant males that courted as one of a pair fathered more poults from fertilized eggs than males that courted alone. The theory behind team-courtship is that less-dominant males have greater chances of passing genetic material on than when courting alone.

After mating, hens build nests in places screened by brush in tall grass. Nests are shallow depressions in the ground, into which they have scratched dry leaves and grass. Hens lay clutches of eight to twenty eggs, usually one egg per day. Eggs are colored white to cream, with brown speckles. Should something happen to the eggs, hens may lay one or two more clutches.

Hens incubate and turn the eggs for twenty-eight days, and hatching takes place over a twenty-four hour period, during which time the hen remains in the nest. Newly hatched poults can leave the nest in about twelve to twenty-four hours, but, for the first five to six weeks, they roost on the ground, under the hen's wings. They soon learn to leap into the lower branches of trees to roost, but still under their mother's wings. Poults remain with their mothers for about nine months.

Although not as numerous as buffalo, wild turkeys ranged over much of North America and, like buffalo, they played a significant role in the diets and cultures of North American Indigenous peoples. Native Americans caught turkeys with snares and traps, and some eastern tribes burned parts of forests to create open spaces to attract mating birds and provide hunters with clear shots.

Native Americans had differing attitudes about wild turkeys and how they made use of them. Some ate the flesh, others would not, even when faced with starvation. The Cheyenne refused to eat turkeys because they believed them to be cowardly and, eating the flesh, they believed, would make them cowardly as well. The Hopi would not eat the flesh, but they relished the eggs and ate them whenever they could find them. Other groups ate the eggs, but not the flesh. The Navajo ate turkeys, but not their eggs.

Whether they did or did not eat turkeys or their eggs, Native Americans valued wild turkeys for their parts. Bones were used for making tools, musical instruments, spoons, awls, beads, and other decorations. Spurs were used as arrow points. Because an adult turkey can produce a large number of feathers, which grow back when plucked, Native Americans valued wild turkeys for their feathers, even if they refused to eat the flesh or the eggs.

Some fletched their arrows with wild turkey feathers. Others, believing turkeys to be cowardly, did not, fearing that the feathers would distort the arrow's flight. Feathers were also used in rituals, as decorations, and in headgear. Many leaders, such as Catawba chiefs, traditionally wore turkey feather headdresses. Turkey feathers were also used as fans and woven together with strips of hemp to make coats, blankets, and the cloaks that significant individuals of some groups favored. Wild turkey feathers held such a prominent place in some cultures that items made from them were interred with the bodies of the dead.

Wild turkeys are inquisitive and easily captured, and because they congregate readily with humans they were logical candidates for domestication. In North America, Native Americans did not domesticate them, however, and much about the domestication of wild turkeys remains a subject of academic debate.

The issue is complicated in part when the words "tamed" and "domesticated" are used interchangeably. A hunter can tame a wolf to be a companion, for example, but the wolf cannot be said by that to be domesticated. Although tamed, it is still a wild animal. The issue is also complicated because, as Andrew Smith noted, the bones of wild and domestic turkeys differ little from each other. It is, therefore, difficult to determine if the bones found in archaeological sites are those of wild or domesticated birds. Turkey eggshells, suggesting chicks hatched in captivity; turkey pens, suggesting caging; remains of turkeys of different ages, suggesting they were raised at that location—all might be indications of domestication. Feathers with white tips, suggesting a diet of corn, might also be a sign of domestication. Corn does not grow in the wild.

Adding to the debate on the domestication of the wild turkey is that when the Spanish arrived in the western hemisphere, both wild and domesticated turkeys were an important food item for the Indigenous peoples of Mexico. The Aztec empire was at its zenith, and the Europeans naturally assumed that it was the Aztecs who had domesticated turkeys.[1] However, the Aztecs did not arrive in central Mexico until well after turkeys had been domesticated by others and, familiar with wild turkeys in northern Mexico, they adopted domesticated turkeys as a matter of course when they encountered them.

Scholars—such as the Canadian archaeologist Richard MacNeish, working in Mexico—have found evidence of domesticated turkeys at sites in the Tehuacan Valley, with remains dating to 200 B.C.E. to 700 C.E. Domesticated turkey bones, dating to circa 700 C.E., have been found at a site in Guatemala, far beyond the range of wild turkeys in pre-Columbian times, suggesting that domesticated turkeys were an item of trade in Central America.

The southern Mexican wild turkey, *M. g. gallopavo*, is considered to be the nominate subspecies of wild turkeys, and tamed members of this subspecies were taken to Spain in the sixteenth century. Bones of this subspecies have been identified at sites dating from 800 to 100 B.C.E., but it

---

[1] The Aztecs did not develop a writing system, however, so what is known about turkeys among the Aztecs must be surmised from archaeological artifacts and from the writings of Europeans arriving after the conquest in 1519.

is uncertain whether the bones are from wild or domestic birds. Turkey remains from other archaeological sites indicate that birds of the Mexican subspecies were exported to the ancient Maya world.[2]

Because uncertainty surrounds the domestication of the wild turkey, Smith is content to accept that domestication had occurred before Europeans reached the Western Hemisphere. He believes, however, that turkeys had been domesticated for only a relatively short period by the time Europeans arrived. Early in the process of domestication, he wrote, domesticated animals are smaller than wild ones, and pre-Columbian domesticated turkeys were much smaller than wild turkeys, suggesting a relatively short period of domestication by the time Europeans were introduced to them.

That domesticated turkeys were found in a relatively small area in Central America also suggests that domestication had not occurred very long before the arrival of Europeans. Other domesticated foods such as beans, squash, peanuts, and sweet potatoes had been widely disseminated in pre-Columbian times. Not so domesticated turkeys.

Whether they did or did not domesticate turkeys, the Aztecs made them an integral part of their diets, culture, and religious practices. They prepared turkeys in a number of ways, roasting them over fires and incorporating them in soups and stews.

Europeans also found turkeys to be tasty. Bernardino de Sahagún, a Franciscan priest, arrived in Mexico in 1529, and for more than forty years recorded what the Aztecs reported to him about their pre-Columbian customs, religion, diet, and daily life. Of the turkey, Father Bernardino wrote, it "leads the meats; it is the master. It is tasty, fat, savory." Hens were particularly "tasty, healthful, fat, full of fat, fleshy, fleshy breasted, heavy-fleshed."

Poultry industry expert Karen Davis noted that the Aztecs required large numbers of turkeys for food, but also for tribute and sacrifices. In 1430, the lord of Texcoco required 1000 turkeys per day for his household, 365,000 turkeys per year. Montezuma's household also required a

[2] Karen Davis, noted for her work on the modern poultry industry, in 2001 published her study on turkeys titled *More Than a Meal: The Turkey in History, Myth, Ritual, and Reality*. She is among those who credit the Aztecs with domesticating turkeys.

prodigious number of turkeys, 1000 per day for a total of 365,000 turkeys per year. Three hundred guests, plus a thousand guards and attendants, were often served turkeys at meals.

As many as 1,400 to 1,600 turkeys per day were required for each lord of the fiesta of the Tlaxcalan god Camaxtli. Meals at which between 1,000 and 1,500 turkeys were consumed were not uncommon. Turkeys were not consumed only by humans. Montezuma's menagerie of raptors and carnivores required large numbers of them. The raptors ate only turkeys (eagles, one per day), as many as five hundred per day.

Client cities were required to give large numbers of turkeys to the Aztec emperor as tribute, and turkeys were also sacrificed in religious rites. In Mexico and Central America, turkeys were regarded as symbolic manifestations of earth, rain, and fertility, and they were ritually sacrificed to propitiate these elements. The Zapotec people of Central America ritually sprinkled turkey blood on their newly planted fields. Merchants offered turkey heads as thank offerings when granted a safe return from their travels. The Aztecs deified the turkey as the Jeweled Bird, believing that the deified turkey could cleanse humans of contamination and absolve guilt. Because turkey wattles were believed to cause impotence, they were ground, combined with chocolate, and served to enemies.

Andrew Smith noted that for all their importance to the Indigenous peoples of Mexico and Central America, turkeys were seldom incorporated into their art. Some representations of turkeys have been found on pottery and on jewelry. The Aztec emperor Montezuma sent Hernando Cortés six turkeys made of gold, intending to impress and appease him. The gift did not serve its intended purpose. The gold objects only whetted Cortés's appetite for gold and made him the more desirous of capturing Mexico City and seizing its treasure.

Although Spanish explorers, conquistadors, and priests reported seeing domesticated turkeys in Mexico and Central America, there is as much uncertainty about when and by whom they were taken to Europe as there is about when and by whom they were domesticated. That turkeys were taken to Spain is certain. They were being raised there in the 1520s, and pairs were being given as gifts to prominent officials in church and government, with the admonition that they were only to be admired, not eaten. Thus begins, as was befitting this aristocratic fowl, the turkey's Grand Tour of Europe, during which its domestication, having begun in the Western Hemisphere, was further refined, as were the social graces of upper-class European young men of means. Four hundred years later, turkeys, now defined as strains or varieties, strutted and preened and were brought to perfection at the All-American Turkey Show held in Grand Forks, North Dakota, the Turkey Capital of the World.

CHAPTER 3

An American Turkey in European Courts

(with apologies to Mark Twain)

IN THE EARLY SIXTEENTH century, Spanish explorers and conquistadors began bringing an exotic bird with them on their return to Spain from the Western Hemisphere. But, what was this bird? What was its name? Spaniards, and many other Europeans, were familiar with guinea fowl and peafowl or peacocks, and this New World bird was similar to, but not exactly like, the birds they were accustomed to seeing. Only later was it determined what this new bird was, and, eventually, it became known as a turkey.

During the age of discovery, when Europeans first encountered turkeys, explorers lacked detailed knowledge of the world's geography. Exotic plants and animals, often with unknown provenance, were taken to Europe from areas of the world that were not shown on the maps of the time, areas that had no names, at least no names that were familiar to Europeans. Indisputably, the turkey originated in the New World and was introduced into Europe from Mexico, but, for years after its arrival in Europe, it was confused with the peacock and the guinea fowl. Like the turkey, neither were native to Europe.

Until trade routes were discovered around the southern tips of Africa and South America, most things exotic reached Europe through Turkey, many of them from places as exotic to the Europeans as the items themselves, such as India, Arabia, and Africa. The peacock was associated with India, the guinea fowl with Africa, and both were imported into Europe through Turkey. Although the turkey originated in the Western Hemisphere, people in Spain, Italy, and elsewhere assumed that it also had originated in India and that it also had arrived in Europe through Turkey.

Indeed, the French name for turkey, *dinde*, can be translated into English as "from India." Many other countries also have names for turkeys that assume an Indian origin—*diiq* (Indian rooster) in Arabian countries, Indjushka (bird of India) in Russia, and Hindi (India) in Turkey. Given their lack of knowledge of the world's geography and their scant appreciation for nomenclature, Europeans can be forgiven for assuming that the turkey had come from India. It was known that Columbus, sailing for Spain, intended to find a shorter water route to India by sailing west, and he believed that he had done so. When he returned from the Caribbean, he reported that he had reached "The Indies."

Even Carl Linnaeus, the eighteenth-century Swedish botanist, in 1758 gave the turkey a scientific name reminiscent of the commonly held beliefs of where the bird was from. By the mid-seventeenth century, it was common knowledge that the peacock, the guinea hen, and the turkey had originated in geographic locations that were far from one another. Linnaeus had to have known that the turkey had originated in America, yet he gave the bird the generic name *Meleagris*, the Greco-Roman name for guinea fowl. Even less understandable is why he chose *gallopavo* for the species name. *Gallopavo* is Latin for "peafowl" of Asia (*gallus* for cock and *pavo* for chickenlike).

Europeans can be forgiven too for confusing the turkey, the peacock, and the guinea fowl because the generic terms used when referring to them meant unfamiliar or showy birds, which all of them were. The three birds have similar physical characteristics, and explorers and others described them using similar terms. Even the best of sixteenth-century naturalists would have had difficulty distinguishing one bird from the others, based solely on the confusing and often vague descriptions brought back by those who had encountered them.

It is unclear precisely when generic terms used to refer to the turkey were changed to a term that specifically meant the turkey. A. W. Bryant, in his "Brief History of the Turkey," noted that there is no agreement on how or when the turkey got its name and that the nomenclature "followed a tortuous path." Even its scientific classification was, in his words, an "ill-chosen scientific name."[1]

[1] A. W. Bryant, "Brief History of the Turkey," *World's Poultry Science Journal* (1998).

Suffice it to say, the turkey did not lack for names, including those ill-chosen. Smith believes that many of the explanations for how the turkey got its name are fanciful at best. Other explanations, although plausible, lack documentation. One explanation for the turkey's name, which Smith identified as an "onomatopoetic derivation," is that the turkey named itself, because it makes a "turk, turk" sound.

But, Smith noted, neither wild nor domesticated turkeys make the sound. Another fanciful explanation is that the turkey's head resembles a fez, headgear worn in Turkey, hence the name. Yet another explanation is that the word turkey is a mispronunciation of Mexican Indian names for the turkey.

Plausible, but improbable, Smith believes, is that the word turkey came from the Hebrew word *tukki*, meaning peacock. The peacock came from India, and in the Tamil language of India peacock is *toka*. The word's primitive meaning refers to a train or trailing skirt. The word *toka*, adopted into Hebrew, became *tukki*, which, by the "genius of the English language," says Smith, became turkey.

Jewish traders also began to deal in foreign birds, curiosities, and exotics that could reap exceptional profits. Importing turkeys provided such an opportunity, and Jewish traders may have designated the turkey as the "American" peacock, utilizing the Hebrew name for peacock—*tukki*—in market places. Because the word was heard so often, it was anglicized into turkey, formerly spelled *turky*, the word used by the Bishop of Oxford when writing to the Earl of Buckingham.

Mario Pei, Professor of Romance Languages at Columbia University, offers two reasonable, if not definitive, theories for how the turkey got its name. The first is that when Europeans saw turkeys in America, they identified them as a type of guinea fowl. Guinea fowl were already being imported into Europe through Turkey and Constantinople by merchants who were nicknamed Turkey *coqs* or roosters. The North American bird became known as "turkey fowl" or "Indian turkeys." The phrase was eventually shortened to "turkey."

The second theory derives from turkeys from America being brought to England in merchant ships from the Middle East, much of which was part of the Ottoman Empire. The importers were known as "Turkey merchants" so the birds became known as "Turkey birds." In time, the name became just "turkeys."

Suffice it to say, there is no lack of theories explaining how a bird native to America got a name that associates it with Turkey, a country that had nothing to do with the bird's origin. "Illogical and inappropriate as it may be in the English language," A. W. Bryant noted, "this North American bird has become stuck with the name turkey." And, it is too late now to change it.

Also illogical is that Europeans accepted turkeys as food almost from the moment it arrived in Europe. That Europeans only slowly adopted such New World foods as corn, potatoes, and tomatoes—believing them to be strange if not poisonous—is typical of the process of culinary change in early modern society.[2]

Corn, or maize, was grown in Spain within fewer than fifty years after the Spanish reached the New World and became familiar with the crop. Not for a century and a half, however, did it become commonplace in Spanish diets. Many Europeans considered corn to be feed for animals, but not for humans. Potatoes were adopted even more slowly than corn, when the Spanish became familiar with them after conquering Peru, where they were a major part of the Indigenous diet. Europeans were slow to adopt the potato, in part, because cultivating them requires a climate more temperate than that of Spain and the Mediterranean. Only in the eighteenth century did potatoes become a major food for peasants in Scotland, Ireland, and Germany; potatoes were not cultivated in northern Europe until new and more hardy strains were developed. Even then, they continued to be regarded as food suitable only for the poor.

Tomatoes were introduced into Europe from Chile, Bolivia, Colombia, and Peru early in the sixteenth century. The Aztec word for them was *tomatl*, which gave rise to the Spanish word *tomate*, from which came the English word "tomato." For hundreds of years, however, many Europeans refused to eat tomatoes, believing them to be sinful, seductive, and poisonous, beliefs reflected in the names given the bright, shiny fruit.

---

[2] See Jean-Louis Flandrin and Massimo Montanari, *Food: A Culinary History from Antiquity to the Present* (Columbia University Press, 1999).

Some, believing tomatoes to be an aphrodisiac, called them "love apples." Others, calling them "apples of paradise," considered tomatoes sinful. Believing them to be poisonous, because tomatoes are members of the nightshade family, the most notorious being the *bella donna*, people refused to eat them. Until the late nineteenth century, tomatoes were cultivated in Europe almost exclusively as curiosities or for decorative purposes. They were grown in Italy mainly as ornamentals.

In sharp contrast to their reluctance to accept foods from the New World that were strange to them, Europeans quickly began raising and eating turkeys. By the 1520s, turkeys were being raised in Spain, and they were plentiful enough that they were an important food for the wealthy. That they were referred to in Spanish manuscripts in the 1530s suggests they had become common by that time. From Spain, turkeys spread to Italy, Germany, France, England, and the Scandinavian countries. Turkeys had arrived in Germany by 1530, and they quickly became an important food. In cookbooks, they were referred to as Indianischen Hanen, the German word for turkey. The first depiction of a turkey to appear in any cookbook was that of Marx Rumpolt. *Ein New Kochbuch*, published in 1581, contained an image of a turkey engraved by a German printmaker named Virgil Solis. Rumpolt's recipes for banquets all recommended a turkey dish, suggesting that turkeys were plentiful in Germany by the 1580s.

Shortly after being introduced into Spain, the turkey arrived in France. Barbara Ketcham Wheaton, in her *Savoring the Past: The French Kitchen and Table from 1300 to 1789*, noted that the first reference to turkeys appeared in 1528. Fifteen years later, turkeys were an important food in France, and, at a banquet in 1549 given by Catherine de Medici, seventy turkeys were served, likely because by that time they were cheaper than peacocks, pheasants, bustards, and cranes.[3]

When Charles IX married Elisabeth of Austria in 1570, turkeys were served at the wedding banquet. One hundred years later, historian Andrew F. Smith noted that Cardinal Jacques Davey du Perron wrote that turkeys had "increased wonderfully in a short time." They had become "a

[3] Barbara Ketcham Wheaton, *Savoring the Past: The French Kitchen and Table from 1300 to 1789* (University of Pennsylvania Press, 1983).

very good asset," and they were being driven from Languedoc to Spain "in flocks like sheep." By the eighteenth century, turkeys had replaced geese as the birds of choice at banquets, and they were being eaten by people in every social class. Half a million turkeys per year were being consumed in Paris alone.

That the turkey arrived in Europe early in the sixteenth century was fortunate. Traditional farming methods had exhausted large tracts of arable land, and agricultural production was not keeping pace with increases in population. Much of Europe was on the brink of widespread malnutrition and potential starvation, and, observed Smith, turkeys, providing sustenance for "protein-starved" Europeans, quickly became a widely available food "for all but the poorest of the poor."

Helping to explain why Europeans so readily adapted to raising and eating turkeys is that—unlike New World foods such as corn, potatoes, and tomatoes, with which Europeans were unfamiliar—turkeys were fowls with which Europeans were familiar: chickens, ducks, geese, peacocks, and guinea fowls. Because turkeys were larger than chickens, they supplied more meat and, although they were smaller than peacocks, they tasted better. Peacocks were also difficult to raise, but any farmer could raise turkeys.

Turkeys could also be prepared using existing recipes, and they did not require special cooking skills or equipment. And, as turkey numbers increased, prices dropped. By mid-sixteenth century, they cost less than chickens, and, according to *An Encyclopedia of Domestic Economy* (1848), turkey flesh was "tender, delicate, nutritive, and restorative, of excellent flavor, and more dense and substantial than that of chicken."

From Spain and Italy, turkeys were rapidly disseminated throughout Europe, reaching Scandinavia by the 1550s. When and by whom turkeys were introduced into England is uncertain. William Strickland, the son of a Yorkshire gentleman, sailed to the New World with Sebastian Cabot in 1526, and he is often credited with introducing turkeys to England. Strickland, dealing with Native American traders, is supposed to have acquired six turkeys, which he sold in Bristol for a *tuppence* each. That this is how turkeys were introduced into England is unlikely, noted Andrew Smith, because English explorers along the North America's east coast would have encountered only eastern wild turkeys, which did not reproduce in captivity.

Whether the enterprising William Strickland introduced turkeys to England may be uncertain, but, what is certain is that he was taken by turkeys. With the proceeds of his voyages and his other business ventures, Strickland acquired a number of estates, including Wintringham and Boynton in the East Riding of Yorkshire, and in 1550 Edward VI granted him a coat of arms illustrated with what is described as a "turkey-cock in his pride proper." The official record of his crest in the archives of the College of Arms is thought to be the oldest surviving European drawing of a turkey, and the church at Boynton is decorated with the family's turkey crest. There are stone sculptures of turkeys on the walls, images of turkeys in the stained-glass windows, and a lectern carved in the form of a turkey. The Bible is supported by a turkey cock's outspread tail feathers.[4]

Whether it was Strickland or someone who is lost to history who introduced turkeys to England, the turkeys were instantly successful among those in the upper classes. Turkeys had more meat on them than the small Tudor chickens, and they were far tastier than swans or peacocks. Turkeys must have been exceedingly plentiful by 1541, and ecclesiastics must have been indulging themselves by eating too many of them, because Thomas Cranmer, Archbishop of Canterbury, issued an injunction limiting the number of turkey cocks that could be served at ecclesiastical dinners.

Henry VIII was the first king to enjoy eating turkeys, and Edward VII made it fashionable to serve turkeys rather than peacocks at Christmas. Thomas Tusser (c. 1524–1580) was an English poet and farmer in Suffolk, near the Stour River, where he wrote his *Hundred Good Pointes of Husbandrie*, which appeared in numerous editions. By 1573, Tusser noted, turkeys had become standard fare at Christmas on the tables of those who could afford them.

So common must turkeys have become by the turn of the century, and so current must the name turkey have been, that William Shakespeare could make reference to the bird in a play and not provide either context or explanation. The playwright could assume that members of the audi-

[4] Strickland was elected a member of the House of Commons in 1558 and continued to serve in Parliament for a number of terms. He died in 1598.

ence would know to what he was referring. Shakespeare's *Twelfth Night or What you Will*, among his finest comedies, was written in 1601–1602 as a Twelfth Night entertainment for the close of the Christmas season. The first recorded performance was on February 2, 1602, at Candlemass, the formal end of Christmastide. In Act II, Scene V, Fabian contemptuously compares Malvolio's walk and attire to the strut and plumage of a male turkey: "O, Peace!" he exclaimed, "Contemplation makes a rare turkey-cock of him: how he jets under his advanced plumes!"

With the Protestant Reformation in England, and especially with the Puritans, Christmas was considered to be a "Popish holiday," and its celebration fell out of favor, if its observance was not proscribed altogether. Also giving the event a bad name was that it was often a day of drunken revelry rather than a holy day.

During the Restoration, when Christmas was again being observed, turkeys were once again on the tables of the wealthy as was noted by John Gay, English playwright and poet (1685–1732). Gay is best known for his *Fables* and the *Beggar's Opera* (1728), familiar today in an adaptation, *The Three Penny Opera*. Biographer Phoebe Fenwick Gay, in *John Gay: His Place in the Eighteenth Century*, noted that the *Fables* met with "great, immediate and unqualified" success upon their publication, and they remain among the most reprinted of John Gay's works.[5]

In one fable, titled "The Turkey and the Ant," John Gay expressed the lament of England's turkeys at Christmas time. The fable reads in part:

> How bless'd how envied were our life,
> Could we but 'scape the poult'rer's knife!
> But man, cursed man, on Turkey preys,
> And Christmas shortens all our days:
> Sometimes with oysters we combine,
> Sometimes assist the sav'ry chine.
> From the low peasant to the lord,
> The Turkey smokes on ev'ry board.

By the late eighteenth century, it was the custom for employers to give turkeys to their employees at Christmas. In 1839, Charles Dickens

---

[5] Phoebe Fenwick Gay, *John Gay: His Place in the Eighteenth Century* (Collins, 1938).

received one from his lawyer, Charles Smithson. This may have been the inspiration for an event he included in *A Christmas Carol*, published four years later, in 1843. Ebenezer Scrooge, having been visited by the Spirits and no longer the man he had once been, bought a prize turkey and had it delivered to his clerk Bob Cratchit in time for the family's Christmas dinner. In Scrooge's time, Victorian England, the middle class set the moral tone for society, and its members buttressed the Empire. For them, declared Isabella Beeton in her *Book of Household Management* (1861), a Christmas dinner "would scarcely be a Christmas dinner without its turkey."

Turkey raising as an agricultural enterprise began in England, and by the 1570s, large numbers of them were being raised on those farms large enough to produce sufficient feed and large enough to allow the birds to roam freely. Hugh Plat's *Jewell House of Art and Nature* (1594), was among the first works advising farmers on how to raise turkeys profitably and how to get them to fatten faster so their flesh became juicier and more tender.

By the eighteenth century, noted Thomas Pennant in "An Account of the Turkey" in the *Philosophical Transactions of the Royal Society of London* (1781), English farmers were crossbreeding turkeys to produce larger and better tasting birds. Andrew Smith's chapter in *The Turkey* titled "The Well-Bred Turkey; or, How The Turkey Lost Its Flavor," includes examples of such crossbreeding. Smith relates that the Englishman John Mortimer, as early as 1708, wrote of the "Gentleman" who crossed a "Hen-Turkey of the wild kind from Virginia" with an "English Cock" to produce "a very fine Breed."

The crossbreeds bred wild in the fields, and they were much larger than the domesticated turkeys then being raised in England. In 1736, the English agricultural writer Richard Bradley reported on a "Breed" of bird in "the West of England," somewhere "between the Turkey and the Virginia Bustard," that produced the largest bird of that sort he had ever seen. Its flesh was firmer and tastier, he believed, than that of the common turkeys then being raised in England. In 1750, the Englishman William Ellis reported that the turkey varieties then most common in England were the Suffolk or Norfolk turkey and the Blue Virginia turkey.

The Black turkey—often referred to as the Spanish Black or the Norfolk Black and considered to be the oldest turkey variety in England—was developed in Europe from the Aztec turkeys brought from Mexico by the Spanish. Early explorers in the New World mentioned Black turkeys, but they apparently were not the dominant type anywhere, and it may be that, because of their rarity, these were the birds taken to Spain. Breeders selected for this trait until it became predominant and, in Spain and England, the Black turkey was completely black, showing no white in the feathers. Shanks and toes were also black.

Known at first as the Spanish Black, the name was changed to Norfolk Black to reflect that more of them were raised in Norfolk Country than anywhere else. A single-breasted bird with tighter, firmer meat and a gamier flavor, the Black turkey remained the mainstay of turkey production for three hundred years. From the Black turkey, brought to the New World by early colonists and crossed with the wild turkey, were bred the Bronze, Narragansett, and the Slate turkeys that remained predominant strains in the United States until the twentieth century.

By the mid-seventeenth century, turkey raising in England had developed into a commercial business, with production concentrated in Norfolk and Suffolk, counties known for providing London with choice poultry. Naomi Riches, in *The Agricultural Revolution in Norfolk*, noted that sufficient feed for large numbers of turkeys was produced on the flat, drained fen land and that Norfolk and Suffolk turkeys had no equal for flavor, attributed to the dryness of the soil and the fact that the turkeys could range freely over a wide area.[6]

Because of its characteristics, the Norfolk Black was a particular favorite. The breasts were narrower and smaller than on other birds, the flesh had a fine texture, and it had a fuller, gamier flavor. Norfolk turkeys commanded high prices in London markets and, comparing them to diamonds, one writer reported that their value increased in geometric ratio to their weight.

Turkey's may have been produced in what novelist, journalist, and spy Daniel Defoe described as "infinite" and "prodigious" numbers in Norfolk and Suffolk Counties, and there may have been an almost in-

---

[6] Naomi Riches, *The Agricultural Revolution in Norfolk* (Routledge, 1967).

satiable demand for them among wealthy Londoners, but such was not the case in other parts of England, where raising turkeys caught on, but slowly. For small, self-sufficient farmers and small-holders, turkeys were known as a difficult bird to raise and on which to realize a profit. Farmers, on their small holdings, did not raise sufficient feed for a flock of turkeys, and the high mortality of poults because of weather, disease, and predators discouraged farmers from raising more than one or two turkeys for their own use.

For these families, chickens remained the bird of choice. Chickens required less feed, less space, and less care than turkeys. They were cheaper to buy and breed from, more chicks survived, and chickens provided more meat because they matured faster. Turkey hens do not begin laying eggs until late in the spring, and they produce a limited number of clutches. Chicken hens lay eggs throughout the year, eggs that added to the income and diets of self-sufficient tenant farmers.

It mattered little that turkeys were being produced in large numbers and that there was a demand for them in large urban centers if they could not be gotten to market. Among the problems was that distances of up to one hundred miles separated the farms where turkeys were raised and the markets where they were sold.

Another problem was that, in the days before refrigeration, turkeys had to be delivered live to markets. Farm produce was transported to market in horse-drawn carts, but even the largest farm carts could not hold turkeys in sufficient numbers to make this an economical way to transport tens of thousands of turkeys to markets each season. The solution was to have the turkeys walk in droves, driven by men known as turkey merchants, the same way cattle, sheep, and other animals were driven to market.

Livestock had been driven to market in England from the time of the Romans, and by the end of the eighteenth century, the practice was well established. Fast-growing urban centers created a demand for meat, and Smithfield Market in London became the largest and best-known cattle market in the world.

Phil Carradice, in his essay "The drovers of Wales," offers an informative treatment of the drover trade. Drovers' roads were wider than packhorse paths in order to accommodate large cattle herds and sheep flocks. Cutting across open country in order to save time and distance and to avoid toll gates, drovers used paths that were later incorporated into modern roads and highways. The roads had frequent sharp turns to right or left, depending on the prevailing winds. On the turns, both men and beasts were afforded protection from wind, driving rain, and sleet.[7]

Drovers were a distinct breed of individuals, trustworthy, charismatic, and hardy. They took pride in their work. They usually walked, four or five men to a drove of two- to three-hundred animals. Drovers were sometimes assisted by well-trained dogs that were sent home on their own at the end of the drives, while the men remained in the market towns to enjoy themselves and spend their hard-earned money. Using their homing instincts, the dogs often followed the same route the drovers had used on the trip to market, perhaps stopping at the same inns for the night.

At a time when travel was costly and difficult, drovers were in charge not only of livestock. They also carried the news and served as messengers, and they carried deeds and other important documents for people. Trusted with people's money, drovers helped promote banking. One bank was eventually incorporated into Lloyd's of London. Driving livestock to market in droves—an established practice—was thought to be the only expedient way to get turkeys to market. Let them walk.

A number of contemporaries documented the large flocks of turkeys being taken on drives of up to one hundred miles. Drives lasting perhaps three months did not escape notice.[8] The comparison to other livestock drives is apt, but only up to a point. As grazing animals, cattle are accustomed to walking, and they have hooves. With care, they can be taken on drives of several hundred miles. Turkeys, however, forage, and their feet are covered with skin, so farmers used a number of means to protect the turkeys' feet on drives of seventy to one hundred miles. Some fitted the feet with boots made of leather or sacking. Others drove the turkeys through warm, sticky tar and then through sand. When the tar dried,

---

[7] Carradice's essay, referred to here and in following paragraphs, can be read at "The drovers of Wales," Wales History, https://www.bbc.co.uk/blogs/waleshistory/2012/03/drovers_of_wales.html.

[8] Their accounts are cited in works such as *The Agrarian History of England and Wales* and Naomi Riches's *Agricultural Revolution in Norfolk*. Karen Davis, in *Turkeys: More Than a Meal*, compared turkey drives with the cattle drives of the American west.

Turkey drovers and their turkeys. Photograph courtesy U.S. Department of Agriculture.

it became like rubber boots on the turkeys' feet, and the sand provided grip and added protection so that the feet did not become blistered or damaged. With feet protected and one wing clipped to discourage flying, tens of thousands of turkeys crowded the roads each year between farms in Norfolk and Suffolk Counties in the weeks before Christmas, causing traffic jams on roads and streets.[9]

Daniel Defoe is best known for his novels *Robinson Crusoe* (1721) and *Moll Flanders* (1722). Both are literary classics. As a writer, Defoe had the rare gift of historical imagination, and his narratives—written with an eye for detail—read like documentaries. His *Journal of the Plague Year* (1721) is written in the first person, one man's experience in 1665, the year the Great Plague struck London. So vividly does Defoe write that readers can identify neighborhoods and streets, even houses in which events took place, and they hear—as he did—the mournful and repeated

[9] The most detailed descriptions of these drives, and also the most interesting because they are written as if they are firsthand accounts, are those of Daniel Defoe in his *Tour Thro' the Whole Island of Great Britain, Divided Into Circuits or Journies* (1724).

cries of "Bring out your dead" by those who were gathering the bodies of victims.[10]

Defoe's *Tour Thro' the Whole Island of Great Britain* (1724), a remarkable guidebook, is no less absorbing than his account of the bubonic plague that gripped London in 1665. Those reading his detailed descriptions of the countryside, the roads, and the villages through which he passed, can imagine themselves sitting beside him on the seat of his carriage, listening to his running commentaries. Reading Defoe's detailed accounts of the turkey drives that occurred every autumn, starting in August and continuing until October, is to stand with him on the roadside, hailing the turkey merchants and counting the turkeys as they crossed the River Stour on the Stratford Bridge.

"Suffolk," noted Defoe, was "particularly famous for furnishing the city of London and all the counties round, with turkeys." Moreover, more turkeys were raised for market in Suffolk and Norfolk Counties, he believed, than in all other counties combined. As many as three hundred droves of turkeys, a total of 150,000 birds, passed every autumn over the Stratford Bridge that spanned the Stour River, forming the boundary between Suffolk and Essex Counties. Stratford Bridge, fortunately, was a toll bridge, so there are records confirming Defoe's figures.

Each drove contained between 300 and 1,000 turkeys, an average of 500 per drive by Defoe's calculations, giving the total of 150,000 turkeys in all. And, Defoe noted, this was "one of the least passages, the numbers which travel by New Market-Heath, and the open country and the forest, and also the numbers that come by Sudbury and Clare, being many more." New Market, Sudbury, and Clare were small market towns in Suffolk County, on or near the River Stour, and some sixty or seventy miles north of London.

Turkey drives started in August and continued until October, when the condition of the roads made walking on them difficult for turkeys. Because most of the fields had been harvested, the turkeys grazed on the stubble and fed on insects. Flocks of one thousand turkeys could be handled by two turkey merchants carrying long poles of willow or hazel with

[10] Based on extensive research, Defoe's account is more detailed and readable than contemporary first-person accounts, such as the one by Samuel Pepys.

red cloth tied to the ends with which to control the birds. Such a drove could travel about one mile per hour, stopping at dusk, when the turkeys went to roost. Using Defoe's references to place names, readers could follow the turkey droves down the Old London Road, from Ipswich, past the village of Berghold, across the Stour River on the Stratford Bridge, six miles to Colchester, then on to Chelmsford and Brentwood, and thus to London.

"Besides these methods of driving these creatures on foot," Defoe noted, "a new method" was also being used to transport turkeys to London markets: "carts form'd on purpose." He provided detailed descriptions of both the new method and the carts.

Farm carts were modified to carry four "stories or stages, to put the creatures in one above another, by which invention one cart will carry a very great number." Another innovation, Defoe noted, again with detailed descriptions, was that the horses pulling the carts were hitched, not in tandem, the traditional way, but abreast, "fastened together by a piece of wood lying crosswise upon their necks, by which they are kept even and together . . . so quartering the road for the ease of the gentry that thus ride."

Readers would have understood Defoe's use of the phrase "so quartering the road" and his reference to the gentry. The word "quartering," used when referring to roads, has several meanings. A quarter is one of the four parts of a road, marked out by horse tracks and wheel ruts. The word also means to pass back and forth across the road, to cross and recross, so as to avoid obstructions, wet spots, potholes, or ruts.

As Defoe used the term, he took pains to note that the horses were hitched abreast, meant to have the cart straddle the ruts, a wheel on either side. Were the horses hitched in tandem and with the cart wheels straddling a rut, the horses would have been forced to walk in the rut. Hitched abreast, the horses could walk, one on either side of a rut just as the wheels straddled it.

The gentry were the landowners, just below the nobility in social rank. Unlike the gentlemen farmers or yeomen just below them in rank, members of the gentry took no part in the actual work of farming. Rather, they were active in local affairs, such as serving as justices of the peace.

They may have been entitled to bearing a coat of arms, as was William Strickland. Because they were to be deferred to by those beneath them on the social scale, in Defoe's reference, they had the right of way on the road and, for their "ease," the loaded carts had to make way for them.

Defoe, after his detailed descriptions, commented that by using this method of transporting turkeys, "they hurry away the creatures alive, and infinite numbers are thus carried to London every year." Young turkeys and poults, "which are valuable, and yield a good price at market," he noted, were transported the same way. With frequent changes of horses and by not stopping at nightfall, turkeys could be transported the one hundred miles in two days and a night.

Those wanting to learn how live turkeys were transported from Norfolk and Suffolk Counties to London markets could desire little more than what Defoe's accounts provided. Dressed turkeys in large numbers were also transported to London, and these not in farm carts, but in style, in horse-drawn passenger coaches. Those especially favored shared space with His Majesty's Mail in the fast-moving Royal mail coaches. Contemporaries, fortunately, noted these impressive shipments as well.

Thomas Kibble Hervey's *The Book of Christmas: Descriptions of the Customs, Ceremonies, Traditions, Superstitions, Fun, Feeling, and Festivities of the Christmas Season* (1836), provides an exhaustive historical account of old English Christmas customs, including feasting on roast beef, turkey, plum pudding, and mince pies, and customs such as singing carols and mumming. Hervey's book is best remembered for its delightful illustrations by Robert Seymour (1798–1836), one of the most successful caricaturists of the period. Seymour, unusually observant and possessing a keen sense of satire, was so prolific that it has been estimated that he drew one-third of all the British political cartoons of his era. He was also commissioned to illustrate the works of Shakespeare, Milton, Cervantes, and Wordsworth, as well as those of Charles Dickens.

Some of his finest caricatures are in Hervey's *Book of Christmas,* and they serve to capture the spirit and joy of the season. Among the best is his illustration of a coach from Norwich nearing London loaded with turkeys. Seymour made the drawing in 1835, at the height of the coaching age and, given its detail, he must have observed the scene and have been inspired by it.

William Hone published a number of *Every Day Books*, two of them in 1825 and 1827. In one, he noted that "the number and weight of turkeys" entering London by passenger coach for the Christmas season, if it could be determined, "would surpass belief." Between a Saturday morning and the night of Sunday December 22, 1793, for example, 1,700 turkeys, weighing over nine tons, were sent from Norwich in Norfolk County to London. Two days later, half as many more were sent. "Within and without, the coaches were crammed with the bird of Turkey," and it was not unusual for proprietors to refuse inside passengers in favor of turkeys, because ferrying turkeys paid better.

As the supply and demand increased, proprietors added coaches to accommodate, not human passengers, but turkeys. The most detailed and interesting account of transporting turkeys from Norfolk and Suffolk Counties in passenger coaches is that of Charles George Harper in *The Norwich Road: An East Anglian Highway*. Harper, an English author and illustrator, wrote many self-illustrated books exploring the regions, roads, coastlines, and old inns of Britain.[11]

Harper's account of transporting turkeys from Norwich to London in passenger coaches deserves to be quoted in full. At Michelmas and Christmas, he wrote,

> it was often difficult to secure a seat on or in any of the "up" coaches from Norwich or Ipswich, for while every available inch of space on the mails was occupied by festoons of dead birds consigned by country cousins to friends in town, the whole of a stage-coach was frequently chartered for the purpose of dispatching heavy consignments of these noble poultry to the London Markets. Christmas provided extraordinary sights along the Norfolk Road, in swaying coaches, with parcels and geese and turkeys mountains high on the roof; with

[11] In all, Charles George Harper produced some 170 topographical works, including *The Norwich Road: An East Anglian Highway* (Chapman and Hall, 1901).

> barrels of Colchester natives [oysters] in the boot, and hampers winging heavily between the axle-trees on a shelf called "the cellar," while from every rail or projection to which they could be either safely or hazardously tied depended other turkeys or fowls, booked at the last moment before starting. It was something in those days to be a turkey or a goose, before whose importance the claims of human passengers faded; but it was a fleeting elevation which the philosophic did not envy, thinking that here indeed the poet was justified in his sounding line— "The paths of glory lead but to the grave."

"The lord of the road was the Mail Coach," wrote Harper, "besides which the stages were very commonplace affairs." Especially favored, therefore, were the turkeys that shared space with His Majesty's Mail on them.

Mail coaches—expensive and exclusive—were constructed to carry four passengers inside and five outside, besides the coachman and the guard. Because they traveled faster than the stages and were associated with pomp and circumstance, fares were more expensive. The guard wore a scarlet coat and, to protect His Majesty's Mails, he was armed.

Everything gave way to the mail. Pikemen swung gates open and did not ask for tolls. Cautious highwaymen left them alone as conviction of robbing a mail coach meant death, so they concentrated on robbing the plebian stages. Mail coaches were gorgeous in appearance, painted in the traditional post office colors of black and red. Door panels bore the royal coat-of-arms, and the insignia of the four principal orders of knighthood were on the quarters.

The farm carts that Defoe described could transport turkeys from Norwich to London in two days and a night. Passenger coaches could cover the same distance in about the same time, perhaps a bit less. As late as 1821, the Norwich Mail made the same trip in fifteen and one-half hours.

Seeing turkeys being transported from Norfolk and Suffolk Counties to London in farm carts and passenger coaches may have been entertaining and a welcome diversion, but turkey drives remained the primary means of getting tens of thousands of them to market until the mid-nineteenth century, when railroads provided a faster, less labor-intensive, way. Only in some areas, did the practice persist into the 1930s. The turkeys on these drives were primarily the White Holland and the Norfolk Black, strains developed in England from the domesticated birds introduced into Europe in the early sixteenth century.

For three hundred years, the Norfolk Black was the most popular strain of turkey raised commercially. And it was the Norfolk Black that served as the foundation strain for the Bronze, Narragansett, Bourbon Red, Black, and Slate turkeys vying for attention and awards at the All-American Turkey Show in Grand Forks, North Dakota, the Turkey Capital of the World in the years before World War II.

CHAPTER 4

Carrying Coals to Newcastle: English Colonists Taking Turkeys to America

(with apologies to James Melville, ***Autobiography*** [1583])

DOMESTICATED TURKEYS WERE A significant food source in western Europe by the end of the sixteenth century, but, according to Andrew F. Smith, wild turkeys were unknown to most Europeans. It follows that most Europeans were also unaware that the turkeys that were providing their much-needed protein had their origin in America. Fascinating for what it reveals about this lack of awareness is a document written in 1584 by Sir George Peckham. In his *Principal Navigations, Voyages, Traffiques and Discoveries of the English Nation*, Peckham advised English colonists heading for North America to take "turkies, male and female," with them for propagation.

If the domesticated turkeys were intended to supplement the colonists' food supply, they need not have bothered. Eastern North America teemed with large flocks of wild turkeys. Not only were wild turkeys plentiful, they were larger than European domesticated turkeys and, as attested to by Thomas Wentworth Higginson, the first minister in Massachusetts Bay Colony, they were "exceeding fat, sweet, and fleshy." Some colonists did not even bestir themselves to hunt wild turkeys. Thomas Morton of Massachusetts could stand on his porch and shoot whatever turkeys were needed for a meal.

Heeding Peckham's advice, colonists in Jamestown were raising turkeys by 1614, and turkeys were being raised in Massachusetts Bay Colony by 1629, if not before. On the eve of the American Revolution, turkey raisers in southern colonies were exporting turkeys to the West Indies, and eventually large numbers were being sent to Europe. Two producers in Massachusetts sent 1,300 live turkeys to London in one month in 1833. Most farmers, however, considered domesticated turkeys a nuisance and uneconomical to raise. To prevent them from destroying crops and causing other damage, they had to be penned, fed, and cared for. Toms, especially, were considered to be mischievous pests and they had to be separated from hens, particularly in the spring when the hens were producing eggs and incubating them.

The turkey industry developed in America much as it had in England, slowly and in fits and starts. Different, however, is that domesticated turkey numbers increased in America for reasons other than to furnish meat, and turkey production in America eventually became commercialized to the point that would have been incomprehensible to those raising turkeys in Norfolk and Suffolk. Also different is that London, although among the world's largest poultry markets, was not known as the turkey capital of the world, and there was no All-British-Isles Turkey Show each year to which producers brought their prize turkeys and competed for prizes and awards.

Different too, as the title of Karen Davis's *Turkeys: More than a Meal* reminds us, turkeys in America are more than a meal. Turkeys figure in the nation's history; they are enmeshed in our mythology and interwoven into our social fabric. In America, the turkey—wild and domestic—has become an icon, a symbol, and a figure of speech. Dictionaries include lists of the many ways the word turkey has entered our vocabularies, from a theatrical production that is a failure to an eponymous dance to talking forthrightly with someone.

Among the most persistent myths associated with turkeys is that the Puritans at Plymouth observed the "first" Thanksgiving in 1621 and that turkeys were the feast's centerpiece. In his history of Plymouth Plantation, however, as Karen Davis notes, Governor William Bradford made no mention of a "thanksgiving" held by Pilgrims in the fall of 1621, and he referred to turkeys only in passing.

Also a deflator of myths, Andrew Smith observed that in the seventeenth and eighteenth centuries ministers and governors in North American colonies proclaimed thousands of thanksgiving days, to offer thanks for a timely rain, a good harvest, or a military victory. Alexander Young was apparently the first to associate the Pilgrims with Thanksgiving. In 1841, he published a copy of a letter, dated December 1621 and written by Edward Winslow, a leader in what was called Plimoth Plantation, to a friend in England. Winslow described a three-day event, but provided no dates. "Our harvest being gotten in," he wrote, "for three days we entertained and feasted" together with "Massasoit with some ninety men."[1]

The Pilgrims made no subsequent mention of the event and there is no indication that the event was remembered or observed in later years. Moreover, for religious reasons, Puritans eschewed traditional English holidays, considering them to be frivolous and associated with revelry and drunkenness.

Just as persistent as the myth of the first Thanksgiving is the myth that Benjamin Franklin wanted the wild turkey to be the national bird, and that he wanted the wild turkey—not the bald eagle—depicted on the Great Seal of the United States. Franklin, himself an American icon, had a significant hand in the turkey becoming an American symbol, but this myth originated in a now-celebrated letter (dated January 26, 1784), that Franklin, then in France, wrote to his daughter Sarah Bache in Philadelphia. To his daughter, Franklin confided that he would have preferred having the wild turkey on the Great Seal, because, he noted, the bald eagle

> is a Bird of bad moral Character; like those among men who live by sharping and robbing, he is generally poor, and often very lousy. He does not get his living honestly. . . . For in truth, the Turk'y is in comparison a much more respectable Bird, and withal a true original Native of America. Eagles have been found in all countries, but the turkey was peculiar to ours. . . . He is besides, though a little vain and silly, a Bird of Courage, and would not hesitate to attack a Grenadier of the British Guards, who should presume to invade his Farm Yard with a red Coat on.[2]

---

[1]Andrew F. Smith, "The First Thanksgiving," *Gastronomica* 3, no. 4 (Fall 2003):79–85.

[2]"From Benjamin Franklin to Sarah Bache, 26 January 1784," https://founders.archives.gov.

Franklin's last line revealed a commonly held belief, that the color red provoked turkeys and that they would charge anything with red on it. Franklin's letter, made public, has often been quoted, and, just as often, misinterpreted.

Franklin's role in the turkey's becoming an American symbol is significant to be sure, but not because he wanted its image on the Great Seal. Before 1776 and the American War for Independence, turkeys were just large birds that could be hunted or raised and then eaten. With independence, however, the new nation had to find ways to differentiate itself from England. One way was to develop things distinctively "American," things with nationalistic values, such as American foods and American ways of preparing and eating them.

Franklin liked to eat turkeys, and among his recipes was one for an oyster sauce for a boiled turkey. Turkeys also figured in Franklin's scientific experiments. In 1749, he proposed that turkeys could be killed by electric shock and roasted by an electrical jack in front of a fire kindled by an electrified bottle. A year later, attempting to carry out the experiment, he nearly electrocuted himself.

In 1743, Franklin had become intrigued with electricity after observing parlor tricks performed with static electricity. Wanting to understand electricity and believing that it had practical application, Franklin began experimenting on his own. In time, he utilized the Leyden jar, invented by Pieter van Musschenbroek in 1745, and named for the Dutch university where he taught.[3]

Among Franklin's experiments with Leyden jars was using an electric shock to kill chickens. Turkeys, however, although stunned, were not killed—until Franklin succeeded in killing one by joining several Leyden jars. Doing so, he accidently discovered a tenderizing technique, one still used in the meat industry. Sending an electric current through an animal causes the muscles to relax, thus modifying the effects of rigor mortis.

---

[3] The Leyden jar, an early capacitor, could hold and store electricity. William Watson, an Englishman, improved the jar by lining it with foil; with this improvement, he sent an electrical charge across the River Thames. Others began sending electrical jolts through human chains, people holding hands. The electric current made all of them jump simultaneously.

In a letter written to his brother John in Boston on December 25, 1750, Franklin described how he nearly electrocuted himself in a demonstration.[4] The description begs to be quoted. "I have lately made an experiment in electricity," Franklin wrote,

> that I desire never to repeat. Two nights ago being about to kill a turkey by the shock from two large glass [Leyden] jars, containing as much electrical fire as forty common phials, I inadvertently took the whole through my own arms and body, by receiving the fire from the united top wires with one hand, while the other held a chain connected with the outside of both jars. The company present (whose talking to me, and to one another, I suppose occasioned my inattention to what I was about) say that the flash was very great and the crack as loud as a pistol; yet, my senses being instantly gone, I neither saw the one nor heard the other; nor did I feel the stroke on my hand. . . . I then felt what I know not well how to describe; a universal blow throughout my whole body from head to foot which seemed within as well as without; after which the first thing I took notice of was a violent quick shaking of my body which gradually remitting, my senses as gradually returned.[5]

Turkeys being interwoven into our social fabric, Americans sing of those sitting in the straw, and turkeys appear in folk and formal poetry and literature. Popular throughout the mid-nineteenth century, the patriotic protest ballad "American Taxation: A Song of Seventy-Nine," probably dating from 1779 as the title suggests, mentions the wealthy feasting "on turkeys, fowls, and fishes."

---

[4] Two years later, Franklin, having failed to kill himself while electrocuting a turkey, could have succeeded in doing so by another means, if that was his intent. Flying a kite in a thunderstorm in a vacant field near Philadelphia, Franklin proved his hypotheses that lightning and electricity were one and the same, when a bolt of lightning followed the wet kite string to the ground. The lightning charge, had Franklin taken it through his body, would have been many times more powerful than the charge he had generated with his two Leyden jars.

[5] "From Benjamin Franklin to Sarah Bache, 26 January 1784," https://founders.archives.gov.

Ebenezer Cooke, a Maryland lawyer and writer, referred to turkeys in his epic poem *The Sot-weed Factor; or, a Voyage to Maryland* (1707) and, in a footnote, he commented that Maryland's wild turkeys were not only large, their meat was also "very good." In Henry Wadsworth Longfellow's epic *Song of Hiawatha* (1855), the merry Pau-Puk-Keewis danced his Beggar's Dance to please the guests gathered for Hiawatha's wedding, then sat down and cooled himself with his fan of turkey feathers. In his poem, "The Turkey Cock" (1922), D. H. Lawrence referred to the turkey as a noble icon of pre-Columbian America, and in *The Plumed Serpent* (1926) he wrote, "No sound on the morning save a faint touching of water, and the occasional powerful yelping of the turkey-cock."

Novelists described turkey shoots, an important Christmas tradition in America, and painters frequently depicted them on canvas. In one variation of a turkey shoot, turkeys were tied in an open area, and marksmen paid to take shots at a turkey's head. A hit allowed the shooter to claim the bird. In a variation of this, a turkey's head was thrust through a hole in a plank, and its body was secured behind the plank. Shooters paid a proprietor for a shot, and at a distance of one hundred yards or more, they attempted to hit the turkey's head. If his shot drew blood, a marksman was entitled to the turkey.

James Fenimore Cooper in his novel *The Pioneers* (1823) devoted an entire chapter to a description of what he believed was "one of the few sports that the settlers of a new country seldom or never neglect to observe." In Cooper's description, a turkey was tied to a tree with a string, its body buried in snow and only the bird's head and neck showing. At one hundred yards, marksmen took turns shooting at the bird's head. The first one to score a hit got the turkey.

Turkey shoots fell out of favor after the Civil War, and in 1866 a writer in *Harper's New Monthly Magazine* refused to describe one, because, his words, they were "cruel and unworthy of Christian men."[6] Henry Bergh, who founded the American Society for the Prevention of Cruelty to Animals in 1866, believed that turkey shoots were villainous, and he insisted that targets be substituted for live turkeys. By 1879, as reported in *Poultry*

---

[6] *Harper's New Monthly Magazine* (January 1866).

World, turkey shoots—described as "turkey manglers"—had disappeared (except for the clandestine ones).

Because many farmers believed that domesticated turkeys were a nuisance, raising them would have remained a marginal agricultural enterprise had they been used only as targets in turkey shoots and as food. The increase in domesticated turkey numbers was owed, wrote Andrew Smith, to one thing: tobacco, the "stincking weed of America." Not only was tobacco by far the most important reason for the increase in turkey numbers, this noxious weed was also significant in determining the course of American history.[7]

People in England were using tobacco by the end of the sixteenth century, although King James I, disliking the smell, had attempted to prevent its use. Sir Walter Raleigh popularized pipe smoking, and in 1611, John Rolfe—known as an ardent smoker—began cultivating the crop in Jamestown Colony with seeds he had obtained from Trinidad and Caracas, Venezuela. By 1612, Rolfe was growing Spanish tobacco, and he sent the first shipment of tobacco to England in 1613. From that date on, growing tobacco and manufacturing its products have been among the leading industries in the United States.[8]

So ideal were the soil and climate of the Chesapeake for tobacco that production exploded. By the 1670s, ten million pounds were being exported, giving the region the name The Tobacco Coast.[9] Tobacco was so

---

[7] Tobacco was America's first agribusiness, and its export marked the country's entry into the global market. Because repeated tobacco cropping depleted soil fertility, tobacco was also responsible for the movement, ever westward into the interior, from the tidewater. Tobacco is a notoriously labor-intensive crop, because the steps in its cultivation do not lend themselves to mechanization. To deal with the perennial labor shortage, Africans were enslaved and forced to work on the plantations. One development followed on another—all related to tobacco—and one need only allow his imagination to run to understand how tobacco was instrumental in determining the course of American history.

[8] By 1617, twenty thousand pounds of Virginia tobacco were being shipped to England annually, and the amount doubled in 1618. By 1622, sixty thousand pounds were being shipped annually to England, and by 1629 tobacco was being grown on two thousand acres in Virginia.

[9] The work of tobacco farming began in February and March. Young tobacco plants were planted in May and June. Harvesting was done in August and September. Caring for the crop required constant attention and backbreaking labor—twelve hours per day, six days per week, during the heat of the summer. Curing was finished in November and December, and the cured tobacco leaves were packed in hogsheads for shipment, ending the annual cycle, ready for the Tobacco Fleet from England that brought manufactured goods, foods, and alcoholic beverages to exchange for the crop. The soil rested in December and January before the cycle began anew.

important in the economies of the Chesapeake colonies that it became the medium of exchange. Items were bought, debts were settled, innkeepers were paid, and drinks were purchased with bundles of tobacco. That tobacco was designated the official medium of exchange in Maryland in 1637 was only one indication of the degree to which tobacco figured in the economies of the Chesapeake colonies.

A major problem involved with producing tobacco, however, was how to control tobacco hornworms (*Sphinx carolina*). Tobacco hornworms are large, meaty, juicy, green caterpillars from three to four inches long that develop in mid- to late summer. Because they are well camouflaged and blend in with the foliage, they are often undetected until they have caused severe damage to the plants.

Voracious creatures, each hornworm can consume several leaves, and an infestation of hornworms can destroy up to 90 percent of a crop in a short time. The damage is typically in the upper leaves of the tobacco plant, the leaves of highest quality. Before the development of pesticides, the major way of controlling tobacco worms was by hand. Each day, workers (typically slaves) went through the field, picking the worms from the tobacco plants. Even then, half the crop could be lost.

Hornworms met their match, however, with turkeys. Omnivores, turkeys eat insects, bugs, and worms and, reported British traveler John F. D. Smyth in 1784, turkeys are particularly dexterous at finding the large hornworms. Turkeys, observed Smyth, could keep tobacco plants more free from hornworms than all the workers on a tobacco farm put together, even if they did nothing else but pick hornworms from the leaves. Over time, a single turkey could remove the hornworms from an estimated one thousand tobacco plants, but—if properly managed—a flock of fifty turkeys could pick the worms from an estimated one hundred thousand tobacco plants.

Both George Washington and Thomas Jefferson described how turkeys were driven through tobacco fields to feast on tobacco hornworms. In 1819, John Skinner founded the *American Farmer*, and among the first articles published in the periodical was one on the practice of using turkeys to control tobacco hornworms. Having it "on good authority," Skinner wrote, several thousand turkeys could be hired out in Prince George's

County Maryland, the next summer, at the rate of twenty-five cents per bird per month. If any turkeys died from overwork or from other causes, they were to be paid for at the rate of seventy-five cents each.

By the mid-eighteenth century, virtually every tobacco planter maintained a large flock of turkeys expressly for the purpose of controlling the hornworms in his tobacco fields. Each day, workers drove the turkeys through the tobacco field, the turkeys keeping the leaves more stripped of hornworms than humans could. Turkeys, like sheep and cows, will follow a person and—also like sheep and cattle—turkeys can more easily be herded if the flock contains a few mature hens that know the routine. In pre-World War II Virginia, tobacco planters still kept flocks of turkeys for the purpose of controlling hornworms. When tomatoes became a popular food, turkeys were used to control the tomato hornworm, a worm similar to the tobacco hornworm, and turkeys were used in the west to control grasshoppers.

So familiar were people with using turkeys to control tobacco hornworms that James Fenimore Cooper could refer to the practice in his novel *The Spy* (1821). "If I had you on a Virginia plantation for a quarter of an hour," challenged one of his characters, "I'd teach you to worm the tobacco with the turkeys."

Ebenezer Cooke (1665–1732), London-born poet, referred to both "turkies" and tobacco in his epic poem "The Sotweed Factor, or a Voyage to Maryland, A Satyr" (1708). In a footnote, Cooke added that "wild turkies are very good Meat, and prodigiously large in Maryland." Tobacco was also known as sotweed, and a "factor" was someone who bought and sold, serving as a middleman, for example, between tobacco producers and those who purchased the tobacco leaves. Set in colonial Maryland, Cooke's poem was a satirical description in "Burlesque Verse" of the laws, government, courts, constitutions, buildings, feasts, "Frolicks, Entertainments and Drunken Humours of the Inhabitants of that part of America."[10]

---

[10] The American writer John Barth used the title of Cooke's poem for his novel, published in 1960. Barth's novel, with its Tom Jones-like plot and written in the style of eighteenth-century novelists, is a satirical description of the colonization of Maryland. In Barth's satire, set in 1680s and 1690s London and Maryland, a fictionalized Ebenezer Cooke was given the title "the Poet Laureate of Maryland" by Charles Calvert, Third Baron Baltimore. Ebenezer Cooke, in the novel, was the son of Andrew Cooke, an English merchant who owned a tobacco (sotweed) plantation.

Turkeys served another purpose besides figuring in literature and being used to rid tobacco fields of hornworms. After the tobacco leaves were harvested, their primary purpose fulfilled, turkeys were penned, fattened, and sold to serve as the centerpiece for Christmas dinners.

Ah, but there was a problem. In historian Smith's words, tobacco was, by far, "the most important reason for the growth of the domesticated turkey population in America." But, in the days before refrigeration, turkeys could not be slaughtered on the farm, dressed, and shipped to market. And, transporting large numbers of live turkeys was not as easy as packing cured tobacco leaves in hogsheads, loading the hogsheads on boats, and sending them downriver. The solution, as it was in Daniel Defoe's England, was to have the turkeys walk to markets. Turkey drives, therefore, became the norm throughout America until the 1930s.

CHAPTER 5

The Turkey Drover's Lot Is Not a Happy One

(with apologies to W. S. Gilbert)

THE TURKEYS MAKING GOOD the boast that Grand Forks, North Dakota, was the Turkey Capital of the World did not reach the city's markets on foot and in droves as they had in Defoe's England. Elsewhere, however, as historian Andrew Smith noted in *The Turkey*, turkey drives "became the norm throughout America" until the 1930s. The reasons, as in England, were that roads were poor and few in number; farm wagons held few birds; and, because of the lack of refrigeration, turkeys could not be dressed on farms and then transported to market.

Karen Davis, in *Turkeys: More Than a Meal*, compared turkey drives to the cattle drives of the American West. Perhaps a helpful comparison, but only with a number of provisos. Turkey drives in America started long before and continued long after the cattle drives from Texas to the cow towns of Kansas, but it is the cattle drives, steeped in legend, commemorated in word, and celebrated in song that remain fixed in the popular imagination.

In his *The Day of the Cattleman*, a classic treatment of cattle ranching on the open range of the American West, Ernest Staples Osgood wrote that "in all the varied history" of the western frontier,

> no single activity attracted more attention from contemporaries nor called forth a greater flood of reminiscence, story, and song after it had passed than did the Texas drive. . . . In after years, the drive of the Texas men became little short of an American saga. To all those who saw that long line of Texas cattle come up over a rise in the prairie,

> nostrils wide for the smell of water, dust-caked and gaunt, so ready to break from the nervous control of the riders strung out along the flanks of the herd, there came the feeling that in this spectacle there was something elemental, something resistless, something perfectly in keeping with the unconquered land about them.

The "Texas cattle," to which Osgood referred, were the Texas Longhorns, a breed that made more history than any other breed of cattle the world has ever known. Being feral, Longhorns were strong, hardy, resistant to disease, and immune to Texas fever. Given their fertility and ease of calving, Longhorns multiplied at double the rate of other breeds, and, by the 1880s, they roamed the Texas frontier by the millions, most of them unbranded.

Before the Civil War, a Texas Longhorn, worth at best two dollars, was slaughtered only for its hide and tallow. The carcass was left to rot. Circumstances changed with the end of the Civil War. A major transition in meat consumption patterns occurred. A national preference for pork abruptly gave way to a preference for beef.

A steer worth two dollars in Texas might be worth forty dollars in Kansas, Missouri, and other northern states. The Kansas Pacific railroad had expanded into Kansas and cow towns such as Abilene and Dodge City provided shipping points. There was now a market for Texas Longhorns. Taking note, Texas cattle owners began rounding them up and branding them and sending them north on drives, initiating, noted J. Frank Dobie in *The Longhorns*, "the most fantastic and fabulous migration of animals controlled by man that the world has ever known or can ever know." Between 1866 and 1886, cattle drives were a major economic activity, and an estimated twenty million cattle were driven north over cattle trails, so many, said one trail driver, that in places the dust in the trail was knee deep.

Excellent swimmers, Longhorns—with their long legs and hard hooves—were the preferred trail-herd breed for cattle drives until the late 1880s. Needing less water than other breeds and thriving on buffalo grass, they even gained weight on drives, but, wrote Dobie, "no other cattle known to history had such a disposition to stampede."

And it did not take much to cause Longhorns to stampede—a coyote's yelp, the rattle of a chuck wagon's pans, the hiss of a rattlesnake, a cowboy's sneeze, or the flare of a match. The only way to get control of a stampeding herd was for riders to get to the head of the herd and turn the leaders so the herd started to mill and circle around into itself. The cattle stopped running when they became exhausted. On a hot night, a steer that ran ten miles might lose up to forty pounds of its weight.

The men who drove the cattle were given a new name—cowboys—and the trailing era made the cowboy a universal folk hero. The ending of the Civil War provided a ready supply of skilled horsemen—Union and Confederate veterans (especially cavalrymen), former slaves, Mexican gauchos, Native Americans, and a few immigrants from Europe. In the United States census of 1880, hundreds of men listed their occupation as cowboy.

From Texas, trails ran north to Kansas cow towns where the cattle were loaded into freight cars bound for eastern markets. The Kansas Historical Society recorded that the first cattle drive reached Abilene in August 1867 and, between 1867–1871, when Abilene was the main Kansas cow town, more than three million cattle were shipped from its holding pens. It is Dodge City, however, the terminus of the Goodnight Trail, that remains fixed in popular memory; between 1875–1885 hundreds of thousands of cattle were shipped from this legendary cow town.

Charles Goodnight, perhaps the best-known cattle rancher in Texas and sometimes referred to as the Father of the Texas Panhandle, can be said to have invented the American ranching industry. Utilizing the Texas Panhandle and the Public Land Strip, thus avoiding such hindrances as the tolls levied by the Five Civilized Tribes, his trail to Dodge City became one of the most popular cattle trails.[1]

On a trail drive in 1867, Goodnight observed that bulls' testicles became battered, bruised, and swollen from walking day after day. Two bulls died and a third became quite ill from enlarged testicles. Goodnight solved the problem—and added a verb to the English language—by push-

---

[1] Goodnight displayed his intelligence and ingenuity in many ways. He converted a Studebaker army wagon into the first chuck wagon. Versatile and sturdy and drawn by oxen or horses, his kitchen on wheels was widely copied, and it became as synonymous to cattle drives as cattle and cowboys.

ing the bull's testicles up against its belly and cutting off the scrotum. He sewed up the wound, leaving the testicles riding close to the bull's belly, rather than swinging low between his hind legs. The bull recovered, and his breeding ability was not impaired. Goodnight repeated the operation on more bulls and other ranchers began performing the operation too. The practice became known as "goodnighting," surely among the most unusual of words named after a person.

On the trail, cattle did not move in a mass but in a line stretching up to two or more miles. Every trail herd contained a dominant steer, which, by instinct, was at the head of the herd, leading the way, and, at river crossings, the first to enter the water. Charles Goodnight had such a valuable steer in Old Blue. Tall, gunmetal blue, with a massive horn spread, Old Blue had a calming effect on herds. Goodnight put a bell on the steer's neck, and the rest of the herd learned to follow the familiar ringing, each day assuming the same place in the line. At the end of the drive, Old Blue was not sold, but was taken back to Texas to head other herds north. In eight seasons, ten thousand head of cattle followed Old Blue north to Dodge City.[2]

The Chisholm Trail, fifty miles wide in places and a corridor more than a trail, was named for a trader not a cattleman. Possibly the most famous of the cattle trails, it started at the Rio Grande River, near Brownsville, crossed the Red River, and continued north to Abilene and Dodge City. Between 1867 and 1872, 1871 being the peak year, more than three million head of cattle were driven up the Chisholm Trail to Abilene. As many as five thousand cowboys were paid off during a single day. Trailing cattle became unnecessary with the coming of the interstate railroads to Texas in the mid-1870s, and the Chisholm Trail was virtually shut down by the 1884 season.

Because the Chisholm Trail has captured the popular imagination, the best Western writers apparently believed that it was incumbent upon them to mention it in their works. Bill Canavan, in Louis L'Amour's *Where the Long Grass Blows*, rode a horse strangely marked, a true leopard Ap-

---

[2] When Old Blue was retired, according to Martha Deering in "Old Blue: Top Hand on the Trail," he was put out to pasture and lived until he was twenty years old. His impressive horns were mounted in the headquarters of Goodnight's ranch, and today they are displayed at the Panhandle-Plains Historical Museum in Canyon, Texas.

paloosa. Tall, narrow of hip and broad of shoulder, with blunt and rugged features, Bill Canavan had a tough, confident look about him. At twenty-seven years of age, he had made his decision. Wanting not wealth but a ranch of his own, Canavan was going to ride and fight for himself. He had paid his dues. He had taken a herd over the Chisholm Trail.

Luke Short, possibly the best Western writer of all time, wrote powerful, authentic tales of the American West, a West he lived in and traveled through. Among his fifty books, *War on the Cimarron* is considered by many to be his best. The protagonist is Frank Christian. Christian's partner had been murdered, his land had been stolen, and he had been accused of being a whiskey runner. But Frank Christian had never backed away from trouble, and now he was fighting mad. It was his turn to get back what was rightfully his. He too had paid his dues. He had driven his herd all the bitter way up the Chisholm Trail to the rich grasslands of the Cheyenne territory.

Zane Grey, among the most popular of Western novelists, traveled the West and documented the places he visited. He displayed his powers of observation in *The Trail Driver*, a novel that follows Adam Brite's herd of 4,500 Longhorns up the Chisholm Trail in 1871 to Dodge City. The steer leading the herd was Old Mossyhorns, Brite's Old Blue. Zane Grey's descriptions put readers in the midst of a cattle drive.

"Like a colossal triangle," Grey wrote, "the wedge-shaped herd, with the apex to the fore, laboriously worked up out of the valley." The herd was the wildest bunch of Longhorns Brite had ever seen. "Their widespread horns, gray and white and black, resembled an endless mass of uprooted stumps of trees, milling, eddying, streaming across the flat and up the green slope." To Brite, the herd's movement "was processional, rhythmic, steady as a whole, though irregular in spots, and gave an impression of irresistible power."

Zane Grey's *The Trail Driver* has it all—rustlers, hostile Indians, stampedes, hailstorms, lashing rains, bitterly cold winds, harrowing river crossings, and, as if another stressful situation was needed, the most daring rider turned out to be a young woman posing as a man. Zane Grey's description of riders and cattle caught in an electric storm is the stuff of which nightmares are made.

To quiet the "wiry intractable" Longhorns, night herders sang to them after the herd was bedded down after the day's drive. One night, when Adam Brite's herd was particularly restless because of an approaching storm, Reddie, the young woman who had posed as a man, heard a "weird chanting music" coming on the night air. It was San Sabe singing the Spanish song of a vaquero. Then, "from another quarter came a quaint cowboy song, and when that ceased a faint mellow voice pealed from far over the herd." That, wrote Zane Grey, "was the magic by which the trail drivers soothed the restless long-horns."

When Brite's riders and herd were caught in an electric storm and the Longhorns threatened to stampede, Brite asked Reddie to "sing them quiet." In low, quavering tones, Reddie began singing "La Paloma," and, as she sang, "her sweet and plaintive voice grew stronger." By the end of the song, "she was singing with a power and beauty that entranced the listening cattlemen." When she had finished, Texas Joe, who seldom sang, "burst out with his wild and piercing tenor, and then the others chimed in to sing a wonderful medley down that lonely valley." "At length," wrote Zane Grey, Brite's "great herd" of Longhorns quieted, "chained to music."

When describing the Cattle Kingdom in *The Great Plains*, Walter Prescott Webb commented on the commonly held belief that cowboys sang because the singing soothed the cattle. Webb acknowledged that a cowboy may have sung to the cattle, but, he believed, a cowboy sang mostly "to himself in his own loneliness." Night herders worked in pairs, so a cowboy also sang to let his partner and the cattle know where he was and to reassure both partner and cattle that all was well. Cowboys with clear, gentle, singing voices that carried well in the night were often preferred as night guards, and some trail bosses refused to hire cowboys who could not sing.

Cattle seemed to prefer long, slow, sad ballads. It was not the words that soothed them, some mournful tunes had no words, but the sound of the cowboys' voices. The men sang "Old Dan Tucker," "Nearer My God to Thee," "In the Sweet By and By," "Green Grow the Lilacs," "Bury Me Not on the Lone Prairie," and "The Dying Cowboy." Cowboys also composed impromptu songs, based on the rhythm of their horse's gait, that expressed their thoughts and described their experiences on the trail. Singing was also an entertaining way to preserve cowboy legends.

Among the most popular songs that cowboys sang was one that commemorates the best known of the cattle trails. "The Old Chisholm Trail" dates from the 1870s, but the song is based on an English lyrical work that dates from about 1640. The words were modified to describe what cowboys experienced when trailing herds from Texas to the Kansas cow towns.

Researchers have found over one thousand verses to the song, and each verse, which is followed by a refrain, documents the troubles that cowboys experienced in their work. Western Writers of America members rank "The Old Chisholm Trail" in the top one hundred western songs of all time. One version of the song contains these verses:

> Come along boys and listen to my tale,
> I'll tell you of my troubles on the old Chisholm trail.
>
> Oh, a ten-dollar hoss and a forty-dollar saddle,
> And I'm goin' to punchin' Texas cattle.
>
> No chaps, no slicker, and it's pourin' down rain,
> And I swear, by gosh, I'll never night-herd again.
>
> Oh, it's bacon and beans most every day,
> I'd as soon be a-eatin' prairie hay.
>
> We rounded 'em up and put 'em on the cars,
> And that was the last of the old Two Bars.
>
> With my knees in the saddle and my seat in the sky,
> I'll quit punchin' cows in the sweet by and by.
>
> *Refrain: Sung after each verse:*
>
> Come a ti yi yippee, come a ti yi yea,
> Come a ti yi yippee, come a ti yi yea.

Just as the best western writers believed it incumbent upon them to write of the Chisholm Trail in their novels, the best-known western singers apparently felt compelled to include "The Old Chisholm Trail" in their repertoires. The list includes Roy Rogers, Tex Ritter, Marty Robbins, and Gene Autry—known as the singing cowboy.

Whether or not cattle drives and turkey drives were comparable, as author Karen Davis observed, there is a parallel that she perhaps did not intend with her comparison. The Longhorns driven north from Texas to markets in the north whetted American appetites for beef. The money realized from the sale of the steers whetted the desire of those who wanted to make even greater profits. Wealthy English and Scottish investors bought American ranches and crossed the Longhorns with their Angus, Herefords, and Shorthorns to produce better tasting and more marketable beef (as a consequence, unfortunately, the Longhorn was nearly bred out of existence). Today's consumers may not appreciate that their sirloin steaks and hamburgers have their origins in the long Texas cattle drives.

Turkey drives are not fixed in the popular imagination as cattle drives are, and turkeys, with features that only mothers could love, do not evoke the same sentiments that cattle do. Turkey drives have not become legendary, and no popular turkey trail has been commemorated in song by a singing turkeyboy.

No Larry McMurtry wrote a Pulitzer Prize winning *Lonesome Turkey*, and no turkey drover added a verb to the English language (at least not one that can be used in polite society). Nevertheless, today's consumers owe their roast turkey, turkey drumsticks, rolled turkey breasts, and turkey hotdogs to the turkey drives that were commonplace throughout America until the 1930s.

During the first decades of our nation's history, it was an annual fall tradition to drive turkeys on foot from New England farms to markets in Boston. According to historian Charles Morrow Wilson in *The Great Turkey Drive*, it took about one thousand turkeys to make a trip of 150 to 350 miles profitable, but droves containing three thousand to four thousand birds were common. Droves were often fifty yards wide and a mile long and slow moving, perhaps covering ten to twelve miles per day. As described in a program first aired on Vermont Public Radio on November 27, 2013, turkey drives were not without their hazards, but they were a source of excitement for the sleepy, isolated villages through which they passed.

Roads were poor, and when there were no roads, drives moved through fields and forests. Losses ranged up to 10 percent from predators, from natural causes, from drowning, and from farmers taking birds from the flock. Because turkeys are prone to crowd together and trample each other when driven, men called "shooers" divided flocks into bunches of one hundred birds and used long poles with a piece of red flannel attached to the end to keep the turkeys separated and moving.

At the first hint of darkness, turkeys roost wherever they are. Hundreds of turkeys roosting in a tree or on the roof of a shed or barn can topple the tree or cause the building to collapse under their weight. A drove of turkeys bound for Boston chose to roost on the roof of a schoolhouse, and their weight caused the roof to cave in. As described by Wilson, the schoolmaster who was working inside escaped "by only a feather length." When a toll keeper on a road leading to Boston was tardy in raising the gate, hundreds of turkeys decided to roost on the roof of the tollhouse. Under their weight, the roof collapsed and the drovers were required to pay for a new roof.

Covered bridges, because their interiors were dark, also presented a problem. Reacting to the darkness inside the bridges when they reached it, the turkeys stopped. Turkeys once held up traffic through a bridge for two days. Drovers had to enter the bridge, pick up the birds, and carry them into the sunlight. As soon as the turkeys saw the sun, they started moving towards Boston again. Experienced drovers sometimes lit lanterns and placed them inside a covered bridge before the turkeys reached it.

Wilson repeated a story of a Father John O'Sheehan, who in 1840 encountered a drove of turkeys heading south as he was heading north towards Brattleboro, Vermont. The turkeys started roosting on the backs of his team and on his surrey. The good Father drove for his life and escaped, but not before he, his team, and his surrey were covered with turkey droppings.

It did not take much to spook turkeys on drives—a barking dog, a rifle shot, a piece of white paper fluttering in the wind, a dry twig snapping, a steam engine, or an unseen sound. Kathy Warnes, on her website "History? Because It's There," recounted the experience of Joe Heinrich, poultry dealer in Mt. Sterling, Kentucky. When a turkey drove arrived in Mt. Sterling and the birds were startled, they flew to the tops of the highest buildings for safety. Some roosted there for days, providing tempting targets for sharpshooters who competed to determine who could kill the most turkeys by shooting them in the head. Heinrich remembered that few turkeys were killed that way because a turkey's head is not much larger than a silver dollar.

In 1871, according to an article in *Harper's Weekly*, Connecticut was the leading "turkey State of the Union." In its issue of December 2, 1871, the magazine published an article on one turkey operation in the vicinity of Newtown, Fairfield County, Connecticut. Here, in small, free-range flocks, were produced the choicest turkeys sold on New York markets. A Mr. E. M. Peck "scoured" the countryside, buying only choice turkeys, for which he paid from twelve to fifteen cents per pound. In his pens, he readied the turkeys for market by finishing them on a mixture that included corn meal.

When the turkeys were ready to be slaughtered, the turkey butcher made a small cut in the main vein of the turkey's neck. The turkey died instantly and bled freely. Because the skin on the turkey's "drumstick" was too tender to allow the feathers to be scald-picked, the legs were removed first. Workers then immersed the turkeys in water heated almost to the boiling point so the feathers could be removed more easily. A skillful picker could pick the feathers from seventy-five turkeys per day and was paid one and one-half cents per bird. In one week, Peck's operation could prepare up to 1,300 birds weighing fourteen pounds each for New York markets. Peck had shipped up to fifteen tons of frozen turkeys to New York in a single shipment.

There were also turkey drives in mid-America and in the West. In 1863, a fellow and two young boys drove five hundred turkeys from Missouri to Denver, Colorado, where meat was scarce and high in price, a distance of six hundred miles. Their equipment consisted of a wagonload of shelled corn, drawn by six horses and mules. On a good day, with the wind behind them, the turkeys could cover a good many miles. When the wind was against them, progress was slower. The turkeys lived off the land, and when there were no grasshoppers or anything else to eat, the drovers fed them shelled corn. The fellow reported that he had fared well on the deal.

Turkeys were introduced to California by Franciscan missionaries. By the time of the gold rush, turkeys were becoming numerous but expensive. In 1854, toms sold for $7.00 each and breeding pairs for four times that. By 1866, turkeys were plentiful enough that businessman Henry C. Hooker of Placerville bought five hundred of them for resale in Nevada.

Hooker gained both fame, fortune, and a cattle ranch with an epic turkey drive. As a young man, Hooker owned a hardware business in Hangtown, California, which later became Placerville, a gold rush town. When his hardware store burned, Hooker used his last thousand dollars to buy five hundred turkeys. His intention was to drive the turkeys across the High Sierras to Carson City, Nevada, with the help of a drover and a few dogs. Carson City teemed with people hungry for food and meat of any kind. The drive went well until the turkeys reached the snow line and stopped.

When Hooker's dogs nipped at the turkeys to get them to move, they took to the air and disappeared. Assuming that he would never see the turkeys again, Hooker made his way down a steep slope. As he walked, one by one, the turkeys began appearing, and Hooker and his helper were successful in recovering almost the entire flock. In Carson City, Hooker sold the turkeys for $5.00 each, realizing a tidy profit on his investment. With the proceeds from his turkey drive, Hooker purchased cattle and started his cattle ranch, the Sierra Bonita Ranch, in Arizona's San Simon Valley. The ranch became well known and people began visiting the Sierra Bonita Ranch to ride, admire Hooker's cattle, enjoy the fresh air, and soak in the western atmosphere.

Using the turkey drives to Denver, Colorado, and Carson City, Nevada, as inspiration, Kathleen Karr in 1998 wrote *The Great Turkey Walk*, a heart-warming and funny tale of historical fiction that is especially appealing to young readers. The tale of the protagonist, Simon Green, also showed that one's start in life need not dictate one's future.[3]

Simon Green was a large, brawny lad who had finally graduated from the third grade after having attempted the grade four times. He made up for not being very bright by not being anyone's fool. Deciding at the age of fifteen years that it was time for him to make his way in the world, he hatched a plan that would make him a great deal of money. He planned to drive a flock of turkeys from his home in eastern Missouri to Denver, Colorado, where the birds would command a good price.

Borrowing $250.00 from Miss Rogers, his third-grade teacher and the only one who believed he had any potential, Simon Green on June 15, 1860, bought one thousand Bronze turkeys from Uriah Buffey of Union, Missouri. With a few young helpers, a dog named Emmett, four mules, and a wagonload of shelled corn, Simon Green started out on his "wild Wild West adventure."

Karr's book describes the harrowing events the young drovers experienced on the drive, events that mirrored those experienced on the drives on which the tale is based. In Denver, Simon Green sold 930 turkeys to Amos Quinn of Quinn's Provisions and General Store for $5,485.00. He had paid twenty-five cents apiece for the birds in Missouri.

Simon Green sent an account of the transaction to Miss Rogers. Twenty-one turkeys had gotten lost or had been stolen. Sixteen had been shot by United States Cavalrymen, eaten by coyotes, or were given as gifts to peaceful Native Americans. Green saved thirty-three turkeys—three toms and thirty hens—with which to start his own flock on The Great Turkey Five Ranch, named for himself, the drovers, and the dog who had accompanied him on the drive and who wanted to be in on his new enterprise. Miss Rogers was to receive 10 percent of the proceeds—$548.50—plus the $250.00 loan. Green rounded the amount up to $820.00 and sent the money to her by stagecoach.

---

[3] *The Great Turkey Walk* was selected as the Best Book of the Year by both the *School Library Journal* and *Publisher's Weekly*. *Publisher's Weekly* lauded Karr for having "a cheerful, sassy down-home writing style and a perfect pitch for dialogue." The *School Library Journal* described Karr's book as being "full of good humor" and "page-turning quest-style events." *Kirkus Reviews* noted that the book, "a wide-open western epic," was "inspired by actual events," and it appealed because it taught lessons of perseverance, kindness, acceptance of others, and honesty.

There was no Chisholm Turkey Trail, nor was any turkey trail commemorated in a song that is listed among the top one hundred best songs about turkeys. The late Margaret MacArthur and some fifth-grade students from Newbury, Vermont, did, however, compose a delightful song about a turkey drive from Newbury to Boston. The drive crossed the Connecticut River on the Bedell Bridge and went down the Coos Turnpike.[4]

Margaret MacArthur's long residence in Vermont and her familiarity with its folkways and history explain why the song of the turkey drive included mention of the Bedell Bridge and the Coos Turnpike. The Bedell Bridge, constructed by Moody Bedell in 1866 and spanning the Connecticut River between Newbury, Vermont, and Haverhill, Vermont, was 382 feet long and 23 feet wide. The roadway width was 18.5 feet. Before it was destroyed by a windstorm on September 14, 1979, it was the second longest covered bridge in the United States and the longest two-span Burr Arch truss and timber bridge left in America. The Coos Turnpike, incorporated December 29, 1803, ran from Haverhill through Piermont and Warren. Twelve miles in length, the Coos Turnpike was among the most successful of New Hampshire turnpikes.

The song, only four short verses, describes a successful turkey drive by a boy named Murphy.

> Long ago there was no money
> In the town we live in now.
> They sent flocks and herds to Boston
> of turkeys, geese, sheep and cows.
>
> Across the Bedell Bridge
> A drover boy named Murphy
> Drove on foot to Boston
> One hundred fifty turkey.

---

[4] Margaret MacArthur was an internationally recognized ballad singer and an authority on the folk music of New England and New York state. Her husband restored and customized a harp-zither, and Margaret MacArthur became an expert performer on it. The instrument was copied, manufactured, and sold as the MacArthur harp. MacArthur was also an expert on the Appalachian dulcimer, and she taught the lap dulcimer.

Over the course of three decades, Margaret MacArthur made 4,500 field recordings and transcriptions of folk songs. She was a member of the Committee of the National Folk Association and vice president of the Folk Song Society of the Northeast. A recipient of many awards and honorary degrees, Margaret MacArthur died in 2006, and her extensive collection, which included recordings, transcriptions, and analyses, was placed in the Vermont Folklife Center in Middlebury, Vermont.

To get turkeys into Boston,
They had to travel far
so farmers spread upon their feet
Heavy coats of tar.

Along the Coos Turnpike,
Many went astray,
'til he sprinkled corn and gathered
Many more along the way.

The turkey drive in the song is reminiscent of one described by a Frederic J. Wood in *The Granite Monthly* in 1920. A boy in St. Johnsburg, Vermont, was a helper in a drive of five hundred turkeys from St. Johnsburg to Lowell, Massachusetts. Travelers overtaking the drive alerted residents of villages along the way and crowds of people turned out to see the procession.

When the turkeys became accustomed to the road and to being moved in a flock, the boy discovered that he could lead the way by walking in front of them. As described by Wood, "A gobbler of especial dignity" soon assumed a position beside the boy "and thus the procession advanced" at a good rate. The drove reached its destination without the loss of a single turkey.[5]

Turkey drives do not have Pulitzer Prize authors, but they do have a tortoise and hare story in the "Fable of the Waddling Ducks and the Roosting Turkey." Kathy Warnes recounted the fable on her website. A turkey farmer in the small town of Denver, Arkansas, issued a challenge to a duck farmer. Their birds would race to the poultry market located in Springfield, Missouri, some sixty miles away. The duck farmer accepted the challenge, and the two men arranged the details of the race. Townspeople wagered on its outcome.

On the day of the race, the two men lined up their flocks at the starting point. When the starting gun was fired, the startled turkeys flew out of sight, leaving the slow-moving ducks behind. At sunset, the turkeys went to roost in trees, but the ducks kept walking. The duck farmer walked ahead of the drove, lighting the way with a lantern. His son walked behind

---

[5] Frederic J. Wood, *The* [New Hampshire] *Granite Monthly*, Vol. 52 (1920).

the ducks with another lantern, lighting the way for the stragglers. Walking all night, the ducks arrived in Springfield the next morning before the turkey drover was able to get his turkeys out of the trees.

Roberta Simpson Brown and Lonnie E. Brown, authors of *Kentucky Hauntings: Homespun Ghost Stories and Unexplained History*, shared stories of turkey drives in which their grandfathers participated. In Kentucky and Tennessee, turkey drives continued into the 1920s, and as late as 1930 turkeys were driven to markets in Texas, Colorado, and midwestern states.

In Texas, drives of up to twenty thousand turkeys, requiring two drovers for each one thousand birds, were not unusual. The drovers carried long whips with strips of red flannel tied to the ends with which to flick the birds to keep them in line and moving. Turkeys, it was believed, have an aversion to the color red. The *National Geographic* magazine in March 1930 had an article describing these so-called "turkey trots." Buyers went through the country gathering up bunches of turkeys. They stopped at farms, gathered the turkeys that were for sale, weighed the birds, paid for them, and added the turkeys to the drive.[6]

The Texas State Historical Association has documented some of the drives. In the early 1900s, ranchers in the area around Cuero, Texas, organized their small, family turkey-raising operations and began driving their turkeys to Cuero to be processed in a plant built in 1908. Thousands of turkeys on foot present a spectacle, and people from miles around began gathering in Cuero each November to see it. Taking advantage of the opportunity, enterprising merchants opened up rooms for board, offered hot meals, and held sales. Supposedly, visiting salesmen advised the townspeople to make the event a tourist attraction.

The first Cuero "turkey trot" was held November 25–27, 1912. An estimated thirty thousand people gathered to watch a drove of eighteen thousand turkeys strut through the streets. There were agricultural shows, big name bands, dances, a parade with floats decorated with turkey feathers, a carnival, and a football game. Governor Oscar Colquitt attended, and Minnie Lee Mangham was crowned turkey queen. Fourteen turkey trots were held in Cuero between 1912 and 1972, each larger and

[6] *National Geographic*, March 1930.

more spectacular than the previous one. Pageants with a Turkish theme were featured, and a king and queen were crowned as Sultan Yekrut (turkey spelled backward) and Sultana Oreuc (Cuero spelled backward). The festival was discontinued after 1972 when Cuero no longer had a turkey industry.[7]

Except for commemorative ones, turkey drives are no more. Turkeys lost too much weight on drives, unlike Longhorns that usually gained weight, and after 1930 they could be shipped to market by truck or train. Turkeys were also more difficult to control on drives than cattle. Longhorns may scatter in a stampede, but they remain on the ground and can be rounded up.

Turkeys, although domesticated, retain many of the instincts and traits of their wild cousins. When startled, they fly into trees or atop buildings; they can be coaxed down, if at all, only with difficulty. Longhorns were herd animals, and each morning they assumed their places in the drive. In all accounts of turkey drives, drovers warned that if turkeys were not gathered at first light when they descended from their roosts, they scattered into the forests or fields and were likely lost.

Turkey drives are not without their ditties, fables, or legends, but they are not fixed in the popular imagination as long Texas cattle drives are. Turkey drives and cattle drives were alike, however, in that both were victims of progress.

When railroads were built into the heart of the Texas cattle country, the Goodnight and Chisholm trails fell into disuse, and cowboys were relegated to stringing barbed wire and putting up hay. With improved farm-to-market roads and large trucks, turkeys no longer needed to strut down turnpikes and drovers no longer needed to coax them through covered bridges, out of trees, and down from the tops of tall buildings.

[7] In 2012, the Turkey Fest Association celebrated a once in a life-time turkey trot, the centennial of the one held in 1912, but with far fewer turkeys. The Cuero, Texas, turkey trots are commemorated by the Turkey Fest Association that continues to collect photos, videos, and other memorabilia of those held between 1912 and 1972.

Trucks replaced turkey drives as a means of transporting turkeys.
Photograph courtesy Nebraska Agricultural Experiment Station.

CHAPTER 6

"Turkeys are a lot of darned hard work."

(with apologies to Ann Marie Low)

THE 1930 *CITY DIRECTORY* of Grand Forks, North Dakota, noted that North Dakota was "one of the leading turkey raising states in the Union" and that this new farming enterprise was "growing faster in this state than anywhere else in the country." In eighth place among turkey-raising states in 1920, by 1930, North Dakota ranked second, and its fine turkeys commanded premium prices on eastern markets.

The sixteenth annual All-American Turkey Show was held in Grand Forks, January 16–21, 1939. Announcing the event, an article in the *Grand Forks Herald* on January 11, 1939, noted that "thousands and thousands of dollars' worth of turkeys are bred and sold in the Red River Valley. In fact, this territory has become nationally known as the 'Turkey Territory' of the United States."

S. A. Macaulay, publicity agent for the Northern Pacific Railroad, attended the All-American Turkey Show in 1924 and spoke at one of the sessions. "Northwest turkey raisers," he commented, were "keeping about one jump ahead of the middle west states" in turkey raising, partly because the area's "natural conditions" were "very favorable for raising turkeys."

The statements beg explanation, an explanation that spans half the continent and a period of two hundred years. Until the early twentieth century, turkeys were raised on family farms for local markets, mostly in the eastern states, and, before the mid-nineteenth century, farmers paid little attention to their breeding. Domesticated turkeys were allowed to run free, eating insects and plants and mating indiscriminately.

Almost always a sideline, turkey raising was considered to be women's work. Poultry yards were close to the house so farmwives could watch over the turkeys, and many people believed that successful turkey raising required "a kind, faithful, experienced woman, who loves poultry, to look after them." And, as the "frugal housewife well knows," observed the *American Agriculturist* in its February 1862 issue, the "pin money" from the sale of turkeys was hers to spend as she chose.

Being allowed to run free, domesticated European turkeys mingled with their wild eastern cousins. Martha Ballard raised turkeys on her Maine farm, and she noted in her diary on April 7, 1792, that she had found five wild turkey eggs. Two days later, she found two more. She placed the seven eggs under one of her turkey hens that was setting on her own clutch.

A few days later, Martha found seventeen wild turkey eggs, which she also placed under her domesticated turkey hens. In all, forty-three turkey chicks hatched. It became a common practice for farmwives to gather wild turkey eggs and place them under domesticated turkey hens to be incubated, some in the mistaken belief that the chicks would thereby inherit the traits of both wild and domesticated birds.

In 1869, sisters Catherine Beecher and Harriet Beecher Stowe published *The American Women's Home: or: Principles of domestic science: being a guide to the formation and maintenance of economical, beautiful, and Christian homes*. The book, a collaboration by two of the era's most influential writers, was among the century's most important handbooks of domestic advice, and it became almost a bible on domestic topics for Victorian women. An instructional guide to homemakers, the book illuminated women's roles at a time when they were unappreciated and relegated to performing menial work for which they were poorly paid.

"The most absorbing part of the 'woman's question,'" noted the sisters, was to provide a "remedy for the varied sufferings" of women who were widows or unmarried "and without means of support." But, many "lucrative" enterprises were well within reach of women "with proper abilities and training for the business." If women wanted to, they were advised, they could find "employment both interesting and profitable," and "few objects of labor" were more profitable than raising turkeys "on a modest scale," small flocks of a few dozen birds. Turkeys, once fledged, were hardy and, allowed to range freely, they almost took care of themselves. Among the benefits to women who raised turkeys was that the income, being theirs, would afford them a measure of independence.

As farm turkey flocks became more numerous, newspapers and agricultural journals began offering advice on raising turkeys. Turkey production was also encouraged with the creation of the Department of Agriculture and the passage in 1862 of the Morrill Land Grant Act that established colleges for agriculture. By 1900, states were establishing experiment stations and offering agricultural extension services to disseminate information and advice on agriculture. Experiment stations conducted research on turkeys beginning in the 1890s, and, in 1893, the Rhode Island Station issued its first publication on turkeys. Much of the information was directed at farmwives.

Until the mid-1880s, most turkeys were raised and marketed in New England and in Central Atlantic states, but, by 1890, the center of turkey production had moved to the Midwest. In 1890, Illinois was first among turkey-producing states, with Iowa and Missouri tied for second place. In 1930, the ranking was Texas in first place, North Dakota in second, and Minnesota in third.

It is significant that the rapid advance of North Dakota and Minnesota in the rankings coincided with the beginning of the All-American Turkey Show in Grand Forks in 1924 that promoted turkey raising. Helping to explain the movement of turkey production from eastern states to the Midwest is the prevalence of diseases that nearly decimated turkey populations. In New England, production dropped from 11 million birds in 1890 to 6.6 million in 1900.

The sisters Beecher and others may have been overly optimistic in their belief that turkeys were hardy and, "once fledged," able to care for themselves. Turkeys are susceptible to a number of diseases, especially blackhead and pullorum. Blackhead is not a disease indigenous to the Americas, so turkeys have neither evolved with it nor developed an immunity to it.

Blackhead is caused by a protozoan parasite, *Histomonas meleagridis*. The parasite is carried by the common poultry caecal worm, *Heterakis gallinarum*, found in the caeca (blind pouches of the large intestine) of chickens. The protozoan has little effect on chickens, but it is deadly to turkeys. Chickens serve as carriers, which explains the frequent transmission of the disease from apparently healthy chickens to turkeys. The parasites can live for long periods in the caecal worm and its eggs, presenting a high risk for turkeys if they range over areas containing chicken feces. Turkeys are infected once they ingest the eggs.[1]

Symptoms of blackhead appear seven to twelve days after infection, although the name is misleading—the turkey's head does not always turn dark or bluish. The heads of infected birds are tilted down or drawn to the body, their feathers become unkempt, their wings droop, and their eyes are partly closed. Sick birds have reduced appetites, increased thirst, and sulphur-yellow diarrhea. Turkey poults, infected with blackhead, die within a few days. Older birds may be sick for some time before becoming emaciated and then dying.

Pullorum disease, discovered in 1899, is a world-wide, egg-transmitted disease of poultry, caused by *Salmonella pullorum*. The main reservoirs of infection are the egg-producing organs of infected hens. Chicks from diseased hens are infected at conception inside the egg. *Salmonella pullorum* rarely affects humans, but pullorum disease affects most poultry.[2]

Infection may spread by breathing contaminated dust or coming into contact with down from infected poultry or with other material in the incubator, the shipping box, or the pen that has been used by infected birds. Carrier hens lay infected eggs, and the hatchlings are infected. Hens can be infected by eating infected feed, eggs, litter, manure, or by drinking infected water. If young birds, such as turkey poults, have other diseases, pullorum tends to spread faster.

---

[1] Blackhead occurs when the parasites enter the caeca of turkeys and are able to multiply in the caeca wall and cavity. Turkeys can ingest the organisms in contaminated feed or water, while picking gravel, or preening. Often, a second parasite, the caecal worm (*Heterakis gallinarum*) is involved, a worm that is one-third to one-half inch long and as thick as a thread. The worm lives in the caeca of chickens and other birds. The worm, or its microscopic eggs, can harbor blackhead organisms and carry them from one bird to another. Blackhead organisms are fragile and cannot live alone outside a bird host for more than a few hours. In the eggs of highly resistant caecal worms, however, they may remain infective for four or more years. Infected birds eliminate the organisms in their feces; these are then ingested by susceptible turkeys and they become infected.

[2] Once called bacillary white diarrhea, pullorum is a devastating disease of turkeys. Young birds die during their first three weeks and losses may be as high as 80–90 percent. One poult may infect an entire hatch. Death from pullorum rarely occurs after four weeks of age, and infected adult birds usually show no outward evidence of infection, but they are disease carriers for life.

There being no therapeutic drugs for blackhead or pullorum, prevention was the sole method of treatment (breaking the disease cycle by keeping turkeys separate from chickens and frequently moving them to disease-free ground), explaining in part why turkey production moved from New England and the mid-Atlantic states to the Midwest and justifying the claim that North Dakota in 1930 was one of the leading turkey raising states.

Elwyn B. Robinson wrote in his *History of North Dakota* that the state enjoys a continental climate—cold winters, hot summers, warm days, cool nights, light rainfall, low humidity, and abundant sunshine. The American Guide Series on North Dakota added this: the state had "elbowroom," and its farms provided abundant feed, grains, and forage. In short, noted C. Dyke Page, president of the All-American Turkey Show, the region was particularly well suited for producing quality turkeys that commanded premium prices on eastern markets.

How quickly and the extent to which turkeys were integrated into the Midwest's agricultural economy can be measured by the number of references to them in literature, diaries, letters, farm journals, and agricultural publications.

Turkeys are interesting birds, noted Geri Walton in "Raising Turkeys for Market in the 1800s," but they are neither amiable nor sociable, and there is no great love between them and humans. Nor are they entirely domesticated. No sooner were chicks hatched than they were apt to scurry away at the slightest provocation.

Young turkeys were also difficult to care for, and a heavy rain shower could kill an entire flock. Ann Marie Low, in her *Dust Bowl Diary*, went Geri Walton one better. A particularly sensitive and observant young woman with an "irrepressible spirit," Ann Marie Low wrote of the drought and Depression in the Stony Brook country near Jamestown, North Dakota. "Turkeys," she noted in her diary,

> are a lot of darned hard work. They are as dumb as sheep and must be constantly watched lest they commit suicide in some stupid way. A lot of their feed is grasshoppers they catch in the fields and meadows. We have to keep them shut up in the

> morning until the coyotes are through mousing in the meadows, and then run our legs off in the evenings getting them penned up before the coyotes are out again.
>
> Dampness is death on young turkeys. They must be protected from dew and rain as well as from skunks, weasels, coyotes, and their own general stupidity. Ever since I started raising the feeble-minded creatures, they have kept Bud [Ann Marie's young brother] and me on the run.

Carrie Young, in *Nothing to Do but Stay*, wrote of her mother, Carrine Gafkjen, who at the age of twenty-five staked out a homestead for herself on the windswept North Dakota prairie. Not until she was thirty-four years of age and a successful landowner, did Carrine marry and raise a family.

When Carrie was about eight years old, her mother decided to go into the turkey business, although no farmer in their Norwegian-American community raised turkeys. She started with eight fertile Bronze turkey eggs purchased for fifty cents apiece. Six of the eggs hatched, but only four of the poults survived—a tom and three hens, the beginnings of what Carrine believed would be a thriving business. It did not turn out quite that way. The next spring, the hens produced more than three dozen poults, but as soon as they were hatched, they began to scatter into the neighbor's wheat field.

Thus began what Carrie described as "the most onerous task" of her childhood—herding turkeys. For the entire summer, Carrie and her sister Fran, armed with brooms, had the job of "keeping track of the turkeys." Turkeys, they soon learned, were

> congenitally indisposed to the principle of herding. Neither are they compatible with chasing, shooing, or rounding up. All of our other farm animals had the homing instinct. Our horses could find their way home blindfolded in a snowstorm. Our cows could break out and wander miles from home, but as soon as they saw our car they would lift up their

> heads and run unerringly in the direction of home. Our chickens had the good sense never to leave home in the first place. But once turkeys left home, they have never heard of it; when approached on foreign soil they will panic, break rank, and skitter in every direction except homeward.

Not only did the turkeys resist chasing, Carrie complained, "they feinted this way and that, stretched their necks and retracted them, made garbled noises, and—if all else failed—flew over our heads."

Carrie and Fran prayed that their mother would quit raising turkeys after the first season, but she persisted for the next four summers, during which Carrie and Fran, now teenagers, with sinking hearts assumed that they would still be herding turkeys when they were old and gray. Abruptly, however, "one bright autumn day," Carrie's mother announced that she was "selling out" and quitting the turkey business.

In "a rare burst of hang-the-expense abandon," however, Carrie's mother kept the choicest young turkey for the family's Thanksgiving dinner. "After years of doing battle" with turkeys on the wing, Carrie wrote, she and her sister had their first taste of turkey. It "had a strange, rather dry nutlike taste," and they could not decide whether they cared for it. The girls did confide to each other after the meal, however, that the best part of that Thanksgiving Day was knowing that they would not have to spend any more summers herding turkeys.

Great Plains author Willa Cather wrote of turkeys in her novels and short stories in a way that revealed her familiarity with them. In *Death Comes for the Archbishop*, in which a serving boy dies after being struck on the head, a turkey that was to serve as the feast's main course, was left roasting on the spit when the startled guests fled. In her war novel, *One of Ours*, for which Cather was awarded a Pulitzer Prize in 1923, a troop train had been stopped by a wreck on the tracks ahead, and the fellows had gotten off the train to stretch their legs. Then, when "the locomotive screeched to her scattered passengers, like an old turkey hen calling her brood," the soldiers scampered back to the train.

Cather devoted a section of *My Antonia* to a description of the Cuzak boys. Leo was fawn-like and his eyes were not frank and wide apart like those of the other boys, but deep-set, gold-green in color, and sensitive to the light. Leo, his mother said, got hurt more often than all the other boys put together. He attempted to ride the colts before they were broken, tried to see how much the red bull would stand for, tested the new axe to see how sharp it was, and teased the turkey gobbler.

Among Willa Cather's most telling references to turkeys, however, was in *The Troll Garden*, in the story of "A Wagner's Matinee," a disturbing tale of farm life in Nebraska. Georgiana, a weather-beaten, work-worn farmwife, had been a music teacher at the Boston Conservatory as a young woman, and she had never lost her love for music. She and the "idle and shiftless" Howard Carpenter had taken up a homestead in Red Willow County, Nebraska, fifty miles from the railroad, and, for thirty years, Georgiana had never been farther than fifty miles from the farm. Coming into a small legacy and having to go to Boston to settle the estate, Georgiana arranged to meet her nephew, who had "a reverential affection for her" because he owed her most of the good that had come his way as a boy.

While Georgiana was in Boston, her nephew took her to a symphony concert that featured Wagner's music. At the sound of the beautiful music, Georgiana began to cry. When the concert was over, she sobbed pleadingly, "I don't want to go, . . . I don't want to go!" Her nephew understood. For his aunt, he knew, "just outside the door of the concert hall, lay the black pond with the cattle-tracked bluffs; the tall, unpainted house, with weather-curled boards; naked as a tower, the crook-backed ash seedlings where the dishcloths hung to dry; the gaunt molting turkeys picking up refuse about the kitchen door."

Before World War II, most turkeys were raised on family farms, six hundred thousand of them in 1930, according to an article in the *American Turkey Journal*, with a total annual production of fifteen million birds. Although an important source of income, turkeys were a marginal activity on most farms and, notes historian Barbara Handy-Marchello in *Women of the Northern Plains,* they remained the domain of the farmwife.

Handy-Marchello cites the example of Rosina Reidlinger, a German-Russian in McIntosh County, North Dakota. Because of the vagaries of weather, insects, and disease, noted Rosina, "the poorest farming" was "just relying on crops for all of the income." Better for the farmwife to "have about 200 chickens and 100 turkeys." In the Reidlinger household, income from chickens bought the groceries and income from turkeys paid for fuel and clothing. Poultry produced supplementary income with very little investment.

In *The Prairie Winnows Out Its Own*, historian Paula Nelson provides an economic and social history of South Dakota's west river country in the 1920s and 1930s, years of stress and hardship. On diversified farms, Nelson notes, turkeys supplemented incomes and they were women's work. Agricultural journals such as *The Dakota Farmer* addressed advertisements and articles on turkeys to women. Advertisements for disinfectants or insecticides for poultry houses portrayed women using them, as did ads for incubators. Testimonials for incubators with names such as Sure Hatch, The X-Ray, Successful, and Close-to-Nature were often signed by women. Articles on poultry were addressed to "farm women." Two issues of *The Dakota Farmer* contained such articles. One article, for example, titled "Poultry will Return Profit for Care," was written by "Survivor of 1911," from Zieback County, South Dakota, and another issue contained four letters from women to the magazine's poultry department on issues concerning poultry.[3]

H. E. H., a woman of Montrail County, North Dakota, asked what had ailed her young turkeys the previous summer. They had hatched fine and looked well and plump until they were eight to ten days old, when they started to look sick. They became light and thin, and then "they died like flies." Their droppings were full of small, round white and pink worms. Out of 500, she had lost over 150.

Judge E. L. Hayes, editor of *The Dakota Farmer*'s Poultry Department section, responded to the question, diagnosed the problem, and offered advice. The poults were infected with internal parasites, likely common round worms. The best treatment for round worms was one teaspoon of turpentine, followed by a tablespoon of olive oil, but the turpentine should only be given after the poult had been starved for twenty-four hours. A 10-grain dose of areca nut was also a good treatment. The areca nut could be mixed with soft feed and fed from a clean trough. A one-grain dose

[3] *The Dakota Farmer*, January 15, 1922, and February 1, 1922.

of thymol was also an excellent worm remedy, as was a two-grain dose of santonin.

Because the worms multiplied from eggs, the range and buildings used by the turkeys would be infected with worm eggs. Turkeys become infected after eating the eggs. Because the eggs live for months, Judge Hayes advised, the writer should discontinue raising turkeys for a year or raise them in different buildings and in a new location.

A springtime 1922 issue of *The Dakota Farmer* includes an article by C. B. Titus of Benson County, North Dakota, titled "How to Raise your Turkeys: Instructions on Handling from Hatching Time to Marketing." The article covers everything a woman needed to know about raising turkeys, from selecting breeding stock to how to kill and pick them "quickly, easily and humanely." The article includes a picture of Mrs. E. V. Trenholm of Codington County, South Dakota, with her flock of White Hollands, raised the previous summer from five turkey hens.[4]

"Lice," warned the article, were "the worst enemies of young turkeys," and the breeding flock had to "be kept perfectly free" of them. Painting the turkey house or barn with crude oil and then white-washing it was among the best methods for killing and controlling lice. If hens were kept free of lice, there would be no problem. If they were not, turkeys would die throughout the entire season.

During the winter, turkeys kept for breeding should be fed corn, wheat, meat and table scraps, chopped raw potatoes, and silage. They also needed sand, gravel, and oyster shell. In the breeding season, hens needed green feed and oyster shells to produce strong-shelled eggs, and the hens should be outside during the day unless the weather was unfit. If more than one tom was used, one should be used one day, the other the next, to prevent them from fighting.

Poults needed no feed for three days after they hatched, and a good starting feed included finely chopped hard-boiled eggs, clabbered cheese sprinkled with black pepper, and crushed egg shells. "A grain of whole black pepper," women were advised, would "often revive a chilled or droopy turkey." By far, "the best growing feed" was scalded cornmeal, but it should not be "mushy or watery."

---

[4] *The Dakota Farmer,* March 15, 1922.

Above all, readers were warned, "Do not overfeed; it is the worst thing you can do. Keep them hungry all the time, as an active, hungry turkey will not get sick very easily." Once established, poults should be allowed to run free because bugs, grasshoppers, and grass or alfalfa were the best food for them. The poults did, however, have to be kept dry.

Turkey raisers were also advised to have fresh water available at all times for growing turkeys, preferably in a water fountain, because poults could tumble into an open trough or pan. A disinfectant should be added to the drinking water. "Permanganate of potash" was the best, as much as could "be put on a dime to each gallon of water."

When the poults were ready for market, they should be fed corn heavily for a month or two before they were dressed, and they would gain more weight if they were allowed to run free. Only the best birds—big-boned, well-feathered, and nicely-colored—should be kept for breeding. If young hens were kept for the next season, to avoid inbreeding, toms from another flock should be used.

Turkeys did more than provide supplemental farm income, however. Although agricultural production declined significantly during the Great Depression, turkey production did not. The reasons are obvious. Turkeys could be raised cheaply, and, pound for pound, turkey meat was the least expensive. In a *National Geographic* article, "How Turkeys Saved Families in the Great Depression," the author notes that turkeys "helped countless families weather the Great Depression—and it was often farmwives who ran the cottage industry."[5]

Illustrating the article is a photograph of one such enterprising farmwife in Idaho, difficult to distinguish as she stood in the center of a large flock of turkeys. The photograph, from the *National Geographic* photo archives, was taken by the noted Boise photographer Ansgar E. Johnson, Sr. And hers was only one example among many—farmwives whose turkeys provided the only income when drought, heat, disease, and grasshoppers destroyed crops or reduced yields.

The *Grand Forks Herald* headline on July 6, 1936, was "104 [degrees] Here as Heat Sears Northwest." The temperature was 114°F at Hillsboro, thirty-five miles south of Grand Forks. It got worse. The headline in the

---

[5] *National Geographic,* November 2016.

Saturday edition of the *Herald* was, "108° Here is All-Time Heat Record." The record-breaking temperature, caused by the "onslaught of a blazing sun and a hot, southerly wind," was 2°F higher than the previous record of 106°F set on July 28, 1917, and Saturday was the fifth consecutive day that temperatures had been over 100°F.[6]

High temperatures and drought reduced the wheat crop in the Grand Forks area in 1936 by one-half, but conditions in the western part of Grand Forks County were worse. Grain fields were parched and crops neared total failure. Counties in central and northwestern North Dakota were even more scorched by the heat. In Sheridan County, 52 percent of farmers were on relief, and cattle herds had been reduced by one-third. In Divide County, 90 percent of farm families were on relief. "Relief," noted Elwyn Robinson in his *History of North Dakota*, became the state's "biggest business."

The years of drought were also years of the Depression. In 1932, North Dakota farmers sold wheat for thirty-six cents per bushel, oats for nine cents, barley for fourteen cents, flaxseed for eighty-seven cents, potatoes for twenty-three cents per hundredweight, and beef cattle for three dollars and thirty cents per hundredweight. The price of wheat recovered only to fifty-three cents per bushel in 1938. Most farm-mortgage loans became delinquent, and the number of forced sales increased. From 1930 to 1944, probably one-third of North Dakota's families lost their farms to foreclosure.

In 1933, the low year, the per capita personal income in the United States was $375.00; in North Dakota it was only $145.00. From 1932 through 1937, the per capita personal income in North Dakota was only 47 percent of the national income. "Plainly," noted Robinson, "the 1930s brought much greater hardship to North Dakota than they did to the nation as a whole."

Rosina Reidlinger, the German-Russian farmwife in McIntosh County, North Dakota, believed it unwise to rely only on crops for income, because, in years of low yields and poor prices, income from chickens and turkeys sustained families that otherwise might have lost their farms. An article in the August 1941 issue of *American Turkey Journal*—without providing documentation—noted that farmers who raised turkeys during the years of drought and Depression were better able to pay their bills,

---

[6] *Grand Forks Herald*, July 6, 1936.

protect their credit, and save their farms than those farmers who relied only on crops for income.[7]

Ann Marie Low, she of the "irrepressible spirit," wanted to continue her studies at Jamestown College. When crops failed and livestock sold for giveaway prices, however, her folks did not have enough money to buy shoes for her and her sister, let alone pay their tuition. During semester break in 1931, Ann Marie and her dad tried to figure out how she and her sister could remain in college when there was no money for expenses. They rejected the idea of Ann Marie taking a stenographic job in Jamestown for the summer because the pay would be too low.

The family would be better off if she stayed at home, helped with farm work, and raised turkeys. Turkeys had been worth "five dollars apiece" in the fall, Ann wrote, and, because they had most of the feed needed, "including lots of grasshoppers, the money was chiefly profit." The turkey hens laid more eggs than they could brood, so the extra eggs were put under setting chicken hens. "With luck," Ann Marie noted in her diary, "the turkeys should bring in about $750.00 next fall."

On June 30, 1931, a Tuesday, Ann Marie wrote in her diary that "It is so hot!" The flax had cooked in the ground and fields and pastures were burned brown. Cattle were starving all over the state, and there was no market for them. Horses were dropping dead in the fields from the heat, and cows had stopped giving milk because they had nothing to eat. Farmers pastured their grain fields, then plowed and summer fallowed them to conserve moisture.

"This," Ann Marie noted, "is our third year of drought, and it is a severe depression." If good rains did not come soon, there would be no pasture, then "the cattle must go to market or starve, and they are not worth shipping. The turkeys, however, have done fine."

Unfortunately, the rains did not come, the cattle were sold, and Ann Marie's dad received only $1,312.00 for them. They harvested four bushels of wheat to the acre, enough for seed and a little to sell. They had seventy-five bushels of barley and three hundred of oats—little more than enough for seed. There were only 125 turkeys to sell in the fall of 1931 and the price for No. 1 turkeys had dropped to $4.00. Their luck had not held.

[7] *American Turkey Journal,* August 1941.

The turkeys did not bring the $750.00 the family was counting on to see them through until the next fall.

Nelson, in her history of South Dakota's west river country, cited examples of families that raised turkeys to supplement farm income during the bad years. Iola Caldwell Anderson's family started raising turkeys with a gift of turkey eggs that they put under a brooding chicken hen. Only a few eggs produced poults the first year. The following spring, the family purchased a tom and started their turkey business. Poults were kept dry and fed chick feed and cottage cheese for the first few weeks. "After a few trials and errors," the family discovered that "turkey farming was not an unprofitable occupation."

Turkeys supplied the family with meat, and those that were sold added to their income. The Marousek family in the west river's Meade County also raised turkeys on their farm, located forty miles from Wasta, the nearest town with a railroad. Helen Marousek said that income from turkeys was a vital part of the family's finances during the years of drought and low grain prices.

Glen C. Bidleman of Kingsley, Kansas, reported to the *American Turkey Journal* in early summer of 1935 that his area had escaped the worst of the dust storms, although it had been "plenty bad." Even with the heat and drought, however, his turkeys were doing well.[8] Another Kansas turkey raiser, writing in August 1940, reported that it was still hot and dry, and farmers were selling their livestock because there was no feed. Under such conditions, it was "a wonder" his turkeys were doing so well. "That only goes to show, once more," he wrote, "under what many adverse conditions turkeys can go through and still thrive."

In the same year, Mrs. William Eddie of Northwood, North Dakota, and Secretary-Treasurer of the Narragansett Turkey Club, reported to the *American Turkey Journal* that there was a "serious" infestation of grain rust in her area, and she was thankful that "turkeys can't rust." She had, however, lost twenty-five poults in two heavy rains. The strong wind had blown the poults off their roosts, and the rain had pelted them so hard they died.

[8] *American Turkey Journal,* June 1935.

"A poult surely gives up easily," she complained. One of hers had been out in the rain, and he "looked like his minutes were numbered." His wings were drooping, he was half sitting, his head was pulled in as far as possible, and his eyes were closed. She set the poult on a box, flapped his wings, stretched his legs, and moved his head from side to side. "He soon started to stretch his neck, open his eyes, and stand on his shaky legs." After stretching his neck, opening his eyes, and standing on his own, the poult "decided it wasn't a bad world to live in after all."

Mrs. Earl Gridder of Barton, North Dakota, started raising turkeys in 1918 with nine eggs that hatched seven poults. She soon had a flock of Giant Bronze, and in "one of the dry years when many farmers did not realize as much from their crops," she sold $1,109.00 worth of turkeys. There was "no royal road" to success with turkeys, she wrote, but if one was unafraid of work, eager to learn, and enjoyed being outdoors, one could be successful.

In May of 1935, Mrs. Roy Vasper of Neche, North Dakota, wrote to the *American Turkey Journal* that "turkeys are the life savers on most of our farms during these trying times, when the wind and frost damage the crops and the drought and 'hoppers' take the rest." If turkeys are given care and attention, she believed, "they never fail us." For the past two years, her family's crops were "nearly a failure," but money from her turkeys had helped pay the bills and buy many of the things her family needed.[9]

Another article in the *American Turkey Journal* almost a year earlier was titled "The Drought and Turkeys." The extended drought "hits everyone and everything," noted the author, but, even under these trying conditions, "nothing on the farms of the Northwest will withstand conditions better than the turkey and nothing would replace it for profitable returns to the farmer in areas of drought."

Turkeys withstood heat, they found feed where other stock starved, they consumed less grain than other stock, and they could be marketed at prices several times higher than other stock. "So," concluded the writer, "turkey folks should not become discouraged. If turkeys won't pay, nothing will."[10]

---

[9] *American Turkey Journal,* May 1935.

[10] *American Turkey Journal,* July 1934.

Author Lois Phillips Hudson, born in Jamestown, North Dakota, grew up on her parents' farm near Cleveland, North Dakota. Ruined by the Great Depression and drought, the family migrated to Washington state in 1935, spending several months as migrant workers along the way.[11] Hudson's novel, *The Bones of Plenty*, set in Stutsman County, North Dakota, encapsulates the experiences of the Lows, Caldwells, Marouseks, and other "reapers of the dust," as they fought drought and the Depression.

The story begins on Friday, February 17, 1933. Hudson's George Armstrong Custer, like Hamlin Garland's Tim Haskins, was under a lion's paw. For nine years, Custer picked rocks from the 320 acres he rented from James T. Vick, and, like Tim Haskins, he had put cash and labor into the farm and hoped one day to buy it. He promised himself that if Vick tried to push him off the farm without making a decent settlement with him, he would "take a few thousand dollars out of the old man's hide."

One disaster after another befell the hapless George Armstrong Custer, from having his best mare abort her colt after falling into an abandoned well to having his crop of Ceres wheat disappear in "smut, rust, drought, grasshopper gizzards, [and] middleman's pockets." Custer's "last hope of the harvest year" were the turkeys. He planned to ship the first batch on November 11, in time to catch the Thanksgiving market.

Before it got too dark the night before, he sharpened the knives he would need and, before going to bed, he filled the boiler and tub with water from the well that "was acting almost normal again." As he walked past the turkey pen with the water, Custer "looked over at the dark masses of them roosting on their poles and wondered what they would bring."

According to Custer, turkeys "were a lot of work," "as brainless as a creature could be, and susceptible to all sorts of diseases," but it was a sight to watch them as they "preened their magnificent white-tipped feathers in the Indian summer sun and strutted about so dignified one minute and so ridiculous the next." He hoped that the turkeys would sell for a decent price and make up for the slump in the cream checks.

---

[11] A prolific writer, Lois Phillips Hudson is best known for her autobiographical novel *The Bones of Plenty* as well as *Reapers of the Dust*, a collection of short stories. The books chronicled the years of the Great Depression and drought in North Dakota.

Custer knew that the proper way to "stick a turkey" was to thrust a knife up through its open mouth into its brain to sever the right nerve. If the thrust had gone true, the muscles in the turkey's skin went limp for about ninety seconds—long enough for an accomplished picker to strip off the feathers while the bird was still not really dead. After those few seconds, the muscles tightened as rigor mortis set in. If the knife thrust missed the nerve and rigor mortis set in immediately, the muscles tightened around the pin feathers and the turkey had to be scalded before it could be picked.

By mid-afternoon, the Custers had twenty-five turkeys picked and ready for packing. Custer weighed each one. The largest tom weighed thirty-six pounds, nearly three times what the young hens weighed. The family packed the turkeys, weighing a total of nearly four hundred pounds, into two barrels that Custer rolled out on the porch where it was cool. He intended to haul them to the depot after the chores were done. Custer anticipated receiving a check for $115.00 for the twenty-five turkeys and another check at least as big for the ones he planned to ship the next week.

Not only did turkeys provide supplemental farm income and help families save their farms during the years of drought and the Depression, they also proved their worth by providing farmers with a way to diversify their production. Diversification, it was expected, would increase incomes and land values and provide a hedge against disasters resulting from relying on a single crop in the event of crop failure. Ah, but, turkeys, versatile birds that they are, did more. Turkeys have long been valued for their ability to forage for most of their food, helping to control insect pests in the process.

Lacking pesticides, tobacco farmers used turkeys to rid their tobacco plants of tobacco hornworms. In the Midwest, farmers used turkeys to rid their fields of grasshoppers. "Raise More Turkeys" was the *Dakota Farmer* directive.[12] All those "interested in breeding these great American birds" were encouraged to adopt and heed the slogan. Every farmer with range sufficient to allow turkeys to roam freely was encouraged to ob-

[12] *Dakota Farmer*, February 1, 1922.

tain breeding stock or order eggs for hatching, because "the Grasshoppers Must Go."[13]

It had long been acknowledged that North Dakota was among the leading turkey-producing states in the country and that its fine turkeys commanded premium prices in eastern markets. As reported in a 1934 issue of the *American Turkey Journal*, however, the state's turkeys were also "achieving eminence in the more lofty sphere of improved mental activity." Instead of merely eating and growing, North Dakota turkeys had added "mental gymnastics" to their routine, and they were "doing some real thinking" for themselves.[14]

An incident in the previous summer, reported by Jay Stevens of Lawton, North Dakota, was "advanced in support of this contention." Stevens was in his automobile on his way to town when he came upon a large flock of turkeys on the road. The turkeys gave no indication that they were willing to move out of the way. Indeed, they appeared to be waiting for something. Stevens decided to stop, get out of his car, and shoo the turkeys off the road.

As soon as the car stopped, the entire flock of turkeys "swooped down on it" and proceeded to eat all the grasshoppers that had lodged anywhere on the vehicle. When the birds had finished cleaning the car of grasshoppers, they "calmly dismounted" and moved off to the side of the road so Stevens could pass. They then moved back onto the road to await the next car that came along.

"The traditional impartiality" of the *American Turkey Journal* would not allow asking Stevens what strain the turkeys were, but he insisted that he had had the experience several times and it had been a different strain of turkeys each time. The only conclusion that could be drawn was that all turkeys in North Dakota were "exhibiting this new mental agility." Incidentally, Stevens noted, the turkeys had done a better job of removing grasshoppers from his car's radiator than his garage man had—and he had charged two dollars for the job.

The movement of turkey production from New England and the Central Atlantic states to the Midwest and Northwest does not lack for ex-

---

[13] *The Dakota Farmer*, February 1, 1922.

[14] *American Turkey Journal*, November 1934.

planation. Speaking at the All-American Turkey Show in February 1924, County Agent Langley noted that "the Northwest [was] one of the few sections in the United States where turkeys [could] be successfully raised." Indeed, the Northwest was "especially adapted for turkey raising" with its favorable climate, its prairies, and its abundance of grain and insects for feed. The Northwest was also "remarkably free" of blackhead, the most widely spread turkey disease.

Turkeys were profitable and versatile, but raising them remained women's work. Eighty percent of the turkeys exhibited at the All-American Turkey Show were raised by women. It was the women who attended the educational sessions on turkey raising, those who asked the questions, and those who shared their experiences. At the All-American Turkey Show in 1938, the top prizes in the Bronze, Narragansett, and Bourbon Red divisions were awarded to turkeys that had been raised by women. These were "business women," noted a 1937 *Grand Forks Herald* article, conducting "one of the important industries of the Northwest."[15]

These women could agree with Ann Marie Low that turkeys were a lot of work. For Mrs. Grace Randolph of Donaldson, Minnesota—Minnesota's Turkey Queen—raising turkeys was a full-time job. Mrs. William Eddie of Northwood, North Dakota, also spent most of her time with her turkeys, especially during the months of April, May, and June. She did not mind the work, and she enjoyed being outside, but there were times, she admitted, when "you have to make up your mind to stay on the job if you would succeed."

[15] *Grand Forks Herald,* January 17, 1937.

CHAPTER 7

Fine Feathers Make Fine Birds

(with apologies to Aesop)

ANYTHING PERTAINING TO THE All-American Turkey Show invariably commanded prominent space in the *Grand Forks Herald*, and during the week of the show, its activities dominated the paper's front page. The show's title was often included in the day's headline. On January 17, 1938, the first day of the Fifteenth Annual All-American Turkey Show, the lead article on the paper's front page commented on the four hundred "brilliantly plumaged" turkeys, all "of proud lineage and individual distinction," that strutted, gobbled, and "preened their feathers" while waiting their turns on "the judges' tables."[1]

"Brilliantly plumaged."

"Proud lineage."

"Preened their feathers."

"Judges' tables."

The phrases in small compass explain the purpose and the appeal of the All-American Turkey Show. The phrases were included in accounts of the show, in articles and editorials in the *American Turkey Journal*, and in advertisements of breeding stock. Birds of an excellent type can always win the highest honors because there is always plenty of so-called "good color." Of all varieties of domesticated fowl, few, if any, "possess the magnificence of color" of turkeys.

Feathers tell the true story of the turkey, because a highly colored bird shows good breeding. Claims such as "Large wide feathers show a thrifty bird" and "We each strive to raise our chosen [strain] true to color"

---

[1] *Grand Forks Herald*, January 17, 1938.

were in abundance. As reported, breeders enjoyed walking through their flocks, "stopping here and there to admire feather markings, nicely built blocky birds and prospective show birds." Flock owners took delight in watching potential show birds "blossom out in full beauty at maturity."

"The color standard" indicated "good, pure breeding," and the turkey of best type was sure to win if it had reasonably "good color." Because color represented the strain and type, there was "no excuse for not breeding for color." M. C. Small in his essay "Turkeys" advised breeders that, regardless of market quality, a turkey stood no chance of winning in Exhibition Classes without the "finest of feather characteristics."

In his comments in the *American Turkey Journal*, George Hackett explained why color and feather patterns were critical in maintaining the purity of turkey varieties. "October," Hackett wrote, "is full of thrills for the real turkey breeder" because, as he observed the birds in his flock, he could discern the results of his "untiring effort" to improve the variety. As the Standard breeder went through his flock to select breeders and "those promising youngsters that may bring to him the coveted high honors all progressive turkey breeders strive for," he had two objectives—constant improvement of a better market carcass and "further improvement of standard color."[2]

If, Hackett chided, people think that "standard color is no longer important," let them walk down the aisles in the show room "viewing the fine specimens" in the cages. Yes, they recognized good type, blocky conformation, and symmetry, but few were attracted to a turkey unless it had "smooth and beautiful plumage which is distinctly characteristic of the turkey and always has been." He acknowledged that carcass and type were important, but he did not envy the turkey grower who saw "only dollars as compensation from his turkey flock." Breeders who aspired to taking the top awards with their birds at the All-American Turkey Show would agree. They were breeding for perfection of color.

The American Poultry Association was organized in February 1873 in Buffalo, New York, by representative poultry breeders from the United States and Canada. The Association's primary objective was to "standardize the varieties of domestic poultry" that had become so numerous and

[2] *American Turkey Journal*, October 1941.

so similar "that there was the greatest confusion" in breeding and judging them. Accordingly, the association followed principles that "would be recognized generally as right" and those that would "secure acceptance of its standards as authoritative."[3]

The association issued the *American Standard of Excellence* in 1874 (the word "excellence" was shortly changed to "Perfection"), and, with its publication, "attention turned more and more to the improvement of characters of substantial worth and true beauty." Thereafter, wherever "Standard-bred poultry" were found to be lacking in productive or ornamental value, it was because breeders and judges were not "interpreting and applying the *Standard of Perfection* in accordance with the principles governing the making of the Standard."

Also, the association maintained that "the permanent popularity of a useful breed, or of any variety of that breed," depended on "giving it a finish in every detail of form and color which will make it equally desirable for its beauty." The association believed that "indifference to productive values and neglect of appearances are equally obnoxious to its principles." Clearly, as stated in the third of its principles, the association recognized and emphasized the importance of color in all breeds of poultry.

The *Standard of Perfection* was recognized as the authority in all matters pertaining to every variety of Standard breed of fowl, and it was followed religiously by judges in all poultry shows in the United States and Canada and by all successful breeders. The *Standard of Perfection* went through a number of revisions, but for most of the years of the All-American Turkey Show, the *Standard of Perfection* allowed forty of the one hundred points in judging turkeys to be awarded for color.

Acceding to the demands of breeders who worked to produce birds more for market than for color, the revised *Standard*, used at the All-American Turkey Show in 1941, reduced to thirty-two the number of points awarded for color. This, over the objections of those breeders who wanted the number to remain at forty and those of breeders who insisted

---

[3] Briefly stated, the principles were as follows: First: "In each breed then existing," the most useful type would be the Standard type. Second: "No more breeds should be recognized as having distinctive breed character than could be identified readily by at least one conspicuous character." Third: "Recognition of color varieties in a breed should be limited to plainly distinctive color patterns."

on a more drastic reduction limiting to fifteen out of one hundred the points awarded for color.

Color was a contentious issue, but it was an absolutely vital consideration when breeding and judging turkeys according to the *Standard of Perfection* because, noted Dr. Jacquie Jacob, a Poultry Extension Associate at the University of Kentucky Department of Animal and Food Science, "only one breed [emphasis by the author] of turkey exists." The *Standard* also recognized a single breed of turkey. There are, however, many turkey strains and varieties, and the strain or variety is distinctive because of the color and pattern of its feathers. Dressed turkeys of different strains look very much alike, and their flesh tastes much the same. The most distinctive and readily identifiable feature of a turkey strain or variety is its unique plumage, or, as the American Poultry Association stated as one of its guiding principles, "recognition of color varieties in a breed should be limited to plainly distinctive color patterns."

"The turkey," observed historian A. W. Bryant in his "Brief History of the Turkey," is "known for producing many 'sports' or mutations" that allow varieties to be developed. With most of these strains or varieties, he noted, the focus was "on the plumage and other show characteristics," not on meat or market characteristics. Color, therefore, not size or marketability, is what distinguishes one turkey variety from another.

A "sport" in the plant world is a part of a plant that exhibits unusual or singular deviation from the normal or parent plant. It is a generic mutation that results from a faulty chromosome replication. The result of the sport or mutation is an offspring that is distinctly different from its parent in both appearance and genetics. The genetic change is not the result of unusual growing conditions; it is an accident, a mutation. In many cases, the new trait can be retained and perpetuated by propagation.[4]

Wild turkeys also produce sports. Bill Marchal, an outdoors photographer and columnist living near Brainerd, Minnesota, wrote of wild turkey sports, sometimes identified as "Smoke-phase turkeys," in an article

[4] Sports are common among cultivated plants, and many of the fruits we enjoy eating are the offspring of sports, including Ruby Red grapefruit, Red Anjou pears, and a number of apple varieties such as Royal Gala, Red Delicious, and Fuji. The peach is the most striking instance of a plant producing both bud-sports and seed of commercial value. The peach seed may occasionally produce a nectarine, that, in turn, may produce either a nectarine or a peach.

published in the Minneapolis *Star Tribune*. Smoke-phase turkeys, Marchal wrote, are a naturally occurring "visual delight." Their heads are light blue or gray, and their caruncles are a faint red or pink, similar to those of a normal colored turkey.[5]

As Marchal described, the turkey's feathers, "splendid as they are," are light gray or nearly white and tipped with black, "as if dipped in ink." Tail feathers are only a shade or two lighter than those of their wild cousins. Marchal estimated that as many as one in seventy wild turkeys is a Smoke-phase turkey, capable of producing offspring that are either white, normal colored, or both. Some broods have poults of both colors, but about 95 percent are females.

Mutations in wild turkeys are quickly lost through interbreeding. With domesticated turkeys, however, flock owners can detect mutants or sports as they appear and, by mating selected birds, they can perpetuate the desired features. In this way, the varieties of domestic turkeys have been produced, none of which have counterparts in wild turkeys. The Royal Palm turkey, developed in the 1920s, is an example of a turkey variety developed from a sport.

The Standard of Perfection defines a variety as any bird that has been line-bred for a number of years (some would insist on at least five generations of closed-flock breeding) and which reproduces uniform characteristics with marked regularity. The Royal Palm was exhibited for the first time at the All-American Turkey Show in 1939, but only as a novelty. It could not be entered into competition, judged, or awarded prizes because it was not admitted to the *Standard of Perfection* until 1977.[6]

The turkey, in the *Standard of Perfection*, is considered to be a single breed of poultry, with six recognized and accepted Standard varieties. Most of the *Standard* varieties were admitted to the *Standard of Perfection*

---

[5] *Star Tribune*, February 5, 2011.

[6] The Royal Palm turkey was developed by Enoch E. Carson of Lake Worth, Florida. Carson raised Black, Bronze, Narragansett, and wild turkeys in a mixed flock. The Palm pattern appeared in a male sport, and Carson—by careful selection and breeding—cultivated the pattern in birds until it bred true. The turkey's color pattern is striking. Its feathers are white with black edgings, gradually increasing in proportion until the saddle appears black. The tom's white tail feathers have a black band a few inches from the outer edge, and the base of the tail is also black. The coverts and wings are white with black bands and edgings. The turkey's shanks and toes are pink. Hens have some black on their backs. The Palm pattern is also seen in blue and red colors.

in the 1870s, and these were the varieties that could be judged and entered into competition for awards at the All-American Turkey Show.

Because turkey varieties are produced from sports, it is difficult to keep the variety or strain pure. When selecting birds for breeding in an attempt to improve color, breeders must exercise care, because there is always the tendency for the variety to have some of the coloration of the variety from which it was developed or even for the variety to revert to the wild state. This is especially the case with the Bronze, because its markings are more similar to those of the eastern wild turkey than are those of any of the other varieties of domesticated turkeys.

The difficulty of preserving the genetic diversity, feather pattern, and color of turkey varieties is noted in a number of publications, the Livestock Conservancy's *Introduction to the Heritage Breeds* among them. The difficulty is compounded because of the way that the distinctive feather patterns and colors of each turkey variety are described in the *Standard of Perfection* and other publications.

If beauty is in the eye of the beholder, how should breeders and judges distinguish the "rich, brilliant, copperish bronze" that is not "too bronzy" of the Bronze variety; "the rich, metallic black" of the Narragansett; the "lustrous, greenish-black" of the Black; the "slaty or ashy-blue" of the Slate; or the "deep, brownish-red" of the Bourbon Red?[7]

All varieties of turkeys could be exhibited at the All-American Turkey Show, but those eligible for awards and prizes were those that had been admitted to the *Standard of Perfection*: Bronze, Narraganset, White Holland, Bourbon Red, Black, and Slate. Whether in their viewing pens on the showroom floor or on the judges' tables, there was little besides feather pattern and color to distinguish the birds of one variety from those of another.

Flock owners breeding to meet the exacting standards called for in the *Standard of Perfection* could be excused, therefore, for insisting that forty of the one hundred points in judging be awarded for color. In this, they had the staunch support of George Hackett. Among the poultry in-

[7] Even a hint of green in the bronze of the Bronze is a disqualification, because the bronze is too close to that of the wild turkey. Any slate-color in the black of the Narragansett, too much blue in the ashy-blue of the Slate, or even a hint of orange-red in the brownish-red of the Bourbon Red are enough to disqualify the bird in competition. Small wonder then that turkey experts refer to a variety's feather pattern and color as being "very challenging," "difficult," "almost impossible," or "very difficult" to perfect.

dustry's premier judges and the country's acknowledged authority on turkeys, Hackett was the heart and soul of the All-American Turkey Show.

How Hackett felt about turkeys can be expressed in one word: passionate. Turkeys, those "most majestic of all domesticated fowl," were Hackett's passion. To read his description of a turkey's feathers is to be reminded of how a man might describe the beauty of his lover. Hackett believed that it was an indignity to display turkeys without their plumage, as they were in the All-American Turkey Show's dressed exhibits. For him, feathers made the bird.

When considering the standards to which the *Standard of Perfection* held each turkey variety, it would be well to note the origin and characteristics of each. Originating in the United States, the foundation stock of the Bronze was likely a combination of turkeys imported from Europe and the eastern wild turkey. Running free on farms and in forests, domesticated and wild turkeys turkeys mingled and mated. The Bronze's colors are closer to those of the wild turkey than those of other varieties. Large, hardy, and attractively colored, the Bronze was included in the *Standard of Excellence [Perfection]*, published in 1874.

The Narragansett turkey is named for Narraganset Bay in Rhode Island where it was likely developed. The variety was probably a cross between the Norfolk Black imported from England and the eastern wild turkey. The bird's colors are similar to those of the Bronze, but steel gray or dull black instead of coppery bronze. White wing bars are the result of a genetic mutation that removed the bronze coloration. Known for its calm disposition, maternal abilities, early maturation, egg production, and meat quality, the Narragansett was included in the *Standard of Excellence*, published in 1874. The bird's color is a mixture of metallic black and white, and, noted a poultry writer, it is "an almost impossible one" to perfect.

White turkeys, perhaps part of the wild Mexican stock, were taken to Europe where they were particularly favored in Austria and Holland. Imported to America by colonists, what came to be known as the White Holland was included in the 1874 *Standard of Excellence*. Its relation to European varieties is unknown, but the bird may have been a genetic mutation among Black turkeys. The turkey's pure white color is another that is "difficult to perfect."

Originating in Bourbon County in Kentucky's Bluegrass region and developed by J. F. Barbee, the Bourbon Red was admitted to the *Standard of Perfection* in 1910. Bronze, Buff, and White Hollands were its foundation stock. Handsome birds, the Bourbon Reds have brownish to dark red plumage with white flight and tail feathers. Tail feathers also have soft red bars crossing them near their ends. The tom's breast and neck feathers are chestnut mahogany. Because there is the tendency for Bourbon Reds to have reddish-orange feathers, which is a disqualification in competition, the deep mahogany red color is another that is difficult to perfect.

Included in the *Standard of Excellence* in 1874, the Black variety originated in Europe. European Black turkeys likely derived from the Mexican domesticated turkeys imported into Europe in the sixteenth century that later became known as the Norfolk Black. Black turkeys were standardized in East Anglia, England, before being imported into North America during colonial times. Known for their early maturation and rapid growth, the birds' uniformly black feathers have a greenish-black sheen.

Originating in the United States and recognized by the American Poultry Association in 1874, the Slate or Blue Slate variety is named for its color, ashy blue, with or without a few black flecks. It may also be black or solid blue, however. The blue, another defined as difficult color to perfect, results from either of two different gene mutations (one recessive and the other incompletely dominant). This genetic complexity makes the variety difficult to characterize and difficult to breed, because some color variations breed true and others do not. The color of Slate turkeys can be any number of shades between pure black and white, but these are defects, as are white or rusty brown markings.

Once thought to be a cross of the Black and White varieties, the Slate is a "legitimate mutation," according to the Livestock Conservancy. The variety is one of the most colorful of all Standard turkeys, and a flock of them is "a sight to see." The variety never enjoyed commercial popularity, and fanciers raised them primarily for exhibition.[8]

---

[8] To inform the uninitiated and to remind turkey breeders of how judges assessed and awarded points, the *Standard of Perfection*, the *American Turkey Journal*, and other publications provided helpful descriptions of a turkey's features that figured in judging and definitions of the terms used in evaluations. See Appendix A: Common Terms and Definitions Used in Judging Turkey Features for a sampling of those descriptions and definitions.

When it became apparent that money could be made on turkeys and most of the Standard varieties had been developed, turkey raisers—to foster interest in the varieties—organized clubs, such as the All-American Bronze Turkey Club, the National White Holland Turkey Club, and the National Black Turkey Club. Mrs. Minnie M. B. Brown of Appletown City, Missouri, and Mrs. G. W. Price of Belmont, Ohio, organized the National Bourbon Red Turkey Club in 1907, three years before the variety was admitted to the *Standard of Perfection*.

The listing of turkey varieties in the *Standard of Perfection* and the formation of clubs encouraged the development of turkey shows, the All-American Turkey Show in Grand Forks, North Dakota, among the first and the largest. The shows provided breeders with an opportunity to compete with other breeders to determine whose birds came closest to the descriptions of proper weight, body conformation, symmetry, color, and intricate feather patterns in the *Standard of Perfection*.[9]

In the shows, all turkeys, no matter the variety, were held to the description of "Turkeys" in the *Standard of Perfection*. The words might appear to be straight forward, but judges had to use their discretion when interpreting and applying phrases such as males should have "a very stately appearance" and "the eyes should possess a bold expression."[10]

All breeders of the Standard varieties of turkeys who intended to compete for awards and prizes at the All-American Turkey Show had a common goal. They sought to improve their flocks by producing birds that were as close to perfect as was humanly possible. What constituted "perfection" for Bronze, Narragansett, White Holland, Bourbon Red, Black, and Slate turkeys was explained in the *Standard of Perfection*, and, because the varieties differed from one another primarily in color and weight, these were the standards that were emphasized.

---

[9] Body conformation, symmetry, and quality were supposed to be important, but feather patterns and color were considered to be even more important. The shows, noted M. C. Small in "Turkeys," also provided an opportunity for flock owners to improve their knowledge of breeding, raising, and marketing turkeys. Equally important, the shows allowed breeders to make or renew acquaintances, to socialize, and to enjoy the recognition accorded them by local business owners and members of the public.

[10] Under the heading Shape of Male and Female, toms and hens had to conform to the specifications stated in the *Standard of Perfection*. See Appendix B: Male and Female Turkey Shape Specifications for examples.

At the risk of being tedious, but to indicate the difficulties that breeders of Bronze turkeys faced when selecting breeding stock in their efforts to achieve that "just right shade of bronze" specified in the *Standard of Perfection*, and also to indicate the difficulties judges faced when deciding which bird on the judging table came closest to that just right shade, it would be helpful to know how the *Standard of Perfection* described the color on each part of a Bronze turkey.[11]

The *Standard of Perfection* was almost monotonous in its use of the words "rich, brilliant, copperish bronze" when describing the color of the Bronze turkey. But, arguments without end ensued when attempting to explain when "rich" was "too rich," or the "brilliant" was "too brilliant," or the copperish was "not copperish" enough, or when the bronze was "too bronzy."

The Livestock Conservancy described it as "a very challenging color pattern to perfect." Turkey expert Andrew F. Smith believed that the color was similar to that of the eastern wild turkey, "a light metallic bronze" or "bronze-green." Herbert Myrick, in *Turkeys and How to Grow Them,* described the color as "bronzy brown, with luster in the sunlight," but any "cinnamon" in the color denoted wild blood, for which points were lost in judging. Others used the word "metallic" when explaining "copperish bronze."

Hackett, renowned poultry judge, provided the consummate explanation of the phrase: rich, brilliant copperish bronze. His explanation, which should have laid the issue to rest, bears quoting. "It is always difficult," he editorialized, "to adequately describe delicate shades of color, or to differentiate between the varying shades, so as to give a uniformity of conception as to what is meant. To acquire correct, or generally accepted, ideas can be done only by comparison and with specimens at hand."[12]

Hackett continued, "For this there is nothing equal to the show room, the judging tables and the discussions which ensue." However, there is a generally accepted shade or quality of bronze color, the result of the consensus of opinion of leading Bronze breeders. The *Standard* describes this as "rich, brilliant, copperish" bronze, and as applied to the backs of both

---

[11] See Appendix C: Summary of Standards for Bronze Turkeys for color specifications.

[12] *American Turkey Journal*, April 1937.

male and female, specifies, "the more bronze the better." It is also desirable that the Bronze in all sections be as deep as possible.[13]

When compared to that of the Bronze turkey, the description in the *Standard of Perfection* of each of the other varieties is brevity itself, differing primarily in Standard Weights and Color. For example, Narragansett turkeys were to have light steel-gray wings; rich, metallic black backs; and black saddles. Tails were to be dull black, with each feather "regularly penciled" in parallel lines of light brown, ending in a broad band of metallic black, edged with steel-gray approaching white. Breasts had to be "metallic black . . . with each feather ending in a broad, light steel-gray band edged with black." All aspects were to be free from bronze cast.

Disqualifications for the Narragansett included wings with one or more primary or secondary feathers clear black or brown, or absence of white or gray bars more than one-half the length of primaries. Defects that were to be "severely cut" were the lack of one or more center tail feathers and white or gray bars, other than the terminating wide edging of steel-gray approaching white, showing on the base of the main tail feathers.

White Holland turkeys, both male and female, had to be "pure white," except for the beard, which had to be "deep black." Disqualifications included feathers other than white in any part of the plumage and the color of shanks and toes a color other than white or "pinkish white."

The color of Bourbon Red male turkeys was difficult to describe. Wings had to be a "rich, dark chestnut mahogany" with pure white primaries and secondaries. The back was a "deep, brownish red," and the tail feathers had to be "pure white, with a dimly outlined bar of soft red crossing each main-tail feather near the end." The breast was a "rich, dark chestnut mahogany," with feathers "having a very narrow edging of lustrous black."

Feathers on the body had to be "deep, brownish red," and the fluff had to be "a lighter shade of the same color." The color of the thighs had to be "dark chestnut mahogany" and the toes "reddish pink." The "under-color" of all parts was "red, shading to light salmon at the base." The

[13] For a continued and more in-depth description of Bronze turkey coloring, see the latter part of Appendix C: Summary of Standards for Bronze Turkeys.

color of hens was similar to that of toms, except the breast had to have a "narrow, thread-like edging of white."

Disqualifications included "more than one-third any other color than white showing in either primaries, secondaries, or main-tail feathers." Defects that had to be "cut severely" included "mealy backs, bibs, black barring on body plumage, more than one-third of wing feathers showing pronounced rustiness, black edging on body plumage of females," and "crooked breast-bones." Black turkeys were as briefly described as were the White Hollands. The color of their plumage had to be a "lustrous, greenish black throughout" and the "under-color" had to be a "dull black." Beards were also black. Disqualifications included feathers "other than black in any part of the plumage." A "slight brown tinge" in the tail feathers of hens was not a disqualification.

The requirement as stated in the *Standard of Perfection* that the color of the Slate turkey had to be "slaty blue" was neither more helpful nor more definitive than the requirement that the color of the Bronze variety had to be a "rich, brilliant copperish bronze." The slaty blue could "sometimes" be "dotted with black, but the freer from dots the better." Beards had to be black, and shanks and toes pink or deep pink. Disqualifications included feathers "other than slaty or ashy blue, which may be dotted with black, in any part of the plumage."

Why, when turkey varieties were differentiated primarily by their color; why, when this color was "very challenging," "difficult," "almost impossible," or "very difficult" to perfect; and, why, when Hackett noted that the colors could be traced back to the wild turkeys from which all turkey varieties had been developed and, that unless flock owners exercised good management and care when breeding, the varieties "would likely revert to their wild state;" did flock owners, breeding season after breeding season, persist in their efforts to achieve the "perfection of color" demanded in the *Standard of Perfection*?

George W. Hackett knew why. He explained it when he described "the correct shade of Bronze" of the Bronze turkey. For the word "bronze," insert the predominant color of each of the six varieties of turkeys described in the *Standard of Perfection*, the varieties vying for awards and prizes at the All-American Turkey Show.

"The correct shade of [name of color]," Hackett maintained, "when rightly understood, cannot be forgotten by anyone having a keen color sense, nor can a misinterpretation of the same cause the real [name of breed] fancier to vary his course one iota in his effort to secure that just right shade of [name of color] that is always just a little ahead of human possibilities to obtain."

Standard Bronze Tom
Photo courtesy of Porter's Rare Heritage Turkeys, www.porterturkeys.com.

Standard Narragansett Tom
Photo courtesy of Porter's Rare Heritage Turkeys, www.porterturkeys.com.

Standard Royal Palm Tom
Photo courtesy of Porter's Rare Heritage Turkeys, www.porterturkeys.com.

Standard White Holland Tom
Image used under license from MargoK/shutterstock.com.

Standard Bourbon Red Tom

Image used with permission by Mtshad - Own work, CC BY-SA 3.0, https://commons.wikimedia.org/w/index.php?curid=24543995.

Standard Black Tom
Image used under license from Michele and Tom Grimm/ Alamy Stock/ ID: CRW51C.

Eastern Wild Turkey
Image used under license from Jim Cumming/ Alamy Stock/ ID: 2B86A2M.

Smoke-Phase Turkey
Image used under license from Gina Kelly/Alamy Stock/ ID: 2BCWWBN.

Standard Slate Tom and Hens
Image used under license from Ileana_bt/ shutterstock.com.

The Home Stretch is SOON!

Before you know it your birds will be on the home stretch! Proper feeding is highly important all the time but it becomes increasingly so from now on. Soon the building of large sturdy frames will be completed and then you'll want a plentiful supply of richly flavored turkey meat to top it off and earn more pounds and highest grades.

VITAMIN E ASSURED!

Vitamin E is the essential ingredient in all diets, human, animal and fowl, which assures reproductivity. In turkey feeds it has been supplied by Wheat Germ, which has not been thoroughly satisfactory because of the tendency to become rancid and lose its power. A new discovery

WHEAT GERM OIL

puts this vitally important vitamin into feeds to stay there, without loss of power or deterioration. Always alert for definite improvements, we now use Wheat Germ Oil exclusively in DAKOTA MAID Turkey Mashes, assuring you of a positive and dependable supply of Vitamin E and assuring high fertility and hatchability in your flocks.

DAKOTA MAID

TURKEY MASHES

(In both mash and pellet form)

will bring your birds through to market with the most pounds of quality turkey meat at the lowest possible cost. This is because DAKOTA MAID Turkey Mashes contain every NECESSARY ingredient needed by the bird, with nothing superfluous included to obtain bulk or reduce manufacturing cost. FEEDING FOR PROFIT is sure when you use these splendid feeds

DAKOTA MAID TURKEY STARTER
DAKOTA MAID TURKEY GROWER
DAKOTA MAID LAYING MASH
DAKOTA MAID TURKEY FINISHER
Ask your dealer or write us.

STATE MILL & ELEVATOR

COMMERCIAL FEEDS DIV.

Grand Forks North Dakota

Published Monthly by the PAGE PRINTING CO. at 105 South 3rd Street, Grand Forks, N. D. Subscription 50c per year. Entered as Second-Class Matter May 4, 1932, at the Post Office at Grand Forks, N. D., Under the Act of March 3, 1879.

In 1836, the North Dakota State Mill & Elevator introduced a line of feeds formulated especially for turkeys. *American Turkey Journal*, August 1937.

This Cable turkey vase was awarded to Wallace Jerome of Barron, Wisconsin, at the All-American Turkey Show in 1934 as a sweepstakes prize. Image courtesy of Mary Ella Jerome, Barron, Wisconsin.

Hen Club members held their annual banquet in the Hotel Ryan, located diagonally across from the Hotel Dacotah at Third and First Avenue North. Image courtesy Grand Forks County Historical Society.

American
TURKEY
JOURNAL

Grand Champion at the 1942 All-American. This splendid Yearling Tom Bronze was exhibited by Mr. and Mrs. John Allen, Radium, Minn., from their Standard Bronze flock of many prize winners. The Allen's have won top honors in practically every Northwestern show and the high quality of their strain is recognized throughout the country.

VOL. X
NO. 11

FEBRUARY
1942

PER YEAR
50c

Cover of *American Turkey Journal*, February 1942.

Editor George W. Hackett often used scenes of Hen Club picnics to grace the covers of *American Turkey Journal.* *American Turkey Journal*, June 1936.

One long table was a feature of the Hen Club picnics. Note how well the people are dressed for the occasion. *American Turkey Journal*, August 1940.

CHAPTER 8

A Really Big (Turkey) Shew

(with apologies to Ed Sullivan)

THE FIRST ALL-AMERICAN Turkey Show, "the first exclusive turkey show ever held in the world," opened on a wintry day in February 1924 in Grand Forks, North Dakota, the "center of America's greatest turkey-raising territory." Uncertain about whether it would attract interest and draw visitors, promoters moved the show from the City Auditorium at North Fifth Street and Fifth Avenue North to the Doyle Motor Company at 210 DeMers Avenue. The Doyle Motor Company was located a half block from the intersection of DeMers Avenue and Third Street, the heart of the city's business district, within easy walking distance to hotels, restaurants, and movie theaters.

Promoters' concerns were unfounded. Although it was the "first of its kind ever held," according to follow-up reports the All-American Turkey Show received "considerable attention and support through an unusually large section of the country," and the "large attendance" from the city, country, and nearby towns "taxed the capacity" of the Doyle Motor Company building. A total of 249 turkeys were entered in the show, 110 more than "any turkey show ever held in the United States."[1]

The shows were repeated, and they continued to be successful. By the sixth All-American Turkey Show, held in 1929 on the eve of the Great Depression, promotors, business leaders, exhibitors, judges, county agents, poultry experts, observers, and news reporters—all had exhausted their

[1] The 164 birds in the Bronze class alone outnumbered the turkeys entered in any of the country's major poultry shows. "Success of Turkey Show Warrants Its Repetition in 1925" was the headline of the February 13, 1924, edition of the *Grand Forks Herald*, an opinion shared by E. R. Montgomery, Secretary of the North Dakota State Fair Association, who was in charge of local arrangements for the show. (Source refers to quotes in paragraphs one and two.)

The building with the impressive façade—to the immediate right in the photo—housed the Doyle Motor Company, site of the first All-American Turkey Show, held in February 1924. Photo courtesy Marsha Gunderson, Grand Forks, North Dakota.

store of superlatives in their attempts to describe the show and its success. It was "the greatest turkey show in the world." The All-American Turkey Show was "conceded to be the largest turkey show ever shown on the American continent." The All-American Turkey Show was "the final forum in the selection of the country's best turkeys" and "the leading show in the United States." It was proclaimed "The Premier Turkey Show" in the country and "the Turkey Classic of America." The "World's Finest Turkey Show," the All-American was "the final court of appeal of all Turkeydom." Always the largest, and for a time the only, exclusive turkey show, the All-American Turkey Show served as the model for the turkey shows held all over the country in the years preceding World War II.

The first All-American Turkey Show in 1924 was "splendid and the largest ever held in the Northwest and probably in the entire country," according to Professor O. A. Stevens of the Agricultural College in Fargo,

North Dakota. W. E. Stanfield of St. Paul, Minnesota, on the editorial staff of the *Dakota Farmer* and *Northwest Farmstead* and a poultry judge for twenty-six years, judged the Bourbon Red class at the first All-American Turkey Show. He described the show as "the greatest" he had ever seen.

For six years, Stanfield had been a poultry specialist with the Bushnell Publishing Company, devoted to developing the poultry industry in the Northwest and judging turkeys at the country's premier shows. The 1924 All-American Turkey Show, he said, was "wonderful" and "the greatest show from the standpoint of a turkey show that has ever been brought out in the United States."

The second All-American Turkey Show that opened on February 3, 1925, was the largest of its kind in the United States with a total of 312 turkeys entered in the show by fifty-five exhibitors from twelve states and several Canadian provinces. The 210 turkeys entered in the Bronze class tripled the number of entries in any other Bronze show across the nation." By comparison, at the 1924 show in New York's Madison Square Garden, three exhibitors entered thirty-two Bronze turkeys. At the Boston Fanciers Show in 1924, there were fewer than sixty entries of all turkey varieties and only ten exhibitors. At the Chicago Coliseum Show in 1923, thirty-nine Bronze turkeys were entered by ten exhibitors.

The 1925 All-American Turkey Show was "the finest show I have ever seen," said Dr. J. D. Eastvold of Spooner, Wisconsin, "and I have been in attendance at the Madison Square Garden show as well as at the Coliseum in Chicago, which are considered the best in the world of turkey shows." Eastvold, whose field was tuberculosis in turkeys, had long been associated with the Wisconsin State Department of Agriculture. R. A. Rasmussen, Wisconsin's Washburn County agent, regretted that "all county agents in the country" were not attending the 1925 All-American Turkey Show "to take advantage of the privilege."

Rasmussen was not leaving the show, he insisted, until he had "gotten a great deal of information from these men who have such wonderful turkeys." There were "many wonderful birds" entered in the show, he said, some for which owners had refused offers of $1,000, because the turkeys were desirable breeding stock. "I have judged a lot of shows in this country," he said, "but never saw anything like this before; there are many more birds here than there were at the world's fair in San Francisco in 1915."

The third All-American Turkey Show, "the world's biggest event of this kind" and "the greatest turkey show in the world," was held February 1–5, 1926. "Lauded By Poultry Men Who Have Seen Them All," read a headline in the *Grand Forks Herald*. The show featured 436 turkeys, "some of the most aristocratic birds on the continent," more turkeys than had been exhibited at all the other leading shows combined.

For three weeks before the 1926 All-American Turkey Show opened, notices and articles in the *Grand Forks Herald* had drawn attention to it. "Show Your Turkeys at the All-American," breeders were advised, "Where a Win Means More Than at Any Other Show." "Six of America's Foremost Judges" would judge the entries in this, "The Largest Exclusive Turkey Show in the World," where judging was done "So Everyone Can See."

"The general opinion" was that the fourth All-American Turkey Show that opened on February 1, 1927, was "by far the most successful ever held, in addition to being the largest." O. A. Stevens, of the North Dakota Agricultural College (now North Dakota State University) in Fargo and head of poultry development work in the state, had attended large turkey shows all over the country. "Never before," he said, had he witnessed "the enthusiasm and the spirit that prevails at the All-American Show in Grand Forks." The quality of the turkeys exhibited, he said, was "without question superior to the best ever assembled in a show anywhere." Those 454 turkeys, "representing the best in the industry and from all sections of the country," would compete for a share of the $2,000 premium list.

Exhibitors and visitors braved a blizzard and frigid temperatures to attend the fifth All-American Turkey Show held January 31–February 5, 1928. Hailed as "the greatest exhibit of turkeys ever staged in the world," the show drew 546 turkeys and 134 exhibitors (two-thirds of them women) from every important turkey raising district in North America, from Texas to Saskatchewan and from New York to California. There were over three thousand paid admissions, and 635 people attended the educational lectures. On February 5, the last day of the show, the *Grand Forks Herald* headline read, "All-American Turkey Show Goes Into History as World's Greatest." A lead article the same day was headed, "This was the largest and finest display of high-quality turkeys ever cooped together at one show anywhere in the world."

Ever bigger and better, the sixth annual All-American Turkey Show opened on Monday, January 28, 1929. On the previous day, the temperature was -30°F, and temperatures for almost the entire month of January had been below zero. During the week of the show, temperatures hovered around -12°F, and the mercury rose only to 5°F on the last day of the show, February 1.

The frigid temperatures seemed not to bother the 512 turkeys entered in the show by exhibitors from ten states and two Canadian provinces. The *Grand Forks Herald* Sunday editorial noted that "never before in the history of the world [had] so many fine turkeys been assembled in one place as those that were exhibited at the All-American Turkey show which closed Friday."

The *Grand Forks Herald* front page proclaimed on February 3, 1929, "Officials Declare Turkey Show Was Best World Has Seen" and "Hackett Declares Exposition Has Outdone All Former Shows." George W. Hackett, now manager of the All-American Turkey Show, said of the 1929 show that "there is no question but that it has outdone all past events of the organization by a wide margin."

Begging explanation is why superlatives had to be used to describe the All-American Turkey Shows. They were also used in explanation of the success of the All-American Turkey Shows and in explanations of why they were held annually, even in the worst years of the Great Depression, until 1942, when they were suspended, temporarily it was assumed, during World War II.

Superlatives were also used to describe the interest generated by the All-American Turkey Shows and in explaining why turkey raisers in North Dakota and the Northwest believed it necessary to enter their best birds into competition at an All-American Turkey Show, considered to be the final word on the selection of the country's best turkeys.

Also begging explanation is why, when people in North Dakota and the Northwest were already raising turkeys, did the All-American Turkey Show have its inception? Why was the first show held in 1924, and why was the first All-American Turkey Show held in Grand Forks, North Dakota?

The short answer is, yes, people were raising and selling turkeys, but they wanted something more. It was as if the All-American Turkey Show opened the floodgates on this "something more" that had long been dammed up and pressing for release.

W. E. Stanfield, editor, judge, and poultry specialist, spoke to this "something more" when commenting on the success of the first All-American Turkey Show in 1924. "This," he said,

> is the greatest show from the standpoint of a turkey show that has ever been brought out in the United States. The success of it lies in the fact that it has brought out hundreds of farmer turkey breeders. You can talk your head off, but it doesn't get as far with people as if they can see what you are talking about. And so it is with turkeys. If they can see what is good stuff, they are going to come nearer to raising first class stuff. And this show presents some of the best quality the country has. There are hundred dollar birds shipped in here from other states that are not in the winnings. That shows that somebody around here is already raising pretty high class stock. There are 224 turkeys on exhibit here, and they are some of the finest that ever came into a show room.

From the vantage point of the conclusion of the sixth All-American Turkey Show, the February 3, 1929, *Grand Forks Herald* Sunday editorial also spoke to this "something more." "Now that the show is over," read the editorial, "the exaltation of the winners has subsided somewhat, and the disappointment of the losers is not so keen as at the close of the judging, a fair estimate of the benefits of the show can be made." The sixth All-American Turkey Show had "brought out a fine display of birds," and it had given turkey raisers "their annual opportunity to brush up on points of the industry."

"No greater group of turkey experts" had ever been assembled in the Northwest, the editorial writer noted, than those who had gathered in Grand Forks for the All-American Turkey Show. From these experts, turkey raisers had received advice on how to improve the quality of their flocks. Comparing their birds with those shown by "nationally known authorities on turkeys," turkey raisers had also seen the "glaringly apparent" imperfections in their own turkeys.

But, continued the editorial, "the main advantage" of the All-American Turkey Show had to be measured in the "dollars and cents" it would "bring to the Northwest." Turkey raisers had noted that orders for breeding stock had "fairly poured into the hands" of the owners of prize-winning turkeys. Noting as well that a premium was paid for turkeys "fitted in the North," turkey raisers in North Dakota and throughout the Northwest would be encouraged to increase their incomes by producing better turkeys and selling the best as breeding stock.

"There," concluded the editorial, "is the main value of the All-American Turkey Show, the dollars and cents in additional payments for turkeys that will be forthcoming next year and for many years to come."

Another "something more" that the All-American Turkey Shows provided turkey raisers can be defined only unsatisfactorily, if at all, as fellowship, camaraderie, or togetherness. Inexpressible, this something more had to be experienced by attending the All-American Turkey Show, by mingling with "turkey folk" on the showroom floor amid the hundreds of prize turkeys in their cages, by displaying turkeys and competing good-naturedly for prizes and awards, and by attending the educational sessions. This something more was palpable; it was a hallmark of the All-American Turkey Show, and it was noted by visitors—especially by those from other parts of the country.

J. D. Eastvold, long associated with the Wisconsin State Department of Agriculture, had attended turkey shows in Madison Square Garden in New York and at the Coliseum in Chicago. Never at these shows had he seen, as he observed, "the sportsmanship" that was so much in evidence at the 1925 All-American Turkey Show. What he described as "the spirit of the Westerner" characterized the "average breeder" who had brought turkeys to the show to compete with those of well-known and established breeders from both coasts and points in between.

The camaraderie and cooperation apparent in the 1926 All-American Turkey Show "was considered the most promising" that had ever "been manifested at any turkey show." At the 1928 All-American Turkey

Show, Manville A. Johnson, vice-president of the Northwest Turkey Association, praised "the community spirit and cooperation" that made the shows so successful.

Participant W. E. Stanfield of St. Paul, Minnesota, noted that the 1929 All-American Turkey Show had been successful for two reasons: the quality of the turkeys and "the good, honest, clean-cut fun" that was "always present" at the show. There was, he said, "nothing like it in the U.S.A. and people all over the continent were talking about it." Walter Burton of Dallas, Texas, expected to judge fifty to sixty turkeys at the 1929 All-American Turkey Show, but he was "agreeably surprised" when he saw the large numbers of turkeys competing for prizes and awards. He was even more impressed when the show opened on the bitterly cold morning of January 28.

"The early opening of the show," he said, resembled more "a family party than an exposition." It was "more of a get-together meeting than anything." "Veterans" of previous shows were on hand "to exchange reminiscences of previous shows and to greet new exhibitors." To demonstrate the amiable spirit of cooperation that characterized the All-American Turkey Show, officials began offering a ten-pound pail of clover honey, processed in the local plant, to the most dissatisfied exhibitor at a show. No one ever claimed the honey.

The turkey banquet was the highlight of the All-American Turkey Show, always described as the show's "big event" or "feature event." And it was at the banquets that "the sportsmanship," the "spirit of the Westerner," and "the good, honest, clean-cut fun" were most in evidence. The turkey banquets, always with turkey themes and favors, were held one evening of the show and always at the Hotel Dacotah.

So popular were the banquets that people were encouraged to reserve places well in advance of the opening of the show, and, often, many who wanted to attend were disappointed because the banquet room could accommodate only 375 guests. W. E. Stanfield had been associated with the turkey industry for more than twenty-five years, and he frequently attended the premier turkey shows. The 1928 turkey banquet, he said, was the "largest ever held in connection with that industry."

The evening's program included short addresses by breeders and turkey experts, humorous anecdotes about raising and showing turkeys, turkey calling contests, musical selections, presentations of awards, and the dinner. Then, the highpoint of the evening: banqueters, indulging themselves in a something more that the All-American Turkey Show provided, proceeded to dance the night away.

It remains to explain why the All-American Turkey Show had its inception in the 1920s and why it was held in Grand Forks, North Dakota. The Twenties have earned the appellations "Roaring" and "Golden." Author F. Scott Fitzgerald called the decade "the most expensive orgy in history," and the historicity of his novel *The Great Gatsby* is that it is about money. *The Great Gatsby*, among the classics of twentieth-century literature, brings to life America's Jazz Age, when, as the *New York Times* put it, "gin was the national drink and sex was the national obsession."

The urbanized, industrialized areas of America enjoyed prosperity during the Twenties, but they were unkind to North Dakota and to much of the agricultural Northwest. Here, they were neither "Roaring" nor "Golden." And, given the area's rural character, the low incomes, the lack of electricity, and the paucity of water, there might have been precious few bathtubs in which to make gin, and the obsession was to avoid losing farms to foreclosure.

Elwyn B. Robinson in his *History of North Dakota* described the Twenties as a decade of agricultural depression for the state. Much of the reason for the farmers' plight, Robinson noted, was because of their unwillingness or inability to diversify production. North Dakota is a cool, sub-humid grassland lying in the center of the continent. Climate and topography favored its becoming a wheat-producing state, the land of "Dakota Red" and "No. 1 hard." Wheat brought higher returns than any other crop and, responding readily to labor-saving machinery, its production lent itself to one-crop farming. Wheat became a "mania," and the North Dakota prairie became "one vast ocean of wheat."

Robert P. and Wynona H. Wilkins, in their *North Dakota: A Bicentennial History*, titled one of their chapters "Wheat is King" and noted in the chapter that on North Dakota farms, "wheat was the money maker." That wheat *was* the North Dakota crop is reflected in the index to Robinson's insightful and persuasive history of the state. He has more than thirty entries for wheat, four of them referring to wheat as a "unifying force," and no entries for any other crop. According to Robinson, the state's history, its politics, its institutions, the "Dakota" in the state's name, and his Too-Much Mistake theme—all are written in wheat.

As Eric Sevareid, the nationally-known news commentator from Velva, North Dakota, put it, wheat even helped shape the character of the state's residents. "So far as Velva was concerned," he wrote, "wheat was the sole source and meaning of our lives. . . . We were never its masters, but too frequently its victims. . . . It was rarely long outside a conversation."

It was unwise, of course, for North Dakota farmers to rely on a single crop for their incomes. Concentrating on one crop raised the possibility of a season's income being wiped out by disease, insects, drought, frost, or other natural disasters. Raising wheat on the same land year after year contributed to weed problems and reduced soil fertility.

On the eve of the opening of the first All-American Turkey Show in 1924, yields and quality of wheat were declining, and the Red River Valley—touted as The Nile Valley of the North American Continent—was losing its reputation as the land of the No. 1 hard and being referred to derisively as "the land of No. 2 and 3." Wheat monoculture also continued to keep North Dakota farmers at the mercy of the railroads, terminal elevators, and speculators.

Nevertheless, wrote Robinson, wheat, "still king," produced more than half the state's cash farm income in 1925. This income, divided among 395,000 farm people and almost 78,000 farms, "was simply not large enough." In six of the twelve years between 1919 through 1930, wheat crops were poor and rainfall below average. The price of wheat declined from $2.35 per bushel in 1919 to $1.01 in 1921, before falling to 60¢ in 1930. The price of beef cattle, $8.10 per hundred pounds in 1919, dropped to $4.45 in 1921. The average gross income per North Dakota farm of $2,500 to $3,200 during the 1920s represented, Robinson wrote, "an unspectacular but real farm depression." In 1929, the per capita personal income in North Dakota was $375. For the United States it was $703.

Low farm incomes exacerbated other difficulties. The value of all farm property, according to Robinson, declined by one-third, from

$1,760,000,000 in 1920 to $1,191,000,000 in 1925, "a shattering loss" of over a half billion dollars. Taxes on farm real estate remained high, and heavy taxes and the decline in value of farm property added to farmers' debt burden. In 1920, 70 percent of those working their own farms had mortgage debts. Because taxes on farm real estate remained high and the value of farm property declined, many farmers lost title to their land.

One of every four North Dakota farmers was a tenant in 1920; in 1930, one of every three was a tenant. Robinson quoted from a letter written by William Lemke, whom he described as "an intense, bitter, tenacious fighter for the plain people against the hated interests." Lemke was an attorney and a member of the executive committee of the Nonpartisan League. "The farmers," wrote Lemke, to a friend in 1927, "have never been as hard up as at present. I have more work than I can do trying to save their homes in a legal way and in defending them in many unfortunate lawsuits, due to poverty or rather deflation. It used to be that they were able to pay my carfare and expenses, but it is getting so now they offer me post-dated checks for even actual expenses."

Low farm incomes, partly the result of relying only on wheat, also affected small rural banks. North Dakota had 898 banks in 1920, three for every incorporated place and more in proportion to its population than any other state. Before 1920, good wheat prices, rising land values, and "an almost unlimited" confidence in the state's future, encouraged banks to extend loans to farmers "far beyond the bounds of prudence," an average of 120 percent of deposits, some even higher.

"Dangerously overcommitted," in Robinson's words, "to an unstable, local agricultural economy," banks attempted to collect their loans to farmers when deposits declined. When many of the loans proved uncollectable, the banks failed. "The collapse of North Dakota agriculture in the early 1920s," Robinson concluded, led to the "wholesale collapse" of small, rural banks and the loss of millions of dollars of "hard-earned savings."

Nowhere was there a greater need for farmers to increase their incomes and the value of their farms by diversifying production than in North Dakota. Evangelists, preaching the gospel of diversification under the banner of Better Farming, attempted to get the state's farmers to break off their love affair with wheat and take up other crops, dairying, and beef production. To instruct farmers on how they could make their operations more profitable, the Agricultural College held its first farmers' institute in 1894, and by 1911 there were more than one hundred institutes annually.

The Agricultural College organized the annual Tri-State Grain and Stock Growers convention in 1898 as an agency to educate farmers, and in 1914 it added an extension department. The North Dakota Better Farming Association, organized in 1910, promoted corn production. The association's director, Thomas Cooper, established the first extensive system of county agents and extension workers to advise farm families on how they could increase their incomes and raise their standard of living, and, in 1913, the North Dakota Legislature passed a county agent law.

The work of the North Dakota Better Farming Association was merged with that of the Agricultural College's experiment station in 1914, and Cooper was hired to direct county agent operations. Farm papers such as the *Dakota Farmer* published exegeses of the diversification gospel, and the Great Northern's James J. Hill gave away registered Shorthorn bulls along the line of his railroad in an effort to encourage livestock production.

Henry L. Bolley and John H. Shepperd of the Agricultural College, together with the Oscar Will Seedhouse in Bismarck and the Northrup King Company of Minneapolis, worked to develop crops adapted to North Dakota that could replace wheat. Diversification was also encouraged by the Agricultural Credit Corporation, an agency that extended loans to North Dakota farmers for the purchase of cattle, sheep, and hogs.

In 1924, evangelists of the All-American Turkey Show joined the diversification crusade and began preaching the gospel of diversification. Every speaker at the first All-American Turkey banquet, held on the evening of February 13, 1924, spoke on the theme of diversifying farming operations by raising turkeys. "Diversify and prevent the possibility of a total crop failure that so often overtakes the one-crop farmer," advised state senator Jake Eastgate of Larimore, North Dakota. "The farmer who diversified is almost sure of having part of his crops, if not all of them, bring good returns and the danger of total crop failure from one crop is removed."

"Raising turkeys," Eastgate declared, "seems to be one of the best ways to diversify." So, he advised his listeners, "Raise turkeys." George W. Hackett believed that the first All-American Turkey Show had been a "marked success" not only "from the standpoint of numbers and quality of exhibits," but also because scores of farm men and women had attended the show to learn more about turkeys, how to raise and market them, and how they could diversify their farm operations and increase their incomes by raising them.

The All-American Turkey Show was incorporated in 1928. The second article in the Articles of Incorporation states in part, "The object and business of the corporation shall be to foster, promote and advocate the raising of turkeys, . . . to the end that diversified farming be encouraged and developed."

Thousands of free tickets were distributed to farmers so they could attend the 1925 All-American Turkey Show without charge. Crowds of them did so, to hear talks by N. E. Chapman, of the Minnesota Agricultural College; John Winwoodie, editor of the *Dakota Farmer*, Aberdeen, South Dakota; and O. O. Kemp, president of the Agricultural Credit Corporation of Minneapolis, Minnesota. President Calvin Coolidge had influenced the formation of the Agricultural Credit Corporation, and Kemp explained how the Agricultural Credit Corporation had participated in the diversification crusade by helping banks extend loans to North Dakota farmers.

The rewards of discipleship were attested to by the testimonies of those adhering to the tenets of the diversification gospel. Mr. and Mrs. C. H. Folz farmed a few miles north of Drayton, North Dakota, in Pembina County. Mrs. Folz raised White Holland turkeys. In the fall of 1923, E. H. Cooley, reporter for the *Grand Forks Herald*, visited the farm. "One of the prettiest sights the *Herald* visitor's Kodak took in this fall," he wrote in his column near the end of 1923, "was a yard full of glistening white turkeys . . . These feathered beauties were picking about the farmyard of the C. H. Folz home . . . and the late afternoon sun shining on their gleaming coats dazzled the passerby."[2]

---

[2] *Grand Forks Herald*, December 23, 1923.

ARTICLES OF INCORPORATION.

OF

ALL AMERICAN TURKEY SHOW.

THe undersigned Hereby associate theMselves together for the purpose of forming a corporation under the laws of the State of North Dakota, and agree as follows:

FIRST.

The name of this corporation shall be All American Turkey Show.

SECOND.

The Object and business of the corporation shall be to foster, promote and advocate the raising of turkeys, poultry and other live stock to the end that diversified farming be encouraged and developed. In connection with the work aforesaid this corporation shall conduct the exposition heretofore held at Grand Forks which has been called the "All American Turkey Show" and such otherturkey, poultry and agricultural shows as may be deemed advisable.

THIRD.

The place of the principal business of this corporation shall be In the City of Grand Forks, State of north Dakota.

FOURTH.

This corporation shall exist for the period of twenty years from the date hereof.

FIFTH.

The number of directors or trustees shall be eleven, and the names and residences of those to serve until their successors are elected and qualified are:

Alfred MalMberg, Crookston, Minnesota
Ray Andres, Petersburg, North Dakota
Victor Hartl, New Rockford, NortH dakota
G. S. Mairs, Lisbon, North Dakota.
M. A. Johnson, Michigan, North Dakota.
J. C. Sherlock, East Grand Forks, MinnEsota.
and Julius F. Bacon, R. F. Bridgeman, I. Papermaster,
D. F. McGowen and W. W. Blain, of Grand Forks, North Dakota.

SIXTH.

The amount of the capital stock shall be nothing.

Dated this 19th day of November, 1928.

J. C. Sherlock,
I. Papermaster,
W. W. Blain,

On this 17th day of November, 1928, before me, a notary public in and for said county and state, personally appeared J. C. Sherlock, I. Papermaster and W. W. Blain, known to me to be the persons described in and who executed the foregoing instrument, and they severally acknowledged to me that they executed the same.

Harry P. Bice,
Notary Public, North dakota.
My Commission Expires: OCt 31st, 1930.

N. Seal.

STATE OF NORTH DAKOTA
DEPARTMENT OF STATE.

Filed for record and cerTificate issued the 23rd day of November 23rd, 1928 and recorded in VolumeTwo of Non-Profit Corporations, page 274.

Robert Byrne,
Secretary ofState.

By Charles Liessman,
Deputy.

Charter No. 9932.
Fee $5.00

Articles of Incorporation of the All-American Turkey Show, 1928.

Mrs. Folz, who had raised turkeys for many years, had gotten her start in turkeys with three hens for which she paid $24. Mr. and Mrs. Folz had transformed a run-down, weed-infested farm into a profitable operation by following sound farming practices and by diversifying. They produced wheat, barley, oats, and alfalfa; they milked cows, and they raised beef cattle, hogs, and poultry. Mrs. Folz's turkeys, however, produced more income than any of the crops or other livestock.

G. C. Darrah of Roy, Montana—Montana's Turkey King—raised four to six hundred turkeys per year. In 1925 he sold four hundred birds as breeding stock for $6000, with "practically no expense." "I think turkey raising is about the best bet on the farm," he said. "If more farmers would take it up they would make more money than they do raising wheat. We are from a great wheat state, but we consider turkeys our best crop. Turkeys are our grubstake. If it were not for our turkeys I sometimes think we would be out of business."

Lillian Lohman German raised turkeys on the family's ranch in Montana. She was a graduate of the University of Washington and a member of Kappa Alpha Theta sorority. Not wanting to be bothered by raising turkeys for market, she raised only breeding stock, birds that she sold for $15 to $100 apiece. "Turkey raising," she believed, was "an enjoyable occupation for women. I surely enjoy working with my flock. It is a great thing to find something to do that is pleasurable and profitable as well. Turkey raising is both." German's yearling toms sold for more than her father's yearling steers.

"I can't see why more people don't realize how much money can be made in raising turkeys," said N. A. Nelson of Petersburg, North Dakota, the "Turkey King" of Nelson County. Nelson started raising turkeys in 1919, after returning from military service in World War I, when a neighboring farmer told him how easy and profitable it was. The neighbor had paid $30 for six turkey hens and $10 for a tom. In the spring, after the poults had hatched, the flock left his place, ranged all summer, and returned in the fall, in "fine condition." The farmer kept the hens and tom for breeding and sold the rest for $600. It was like finding the money, he said, because he had paid nothing for the turkeys' feed and care.

In 1924, Nelson sold 435 turkeys for $900, but, because he had "poorly marketed" them by unwisely shipping them to an "unreliable concern," he had lost about $800. Nelson culled his flock of Bronze turkeys and sold 195 culls for $950. He sold 140 turkeys as breeding stock for $1,200, and he had "a nice flock" of 550 poults left. In 1925, Nelson's income from turkeys was over $4,200 with expenses of only $700. Because he could buy feed for his turkeys more cheaply than he could raise it himself, Nelson rented out his quarter section, keeping a field of alfalfa for grazing his turkeys, and concentrated on raising turkeys.

"In the spring," Nelson admitted, "raising turkeys is a crazy man's job." But, "considering the investment and the large returns," he said, "there is nothing that can beat turkey raising. It is better than playing the market." Nelson calculated that his rate of return on his investment in turkeys was "several hundred percent."

Putting it more pithily than the testimonies of those who adhered to the tenets of the diversification gospel, N. E. Chapman, poultry husbandman from the University of Minnesota Agricultural College, advised those attending the All-American Turkey Show banquet in 1925 that "a turkey hen setting for four weeks can do more for the farmer of the Northwest than Congress setting for four months can do."

Turkey kings and sorority sisters were not the only ones making money on turkeys. Many farmers in North Dakota and throughout the Northwest were supplementing their incomes by raising them. According to a December *Grand Forks Herald* article, turkey production in North Dakota was increasing rapidly.[3] The turkey crop in 1924 was 35 percent larger than any previous year, and, the paper noted, "hundreds upon hundreds of turkeys" were being shipped out of North Dakota on the Great Northern Railroad in barrels, the preferred method of shipment.[4]

Each barrel contained two hundred pounds of dressed turkeys, and more than ninety barrels had been shipped out of Towner, North Dakota, on November 16, 1924, alone, and this was only one day's shipment. Turkey raising had reached such proportions in the vicinity of Upham, North Dakota, that turkey shipments rivalled those from any other part of the

[3] *Grand Forks Herald*, December 9, 1923.

[4] *Grand Forks Herald*, November 17, 1924.

state. Prior to Thanksgiving in 1923, 131,000 pounds of dressed turkeys had been shipped from the town, including two railway cars of dressed turkeys consigned to a commission firm in Philadelphia.

Three carloads of dressed turkeys valued at $20,000 had been shipped from Portland, North Dakota, after the "Turkey Days" held in December 1923. "Turkey Day" in Pembina, North Dakota, in December 1923, "brought good results," a carload of dressed turkeys. Farmers "from near at hand and from long distances" had sold their dressed turkeys to creameries and produce houses. Canadian farmers had brought their turkeys to Pembina live and had dressed them in town, to avoid paying duty on them.

There is no gainsaying that large numbers of turkeys were being raised and marketed in North Dakota and throughout the Northwest, but there were no major shows at which flock owners could exhibit their turkeys. New York had its Madison Square Garden Show, Boston had its Fanciers' Show, Chicago had its Coliseum Show, and the west coast had its turkey shows.

Because there were no major turkey shows in the large area between Chicago and the west coast, the region producing a major portion of the country's turkeys, the All-American Turkey Show was a turkey show waiting to happen. Turkey raisers wanted to learn how they could improve the quality of their birds and how they could raise and market them more profitably. Above all, they wanted to mingle with "turkey folks," members of the select brotherhood who spoke the same language, shared similar experiences, and worked for the same end—that of encouraging more people to eat more turkey meat, and to eat it year-round, not just at Thanksgiving and for Christmas.

Were a premier turkey show to have its genesis in the Northwest, then Grand Forks, North Dakota, was the ideal location for it. Grand Forks, declared George W. Hackett, was "the heart of the greatest turkey raising territory in the country." There is no indication that any other city was considered as the location for the All-American Turkey Show, and, apparently, during all the years that the show was held, no city challenged Grand Forks for the honor of holding it.

It had long been acknowledged that the country's best turkeys were those produced in the Northwest, that the Northwest's best turkeys were those produced in North Dakota, and that North Dakota's best turkeys were those produced in the Red River Valley. "The Northwest," declared Ed L. Hayes, was "one of the few sections in the United States where turkeys can be successfully raised." It was the Northwest's climate said W. E. Stanfield, that produced turkeys that were "healthy, strong, and well adapted to conditions."

W. V. Langley, Minnesota's Kittson County agent, believed that the Northwest was "especially adapted to turkey raising" and produced "the best quality of turkeys for eating" because of its "abundance of grain and insects for feed." The Northwest was also "remarkably free" from Blackhead, the widespread turkey disease that had decimated flocks in the east and had forced the turkey industry to move west.

Ed L. Hayes, who conceived the idea of an exclusive turkey show. Image from February 9, 1925; used with permission of *Grand Forks Herald.*

An article on the front page of the *Grand Forks Herald* in January 1926 noted that "the Northwest now leads the world in turkey production and North Dakota leads the Northwest." A week later, the headline on an article in the paper stated that "North Dakota is without question the leading turkey producing state in the United States." "There is," W. E. Stanfield had said in 1924, "nothing like the sunshine of North Dakota and the broad range in the open offered by these prairies to produce turkeys. Conditions here are as nearly ideal as they could well be."[5]

Conditions bid fair for a premier turkey show to be located in Grand Forks, North Dakota, and by December 1923 the headline in the *Grand Forks Herald* was, "Plans Being Made for Big Turkey Show."[6] The lead article on the paper's front page stated that "[a] considerable stimulus to turkey raising in the Northwest" was expected to be the result of "the staging of the first All-American Turkey Show in Grand Forks February 12–14, 1924." The show would open on the morning of February 12, judging would be completed by noon on February 13, and the show would close with "a big turkey banquet" at the Hotel Dacotah on the evening of February 14.

The premium list included cash prizes ranging to $5 for single entries, with more than $100 in cash specialty prizes and a number of silver cups. North Dakota Governor R. A. Nestos and other prominent men had been invited to speak at the banquet, and specialists would provide instructions on how to pack turkeys for shipping. The All-American Turkey Show would be held under the joint direction of the Northwest Turkey Breeders' Association and the Red River Valley Poultry Association.

The All-American Turkey Show owed its inception to Ed L. Hayes, "whose genius conceived of the idea of an exclusive turkey show." The sixth All-American Turkey Show, held in 1929, was dedicated to his memory. A poultry judge who had judged at the North Dakota State Fair for five years, Hayes was editor of the *Northwest Poultry Journal* and president of the Northwest Turkey Breeders' Association.

Visiting Grand Forks late in 1923, Hayes encouraged the city's administrators and business owners to sponsor a turkey show in conjunction

---

[5] *Grand Forks Herald*, January 31, 1926, and February 7, 1926.

[6] *Grand Forks Herald*, December 16, 1923.

with the Red River Valley Poultry Show and Utility Feed Show, scheduled for February 12–14, 1924. He would take general charge of the turkey show, and E. R. Montgomery, Secretary of the North Dakota State Fair Association, would be in charge of arrangements.

Although there are no records indicating where the name originated, the show was known as the All-American Turkey Show from its inception, an apt title. With the success of the first show in 1924, it was indeed *the* premier turkey show in the country and it was truly All-American. Within a short time, the All-American Turkey Show drew exhibitors from every turkey-producing state in the Union and from every turkey-producing province in Canada.

The *Grand Forks Herald* in February 1924, with a full-page announcement and a flair, heralded the opening of "the first All-American Turkey Show ever held in Grand Forks or in the United States." The show's purpose was "to stimulate" the raising of turkeys, to "show conclusively" that raising turkeys was the "most remunerative" enterprise on many farms, and to encourage consumers to eat more turkey because it was "high quality" meat.[7]

Features of the show included lectures by turkey specialists, educational programs on raising and marketing turkeys, judging by judges selected "from among the more competent and best authorities" on turkeys in the country, and the turkey banquet. Entries were open to the world and exhibitors could bring their prized birds to the show, assured that they would be protected from harm and given the best of care. All birds showing symptoms of disease would be removed from the showroom. "A long list of prizes" would be offered, "in order that the raisers of the best specimens" could be rewarded and "re-imbursed for their trouble and pains in making entries."

Turkey specialist W. J. Corcoran of St. Paul, Minnesota, titled his lecture, "The Value of Co-operative Marketing and the Superior Quality of Turkeys Produced in the Northwest." Corcoran was Organization Manager of the Minnesota Poultry Marketing Association. Producers in the Northwest had "just been raising turkeys," chided Corcoran, but "not

[7] *Grand Forks Herald,* February 11, 1924.

marketing them." Turkey raising in the Northwest, he advised, would "come into its own" when producers presented their product to consumers in the "true light" as "high quality" meat and when those raising turkeys adopted "some system" of placing their product on the market in "the most advantageous manner."

Ed L. Hayes spoke on "The Growth of the Turkey Business of Five Years Ago to the Present Time." Turkeys raised in North Dakota in 1919, he said, were valued in the thousands of dollars, but those raised in 1923 were valued in the millions of dollars. The Northwest, he said, was among the few places in the country where turkeys could be raised successfully, and as long as this was the case it was unlikely that prices would ever drop to the point where raising them would be unprofitable. Turkey prices, he believed, were "pretty sure to remain high," but flock owners' profits depended on the way that they raised, fattened, shipped, and marketed their turkeys.[8]

"The Turkey Queen of Minnesota," Grace Randolph of Donaldson, spoke on the "Raising and Care of Turkeys." Randolph, who devoted "her entire time" to raising turkeys, started with turkeys in 1918 with seven hens borrowed from a neighbor. In 1921, she sold her flock for $4,300 and made a new start with Bronze. In 1923, she sold 578 Bronze purebreds as breeders. Because of her reputation for success, she had received letters asking for advice on raising turkeys from turkey raisers in forty-six states and from two Canadian provinces.

Other speakers addressed various aspects of the turkey industry. W. E. Stanfield spoke on the topic of "Raising Turkeys on the Farm," and a representative of the Peter Fox Sons Company of Chicago offered instructions on "How to Properly Pick, Grade, Pack, and Market Turkeys." S. A. Macaulay, publicity agent for the Northern Pacific railroad, noted that Northwest turkey raisers were "keeping about one jump ahead of the Middle West states" in turkey raising, partly because "natural conditions" were "very favorable for raising turkeys," but also because of the care flock owners gave their birds. Basing his estimate on the North Dakota turkey

[8] "Turkey Raising in Kittson County, Minnesota" was addressed by its county agent, W. V. Langley. In 1923, he said, 277,850 pounds of dressed turkeys valued at $72,000 had been shipped from Kittson County. The sale of breeding stock brought the total to $90,000.

crop in 1923, he believed that the 1924 crop would be from one-third to one-half larger.

Figures on the number of turkeys entered in the 1924 All-American Turkey Show vary. Bob Hawes, in his article, "Saving Heritage Turkeys," gave the figure of 210, but E. R. Montgomery, who was in charge of local arrangements for the show, gave the figure of 249. George Hackett of Minneapolis judged the Narragansett and the Bronze varieties, and, he admitted, he had had some difficulty because so many Bronze turkeys were entered in each class. W. E. Stanfield judged the Bourbon Reds and the Slate, and Ed L. Hayes judged the White Hollands and the Black.

Judges followed the exacting standards of the *Standard of Perfection* for each variety of turkeys, and varieties were judged in a number of classes, such as old tom, yearling tom, young tom, old hen, yearling hen, and young hen. Pens of two, three, and five turkeys of the same variety and exhibited by one person were also judged. Exhibitors were provided with Premium Lists that specified the classes and listed the prizes. The 1924 Premium List was mailed to exhibitors early in January.

To those not exhibiting turkeys, Premium Lists would have been boring, if not incomprehensible. Exhibitors, however, pored over them, determining in which classes to compete and calculating their chances of winning. The 1924 Premium List was published in January in the *Grand Forks Herald*; what follows is a portion of it.[9]

Regular Cash Prizes

Twenty entries in class:	1st, \$5; 2nd, \$4; 3rd, \$3; 4th, \$2; 5th, ribbon
Sixteen entries in class:	1st, \$4; 2nd, \$3; 3rd, \$3; 4th, \$1; 5th, ribbon
Twelve entries in class:	1st, \$3; 2nd, \$2; 3rd, \$1; 4th, 50¢; 5th, ribbon
Eight entries in class:	1st, \$3; 2nd, \$1; 3rd, 75¢; 4th, ribbon; 5th, ribbon
Four entries in class:	1st, \$1; 2nd, 75¢; 3rd, 50¢; 4th, ribbon
Two entries in class:	1st, 75¢; 2nd, 50¢
One entry in class:	1st, 50¢

Prizes of fifty cents, seventy-five cents, and one dollar may seem insignificant and hardly work an exhibitor's time and effort. One must, however, appreciate what the exhibitor could purchase with fifty cents or one dollar in February 1924. Perry's Kitchen, at 14 South Fourth Street in Grand Forks, advertised "Every Morning Breakfast Specials." For five

[9] *Grand Forks Herald*, January 13, 1924.

cents, patrons had a choice of two cinnamon rolls, two doughnuts, two cuts of coffee cake, two nut rolls, or two oatmeal Rocks. Breakfast included apple sauce, coffee, and cream, with refills of coffee. The Saturday luncheon special was baked spareribs and sauerkraut, mashed potatoes, bread and butter for twenty-five cents; and the Saturday night supper special was chow mein, rice, noodles, and tea for thirty-five cents.

A few more examples will be helpful in explaining the many categories of awards in this and subsequent All-American Turkey Shows.[10] The Fleischmann Company offered two 2 ½ pound cans of its Special Poultry Yeast, one for the third best exhibit of turkeys and one for the best exhibit of White Holland turkeys.

The Dakota Farmer of Aberdeen, South Dakota, offered a silver loving cup for the best display of turkeys from North Dakota and the same for South Dakota. Silver loving cups were awarded by *The Northwest Farmstead* of Minneapolis for the best display of Bronze turkeys from Minnesota; the *Northwestern Poultry Journal* of Minneapolis for the best display of Bronze turkeys; the *Poultry Tribune* of Mount Morris, Illinois, for the best display of White Holland turkeys; and the *Poultry Item* of Seilerville, Pennsylvania, for the best display of Bourbon Reds.

The Grand Champion in 1924 was a yearling Bronze tom from Alexandria, Minnesota. Its owner, C. J. Cappahahn, refused an offer of $250 for it. As competition at All-American Turkey Shows became keener and as prime breeding stock became more valuable, owners of Grand Champions refused offers of $1000 and trades of new automobiles for their prizewinners, knowing they could realize more from the sale of the champion's progeny as breeding stock.

W. E. Stanfield hoped that cash prizes would be larger in subsequent shows to make it worthwhile for breeders to attend and exhibit their best birds, but it was not the amount of money that was important for exhibitors. It was the sportsmanship of competition, the awareness that a second place at the All-American Turkey Show was the equivalent of a first at any other show, the knowledge that prizewinning turkeys—or their progeny—would command higher prices when sold as breeding stock, and also the good-natured acceptance of not winning any prize at all.

---

[10] For more examples, see Appendix D: Special Prizes, 1924.

CHAPTER 9

Battle of the (Turkey) Titans

(with apologies to Hesiod, "Theogony," c. 700 B.C.E.)

TO ENCOURAGE ATTENDANCE, THE All-American Turkey Shows, starting with the first, included interesting features that would be appealing to visitors, particularly to children, who were not allowed to attend without their parents or teachers, and to those who did not raise turkeys. In 1924, H. M. Eisenlohr of Larimore, North Dakota, displayed Canada geese, a wild duck, and a wild pheasant. He also had several coops of wild fowl, such as brown bantams, game cocks, and white game bantams. Gilbert Johnson of Blabon, North Dakota, displayed a coop of wild turkeys.

Arnold Medlund, who farmed near Driscoll, North Dakota, was—unfortunately—prevented from exhibiting any of his birds. In the spring of 1923, Medlund had crossed his domesticated hens with wild toms. The cross-strain birds were purportedly the wonder of the neighborhood, and produce buyers from nearby Mandan, North Dakota, declared them to be finer than first-grade Bronze turkeys. Medlund sold a few birds, but, intending to have a superior flock in 1924, he kept most of the turkeys for breeding stock.

The turkeys, Medlund said, seemed content to stay on the farm. During the winter, however, with frigid temperatures and heavy snow, "the wild strain surged predominant" and the entire flock "took to the South and West," presumably to the heavy woods along the Missouri River. None were found. The next time, Medlund said, he would clip the wings of his "cross-bred turks."

The 1924 All-American Turkey Show, unequivocally a success, set the standard for those that followed and also for the many turkey shows that sprang up around the country in the years preceding World War II.

Because the 1924 All-American Turkey Show was the first-ever exclusive turkey show, it set the bar high.

Building on the success of the 1924 All-American Turkey Show and anticipating even more exhibitors and even larger crowds in 1925, officials moved the show to the City Auditorium and scheduled it for the first four days in February. As in 1924, the show had three divisions: Turkeys, Poultry, and Utility Seeds. Affirming the distinction of the All-American Turkey Show, however, officials decided at the annual meeting of the Northwest Turkey Breeders' Association on February 14, 1924, that "All-American" would be the trade name for turkeys shipped from the area. Slates of officers and board members were also elected.[1]

Within a year, many of the individuals and their names had become familiar to exhibitors and visitors alike. It is instructive that the four states—Minnesota, Montana, North Dakota, and South Dakota—represented on the Executive Board were those that, in Hackett's opinion, produced the best turkeys. Because they did, he advised, the All-American Turkey Show should accept entries only from those states.

Those attending the meeting set a goal of five hundred turkeys for the 1925 All-American Turkey Show. Reflecting the fledgling reputation after only one show, Robert Kirkwood and William Fox of the Commission House of Peter Fox and Sons of Chicago spoke on the proper way to kill, dress, pack, and ship turkeys. The most important point was to keep the birds clean. The preferred method of shipment was to pack the turkeys in barrels, but only after the carcasses had been thoroughly cooled.

The 1925 All-American Turkey Show opened on Tuesday morning, February 3, to run for four days, and it was well that officials had moved it to a larger venue. Thousands of tickets had been distributed to area farmers, allowing them to be admitted free of charge. Children were also admitted free of charge if accompanied by an adult. Others were charged a small fee, or they could obtain complimentary tickets from the Commer-

---

[1] Officers were Ed L. Hayes, president; W. G. Chapman, Maxbass, North Dakota, first vice president; W. E. Stanfield, second vice president; and Grace Randolph, Secretary-Treasurer. Elected to the Executive Board were Alfred Malmberg, Crookston, Minnesota; Victor E. Hartl, New Rockford, North Dakota; Mrs. Melvin Greenwood, Crystal, North Dakota; H. A. Hartman, Knox, North Dakota; C. J. Kappahahn, Alexandria, Minnesota; and Matthew Wagner, Webster, South Dakota. Board members from Montana would be selected later.

cial Club. Boy Scouts served as ushers for out-of-town visitors, guiding them through the aisles filled with cages on the floor of the showroom.

Making little effort to contain its enthusiasm, the *Grand Forks Herald* announced on the front page the next day that "the biggest turkey show in the United States, by comparative figures, was opened in the Grand Forks auditorium Tuesday evening." At this, the "Largest of its Kind in the Country," a "total of 312 turkeys" had been entered, "210 in the Bronze class, by 55 exhibitors from 12 states and Canada."[2]

According to the article, the 210 in the Bronze class were "over three times the number of entries in any other Bronze show in the country." Alfred Malmburg of Crookston, Minnesota, had twenty-five Bronze turkeys entered, and an exhibitor from New Rockford, North Dakota, had fifteen. At the 1924 Madison Square Garden Show in New York, only thirty-two had been entered by three exhibitors.

Exhibitors from a distance included Cora D. France Forester of Berthoud, Colorado; R. K. Walker of Memphis, Missouri; and Mrs. Rinda Collins of Twodot, Montana. Other exhibitors were from Idaho, Iowa, Illinois, and Wisconsin. Marvin Magnus, of Sterling, North Dakota, exhibited six wild turkeys, two Mexican and four "American." Readers of the *Grand Forks Herald* were reminded that it was from the "American" strain that all domesticated turkeys had been developed.

The evening program on February 4 opened with motion pictures on agricultural subjects and musical selections. C. G. Sterling of the Northwest School of Agriculture in Crookston, Minnesota, spoke on "Cooperation Between Business Men and Farmers." Another address was titled, "Exhibits and Their Importance to Agriculture."

Show manager Ed L. Hayes said that the 1925 All-American Turkey Show was three times larger than the first. The 1926 All-American Turkey Show was larger still, larger than either of the two previous ones. This far removed from the time—and from the times—one can only with difficulty appreciate not only what the All-American Turkey Show had become in the two years since its beginning, but also the enthusiastic support it received from breeders, agricultural experts, state and local officials, towns-

[2] *Grand Forks Herald*, February 4, 1925.

people, and businesses. Were it not for the financial support of businesses and the prizes they donated, the All-American Turkey Show could not have succeeded.

Making liberal use of superlatives, headlines, and leading articles—all of which had become mandatory when referring to the All-American Turkey Show—the *Grand Forks Herald* encouraged participation in the show held February 1–5, 1926. Examples, woven here into a few paragraphs, help to explain why this once-a-year event had made Grand Forks, North Dakota, the Turkey Capital of the World.

Businessmen in Grand Forks and East Grand Forks had raised the money for what they touted as "the world's biggest event of this kind." Final preparations were being made "so that the big five-day international event may get started without delay and be continued without interruption." Preparations went well and the "[p]ounding of hammers and gobbling of hundreds [of turkeys] marked the beginning of the All-American Turkey Show in the Grand Forks city auditorium."

Grand Forks City Auditorium at the corner of North Fifth Street and Fifth Avenue North. From *The Valley and Beyond: More Memories from Grand Forks and Surrounding Areas* (2000), published by (and used with the permission of) Grand Forks Herald.

The large number of prizes on the Premium List had "brought out the greatest display of turkeys ever seen in the world." The turkeys entered in this, "The Largest Exclusive Turkey Show in the World," would be judged by "Six of America's Foremost Judges" who described the All-American Turkey Show as "the greatest turkey show in the world." Their judging would be done "So Everyone Can See," and a win at the All-American Turkey Show meant "More Than at Any Other Show."

The All-American Turkey Show was "Lauded by Poultry Men Who Have Seen Them All." Exhibits at other large shows were "small compared to the exhibits at the All-American Turkey Show held annually at Grand Forks," where the quality of "some of the most aristocratic birds on the continent" was "equal to the quality of any other show." Grand Forks, home of the All-American Turkey Show, by 1926, held "a world position of distinction."

The city was located in North Dakota, the state that was "without question the leading turkey producing state in the United States." At the end of January, these phrases ran along the bottom of the *Grand Forks Herald*'s front page in bold print: "February 1-2-3-4-5," "All-American Turkey Show," "Largest Exclusive Turkey Show in the United States," "More Than 500 Entries," and "Don't Miss It."[3]

So many turkeys were entered in the 1926 All-American Turkey Show that an additional one hundred coops had to be ordered to accommodate them, and so many prizes had been donated that more judging categories had to be invented before all of them could be awarded. Truly international, the All-American Turkey Show drew eighty-six exhibitors—three times as many exhibitors as all the other major shows combined—forty-two of them for the first time, from twelve states, Canada's prairie provinces, Mexico, and from the Island of Jamaica in the British West Indies.[4]

More turkeys, 436, were entered than at all the other leading turkey shows combined, 168 in the Bronze class alone. Four varieties were entered—Bronze, White Holland, Bourbon Red, and Narrangansett—more varieties than at any other major show. "These figures," Hayes said, "are proof that turkey raisers must show and win at the All-American Turkey Show to be supreme in the turkey world."

---

[3] *Grand Forks Herald,* January 30, 1926.

[4] *Grand Forks Herald,* January 31, 1926.

To read accounts of the 1926 All-American Turkey Show in the *Grand Forks Herald* was almost like being in attendance. As described, the entire floor of the city auditorium was taken up with cages in two tiers. "To walk up and down these aisles," according to one account, "was an education in itself, especially so if one stopped to talk with the breeders of these feathered beauties."

Many visitors commented on being delighted with the show and said that they would not have missed it for anything. Readers were reminded that the businessmen of Grand Forks and East Grand Forks deserved praise and thanks for "making this great show possible." The 1926 All-American Turkey Show "was a credit to Grand Forks and worthy of the state of North Dakota."

The "feathered beauties" of which the *Grand Forks Herald* wrote were judged by six of the country's best-known turkey experts, including A. D. Walker, Memphis, Missouri, "generally conceded to be one of the most competent judges of turkeys the country had ever had" and among the country's largest Bronze turkey breeders.[5] Walker had sold turkeys to breeders "in every state in the Union" and in two foreign countries. "This is the best turkey show in the world," Walker said. He had judged for fifteen years, including at two world's fairs, and he had never seen a show as good as the All-American Turkey Show.

Speakers at the afternoon and evening sessions were some of the leading turkey, poultry, and agricultural authorities in the United States and Canada.[6] In his remarks, Oscar Grow, Cedar Falls, Iowa, stressed the size and importance of the All-American Turkey Show when compared with the Fanciers' Show in Boston, the country's second largest show. The Boston show had 175 turkeys; the 1926 All-American Turkey Show had 426.

---

[5] In addition to A. D. Walker, the six were Ed L. Hayes of Minneapolis, "one of the leading authorities on turkeys," who served as dean of the judges; George W. Hackett, Minneapolis, Minnesota, "one of the most popular judges in America"; Oscar Crow, Cedar Falls, Iowa, "another headliner"; W. E. Stanfield, St. Paul, Minnesota, judge for the third year; J. C. Clipp, Saltillo, Indiana, judge and breeder "of the highest quality Bronze turkeys" for over thirty years.

[6] Among the speakers were A. F. Schalk, H. L. Walster, P. F. Trowbridge, and O. J. Weisner from North Dakota's Agricultural College; W. C. Boardman, Secretary of the Aberdeen, South Dakota, Commercial Club; W. C. Allen, publisher of the *Dakota Farmer*; Frank L. Platt, editor of the *American Poultry Journal*; A. A. Yoder, Mount Morris, Illinois, editor of the *Poultry Tribune*; and Charles E. Byrd of Pennsylvania, Secretary of the National Bronze Turkey Club.

After a trip of four thousand miles through twenty-six counties, Ed L. Hayes estimated that the turkeys produced in North Dakota in 1925 contributed $23,504,440 to the state's economy. Touching on one of the reasons for the All-American Turkey Shows, P. F. Trowbridge explained the part that turkeys had in diversification plans for the Northwest. M. C. Herner of the Manitoba Agricultural College spoke on three topics: "The Development of the Turkey Industry in the Prairie Provinces of Canada," "Co-operative Marketing Systems," and "Mating and Breeding."

The 1926 All-American Turkey Show had a number of features in addition to the educational sessions and the judging. All school children with their teachers or parents were admitted free after 4:00 p.m. each day. Among the features important to the Northwest turkey industry were the daily auctions of breeding birds. As reported, the sales offered "an excellent opportunity for present and prospective raisers to make selections from the world's largest and finest collection of breeding stock," which would have "a far-reaching effect on the turkey production" in the territory.

Interest in the 1926 All-American Turkey Show was piqued by a notice in the *Grand Forks Herald* that "several kinds of unusual birds" would be displayed. That may have been an understatement. Several wild turkeys were on display from Mexico, South Dakota, and North Dakota. The largest exhibit was from the Blabon Wild Turkey Farm at Steele, North Dakota.[7]

Perhaps the leading breeder of wild turkeys in the world, the Blabon Wild Turkey Farm supplied wild turkeys for large eastern game preserves. The wild turkeys attracted interest, but curtains had to be placed over their cages for the first few days. Otherwise, in their "wild frenzy" to escape, they threatened to "beat their brains out" on the sides of the cages. Other wild birds included Chinese pheasants and a hen that was the offspring of a pheasant and a turkey.

The most unusual birds were turkhens, the result of experiments carried out in Jamaica. Turkhens were a cross between turkeys and chickens. The *Grand Forks Herald* announced that "[f]ifteen of the feathered freaks"

---

[7] *Grand Forks Herald*, January 27, 1926.

were displayed at the show, where they proved "of much interest to the show visitors" and received "more than a passing glance" from the turkeys that occupied "the real center of interest."

A "remarkable feature" of the "ridiculous-looking" turkhens, the article went on to say, was that they neither gobbled nor crowed. The exhibitors of the "weird birds" explained the value and the benefits of the cross. The wild part of the birds' nature encouraged them to range far afield for their feed, but the hen part of their nature brought them back at night to roost. The size of the chicken hen was slightly increased by the cross, but there was "practically no difference in the flesh."

Cash prizes of $600 were awarded at the 1926 All-American Turkey Show. Special prizes awarded to leading exhibitors included twenty-six silver loving cups of various sizes, a number of silver bread trays, and a chest of silverware.[8]

A number of Special Prizes appear to have been invented. There were prizes for the best-shaped White Holland tom and hen and for the best-colored Bronze tom and hen. The Bronze tom and hen of any age having the best bronze backs received prizes, as did those of any age having the best barred wing. The Bronze pullet and the Bronze cockrel having the best Bronze band on their tails were awarded prizes, and so were the best-shaped Bronze tom and hen.

Taking several minutes for a closely scrutinized comparison, the judges chose a Bronze yearling tom weighing thirty-four pounds as the Grand Champion of the 1926 All-American Turkey Show. The Grand Champion, valued at $1,000, was a North Dakota turkey owned by Victor E. Hartl of New Rockford. Commenting on the Grand Champion, the *Herald* noted that "winning the highest honors of this show practically

---

[8] The *Dakota Farmer*, Aberdeen, South Dakota, provided silver cups for the best display of turkeys from North Dakota and South Dakota. The *Farm Stock and Northwest Farmstead* one for the best display from Minnesota. The Northwest Turkey Breeders' Association provided them for the best displays from Illinois, Oklahoma, Kansas, and Missouri. The Merchants Association of Grand Forks provided silver cups for the best display of Bronze, the best yearling Bronze tom, the best yearling Bronze hen, and the best three young Bronze toms shown by one exhibitor. The American Narragansett Turkey Club awarded a silver bread tray to the best yearling Narragansett tom and silver cups to the best young tom and best young hen. It also awarded $3 to Mrs. Albert Schmidt of Barnard, Kansas, the exhibitor who had come the greatest distance. The Grand Forks Commercial Club provided silver cups for the best Bourbon Red tom and hen, one for the best three young Bourbon Red toms shown by one exhibitor, and one for the Champion Bourbon Red turkey that was exhibited by Mrs. Gladys Honsinger of Lebanon, Missouri.

makes this bird king of the country, for this show outranks all others in size and importance."[9]

Placing second to the Grand Champion was a Bronze tom weighing forty-eight pounds from Roy, Montana, owned by G. C. Darrah, known as Montana's Turkey King. Darrah was another who believed that raising turkeys was "about the best bet on the farm." In the spirit of sportsmanship that was characteristic of those exhibiting turkeys at the All-American Turkey Show, Darrah did not claim the honey and he good-naturedly accepted second place.

Superlatives barely suffice when describing the 1927 All-American Turkey Show. On the first day of February, the *Herald* noted the doors of the Grand Forks city auditorium opened on "the greatest exclusive turkey show in the world" that was "confidently expected to be the greatest ever held."[10]

The "Pride of Northwest Turkeys," worth an estimated $35,000, would be exhibited. Across the bottom of the paper's front page the next day were phrases in bold intended to encourage attendance.[11]

> Don't Miss the All-American Turkey Show
> Take the children today
> See the Greatest Show in the World
> Acquaint yourself so you can tell your friends about it
> Exhibitors are here from New York State, Maryland,
> Texas, Idaho, and Canada
> Hear the High School Band Tonight

"The general opinion," expressed the *Grand Forks Herald* two days later, was that the 1927 All-American Turkey Show was "by far the most successful ever held, in addition to being the largest." O. A. Stevens of the North Dakota Agricultural College and in charge of poultry development work in the state, had attended premier turkey shows all over the country.

---

[9] Hartl farmed 320 acres, but the turkeys he sold as breeding stock provided most of his income. He had shipped turkeys to breeders in seventeen states and had received $12 to $15 apiece for them. He had also sold eggs to breeders in Canada. Hartl was thirty-three years old and had raised turkeys since he was twelve. Raising turkeys, he said, was "one of the most profitable as well as one of the most interesting of occupations."

[10] *Grand Forks Herald*, February 1, 1927.

[11] *Grand Forks Herald*, February 2, 1927.

"Never before," he said, had he witnessed "the enthusiasm and the spirit that prevails at the All-American Turkey Show in Grand Forks."[12]

The welcoming banquet was "one of the most enthusiastic meetings of any kind" Stevens had ever attended, he said, and the quality of the turkeys on exhibit was "without question superior to the best ever assembled in a show anywhere." The All-American Turkey Show, he believed, would be "a credit" to any city and to any group of businessmen "backing it" anywhere. Among the "most gratifying things" about the show for Stevens was that North Dakota turkeys were "well to the front in every class" in which they competed.

Five varieties of turkeys were entered in the 1927 All-American Turkey Show: Bronze, White Holland, Bourbon Red, Narraganset, and Black. Not counting the Special Awards and the Championship Awards, the turkeys competed for 223 prizes and awards, consisting of $2000 in cash, silver cups and trays, and ribbons.[13]

In an effort to acquaint readers with the difficult choices judges had to make when judging such large numbers of turkeys, the *Herald* provided information on the turkey varieties being exhibited, the number of prizes and awards allocated to each variety, the classes of entries, and the number of awards for each class.[14] The number of prizes and awards, the categories, and the number of places for the Bronze variety reflected the large number of Bronze turkeys entered.[15] The categories were the same for the other varieties; the number of places varied.

Any attempt to describe the 1927 All-American Turkey Show would be incomplete were the following features of the show not included. The Bronze variety dominated turkey production, the All-American Turkey Show, and the popular imagination. Bronze turkeys were pictured on

---

[12] *Grand Forks Herald,* February 4, 1927.

[13] The 454 turkeys entered in the 1927 All-American Turkey Show, "representing the best in the industry and from all sections of the country," were judged by six of the industry's most eminent judges. They were George W. Hackett, Minneapolis, Minnesota; A. D. Walker, Memphis, Missouri; Oscar Grow, Cedar Rapids, Iowa; Harry Axtel, Bloomington, Indiana; M. C. Herner, Winnipeg, Manitoba; and Ed L. Hayes, Minneapolis, Minnesota.

[14] *Grand Forks Herald,* February 4, 1927. Understood by exhibitors, but perhaps confusing to others, even a small portion of the material is informative. Prizes and awards: 9 Black, 0 Narragansett, 36 Bourbon Red, 38 White Holland, 110 Bronze.

[15] Adult tom, 7th place, 1–7; Yearling tom, 10th place, 1–10; Young tom, 15th place, 1–15; Adult hen, 8th place, 1–8; Yearling hen, 10th place, 1–10; Young hen, 15th place, 1–15.

greeting cards at Thanksgiving and in depictions of the Pilgrims and the First Thanksgiving. Bronze turkeys were, however, upstaged at the 1927 All-American Turkey Show. For "the first time on record," stated the *Grand Forks Herald*, "a White Holland had won a grand championship."[16]

The Grand Champion, a tom, was owned by Blair Chapman of Brinsmade, North Dakota. Photographs of the tom and one of Chapman's White Holland prize hens were to be published in the *Standard of Perfection* as standards of perfection for the White Holland variety of turkeys. Also at the 1927 All-American Turkey Show, Black turkeys were exhibited for the first time, the fifth of the six varieties of turkeys included in the *Standard of Perfection*. It remained for turkeys of the Slate variety to be entered.[17]

Another outstanding feature of the 1927 All-American Turkey Show was acknowledged in the *Herald*. Many breeders who had exhibited at all the leading turkey shows attended the show in 1927, not to compete, but "to view the best representatives of the several breeds and acquire knowledge of turkey breeding from the educational program in order that they might return home and apply the acquired knowledge disseminated by experts in the betterment of their flocks." These breeders declared the All-American Turkey Show to be "the biggest and best show of its kind in the world," and they regarded it as "supreme in the Turkey World."[18]

Also significant is that during the week of the 1927 All-American Turkey Show, on February 3, the American Turkey Breeders' Association

---

[16] *Grand Forks Herald*, February 4, 1927.

[17] Prices for champions and prize-winning breeders also commanded attention at the 1927 All-American Turkey Show. Mrs. C. H. Folz of Drayton, North Dakota, purchased Chapman's Grand Champion tom for $1,000 and added him to her White Holland flock. Mrs. Lillian Lohman German, known as the "Queen of the Turkey World," was "among the outstanding figures" at the 1927 All-American Turkey Show with her twenty-five Bronze turkeys. Lillian German lived fourteen miles east of Havre, Montana, and, because she regarded the All-American Turkey Show as the most important show in the industry, she would exhibit at no other. Her "beautifully marked tom," Grand Champion at the Madison Square Garden Show in 1926, was prevented from competing in the 1927 show in Grand Forks because his "top knot" froze and turned black when the temperature had dropped to -54°F the week before the show. Havre, Montana, had been known as the coldest place in the United States for the previous forty years, and turkeys raised there had to be hardy to survive. German's "beautifully marked" Grand Champion Bronze tom, raised by Bird Brothers in Pennsylvania, was unaccustomed to Havre winters. Lillian German had refused offers of $500 and $1,000, and the trade of a Ford sedan for her tom.

[18] *Grand Forks Herald*, February 2, 1927.

was organized. Among the founders, officers, and board members were the best known figures in the turkey industry. Ed L. Hayes, Minneapolis, Minnesota, was president; A. D. Walker, Memphis, Missouri, was vice-president; and Ray Andrews, Petersburg, North Dakota, was secretary.[19]

The object of the Association, according to the February announcement in the *Grand Forks Herald*, was "fostering the raising of thorough bred turkeys and the urging of greater turkey consumption."[20] In keeping with the Association's objectives and preceding the All-American Turkey Show, turkey shows would be held during the winter of 1928 in each of the nine districts into which the country was divided.[21] This new national organization was the first of its kind ever to be formed.

Helping to complete the *Grand Forks Herald* description of the 1927 All-American Turkey Show was "the attendance by local people," indicating their "great interest in the turkey industry" and their recognition "of the importance of turkey raising in the Northwest." The Grand Forks Rotary Club and the Y's Men's Club on February 2, for example, adjourned their meetings immediately after lunch and attended the All-American Turkey Show in a body.

On the same day, February 2, a perceptive individual from the *Herald* attended the All-American Turkey Show and shared his waggishly worded observations with readers in a column titled, "Turkeys at All-American Show Many Characteristics." "Probably every characteristic of the human race," noted the witty writer, is exemplified by the birds at the All-American Turkey Show now being held in Grand Forks."

---

[19] *Grand Forks Herald*, February 4, 1927. Members of the Board of Directors were Alfred Malmberg, Crookston, Minnesota; Helen Baker, Chestertown, Maryland; Harry Axtell, Bloomington, Indiana; Mrs. George M. Strech, Greeley, Colorado; Mrs. S. E. Hyde, Boise, Idaho; Charles E. Bird, Myersdale, Pennsylvania; and Mrs. Fiona Horning, Oswego, New York.

[20] *Grand Forks Herald*, February 4, 1927.

[21] The districts were Northwest: North Dakota, South Dakota, Minnesota, Wisconsin, and Montana; Pacific Northwest: Washington, Oregon, Idaho, and Northern California; Western: Colorado, Utah, Wyoming, and Nebraska; Southwestern: Texas, New Mexico, Oklahoma, Nevada, Arizona, and Southern California; Central: Illinois, Iowa, Indiana, Michigan, and Missouri; Southern: Kentucky, Tennessee, Alabama, Louisiana, Florida, and Georgia. Atlantic: Virginia, West Virginia, Maryland, and the District of Columbia; New England: Maine, New Hampshire, Vermont, Massachusetts, Rhode Island, and Connecticut; Eastern: New York, Ohio, Pennsylvania, New Jersey, and Delaware.

> The sourest, most dour looking old individual anybody could ever imagine glummed in the cage fifth from the left in the third row from the wall all day Tuesday and by his general appearance it is likely he will continue glum all the way through life to the roaster that way. Whether or not a prize will cheer him up any is a matter of conjecture only, and so far he has taken the least possible interest in the proceedings. There is that look on his face that is strangely reminiscent of a Democrat the morning after a Republican landslide (or vice-versa) or that of a child who has been subjected to a thorough dose of castor oil.[22]

Given what was said and written about it, the 1928 All-American Turkey Show, held January 31–February 3, 1928, must almost beggar description. For the *Grand Forks Herald*, E. H. Cooley wrote that, "the 1928 All-American held at Grand Forks last week was the greatest exclusive turkey show in the world's history." The paper's headline that day was, "All-American Turkey Show Goes Into History as the World's Greatest."[23]

At the banquet on Wednesday evening, February 1, O. A. Stevens, from the North Dakota Agricultural College, said that the All-American Turkey Show, "was the first exclusive turkey show in the world. Now it is not the only one, but it remains the greatest." By every measure, the 1928 show demonstrated that the All-American Turkey Shows were serving their intended purpose.

Exhibitors from every important turkey-raising area in North America, from Texas to Saskatchewan and from New York to California, entered turkeys in the 1928 All-American Turkey Show, 574 in all, "nearly twice as many birds," noted O. A. Stevens, "as have ever been shown at any other turkey show outside of Grand Forks." Turkeys of all six varieties approved by the *Standard of Perfection* were entered, those of the Slate variety for the first time.

The record number of turkeys, together with the exhibits of suppliers and dealers, taxed the capacity of the Grand Forks City Auditorium and the time and patience of the judges. The large number of turkeys en-

---

[22] See Appendix G: "Turkeys . . . Show Many Characteristics" for more examples from the February 2, 1927, *Grand Forks Herald* column.

[23] *Grand Forks Herald*, February 5, 1928.

ATTEND THE TURKEY SHOW

SAVE BY SPENDING HERE

EVERYONE WILL BE INTERESTED IN THE

ALL-AMERICAN TURKEY SHOW

THIS IS THE LARGEST OF ITS KIND IN THE WORLD AND WILL BE HELD IN GRAND FORKS JAN. 30-31-FEB. 1-2-3.

AND IN KEEPING WITH THIS BIG EVENT

ODELL'S OFFER SPECIAL LOW PRICES IN EVERY DEPT. FOR

TURKEY WEEK!

Both The Show And Our Values Will Be Profitable To You

Regular $2.25 SATIN CHARMEUSE
40 inches wide. A good range of desirable color including black. Special per yard—
$1.59

$3.25 Pure Linen LUNCHEON SETS
50 inch cloth with six napkins to match. Colored damask designs. Set for—
$2.59

81 x 90 PEQUOT SHEETS
The last chance to buy this well known brand sheet at, each—
$1.39

Full Fashioned PURE SILK HOSE
Slight irregulars of a well known $2.00 quality. Big range of colors in all sizes. Pair—
98c

FINAL REDUCTION SALE OF WINTER COATS
VALUES TO $18.50....... $4.95
VALUES TO $25.00....... $6.95
VALUES TO $35.00.... $18.50

NEW SPRING DRESSES
FEATURED AT LOW CASH PRICES
Generous selections as to style, material and color. Big values at
$15.00 - $16.75 $18.50 & $20.00

HOUSE FROCKS
Clever creations made of printed broadcloths and reps—
$2.95

SILK SCARFS
New shapes in gorgeous colorings. $3.50 and $3.95 values at—
$2.95

STAMPED PILLOW CASES
Hemstitched and stamped cases. Assorted designs, priced special at, pair—
89c

CURTAIN PANELS
Full size, lace net panels with border design and fringed ends. Pair:
$1.98

MEN'S HOSE
Novelty rayon mixed. A new shipment of newest patterns at, pr.:
25c

WOOL COAT SWEATERS
For women and children. Prices greatly reduced to—
$1.95 & $3.95

PURE LINEN TOWELING
Steven's pure linen crash. Short lengths of a 35c quality. Sold by the piece in bleached or brown at, yard—
18c & 20c

FANCY ROBE BLANKETS
Part wool, size 66x80. Novel Indian designs. Regular $3.75 value, for—
$2.97

Odell's was located on South Third Street. Stores in Grand Forks featured sales and special prices during the week of the All-American Turkey Show. Image from January 29, 1928; used with permission of Grand Forks Herald.

tered in the 1928 show also prompted the suggestion at the meeting of the Northwest Turkey Breeders' Association on January 29, 1928, that entries for the 1929 show be limited to six hundred and that entries be closed ten days before opening day.

The 574 turkeys entered in the 1928 All-American Turkey Show included 307 Bronze, as many as were shown in all classes combined at any other major show in the country. Of the 307, 102 were young hens, "remarkable pullets, constituting beyond any question a world record for any breed." Most gratifying about this class, according to George W. Hackett, "was its uniformity of quality, there being practically no outstanding specimens. In fact, the difference between the quality of the first and tenth pullet was of minor degree." When judged, very few of the 102 "were given any poor marks on the comparative score cards." What this indicated, Judge Oscar Grow believed, was that breeders were "rapidly agreeing upon a more practical and uniform type" of turkey. "This type," he said, was "more in keeping with the domestic and non-roaming nature."

Of the 307 Bronze, 115 were young toms, believed to be the largest number "ever cooped together at one exposition in the world." Although all of the young toms were good quality breeding stock, only twenty could be placed, and they were sold for up to $150 apiece. Buyers, watching the judges and listening to their comments, had their checks written before the judges had finished their scoring. "It was regarded as absolutely safe," Hackett believed, "to invest these larger values in any bird under the ribbons, as was the case in many instances of transfers during the show."

Thousands of visitors, breaking all records, attended the 1928 All-American Turkey Show, so many they often hampered the judges in their work. There were over 3,000 paid admissions, and 635 attended the lectures. "The big part women have in the turkey raising industry," noted the *Grand Forks Herald*, was "shown by the fact that fully two-thirds of the people attending the lectures were women." The lectures were held in the main section of the auditorium balcony, "in an atmosphere that fitted the subject, the auditorium having a view of the hundreds of birds that filled the lower floor, the stage, and a part of the sides of the balcony. Gobblers, at times, vied with the speakers for attention."

Several hundred people crowded into the auditorium balcony on the afternoon of February 2 to hear an address by Dr. E. Billings, extension veterinary from the University of Minnesota Farm School, who spoke on diseases affecting turkeys. Blackhead, Billings said, caused 90 percent of turkey losses. It could not be cured, only prevented. Other speakers were O. A. Stevens, who spoke on the importance of sanitation and on the breeding, feeding, and marketing of turkeys. More study, he said, was needed on each of the topics. P. F. Trowbridge, director of the North Dakota Experiment Station at the North Dakota Agricultural College, spoke on "A New Era in Agriculture."

A. F. Schalk, veterinarian from the North Dakota Agricultural College, was among the leading authorities on turkey diseases, their spread, and their treatment. "Considerable study of turkey diseases," he said, was being made at the Agricultural College. In 1927, flock owners had sent in 459 specimens for study, compared to 150 in 1926. Of the 459, 150 had "blackhead trouble," 79 were suffering from a disease of the organ "comparable to the human appendix," 43 "suffered from tapeworm, 32 had pneumonia, 29 had a form of cholera, and 24 had tuberculosis, while the remainder of the birds were affected in various other ways." Schalk then proceeded to discuss each of the diseases, explaining the causes and the methods of control or cure. "Profit or loss," was "determined in large measure," he pointed out, "by the health question."

Another large crowd gathered in the program balcony on February 3, the last day of the show, to hear a lecture on raising turkeys in confinement. "Success and profits that were likely to arouse wheat farmers to envy," noted Dr. Williams, were the result of raising turkeys in confinement. Several of the men and women who had "made these records" and had written the letters quoted by the speaker were in attendance to explain how they had done it and to answer questions.

Among them was Ed Brenna of Grand Forks, North Dakota. He had hatched 186 chicks and had raised 182 of them to maturity. The four chicks that died were deformed. His success, he believed, was because his turkeys were raised in confinement. Robert L. Peterson of Bisbee, North Dakota, no longer raised turkeys by the range method of allowing poults to run with the hens. On one acre of land, he had raised 828 turkeys to maturity. The top weight was twenty-seven pounds, compared to the twenty pounds when he had allowed his turkeys to range and forage for feed. His feed bill was only $375. Mrs. Peter Forester, Pisek, North Dakota, raised 231 turkeys in confinement and realized a profit of $550. A. Hoffman, a plumber in Dixfield, Maine, made "a very good return" from the turkeys he raised on a single city lot.

The turkey banquet, held Wednesday evening, February 1, was the largest ever held in connection with the turkey industry. Ed L. Hayes—retiring president of the Northwest Turkey Breeders' Association—presided, and most of the evening's program was broadcast over KFJM, the University of North Dakota's radio station. Manville A. Johnson of Michigan, North Dakota, vice president of the Northwest Turkey Breeders' Association, "praised the community spirit and co-operation that had made the All-American Turkey Show possible and called for "a standing vote of thanks" for Hayes. Responding, Hayes noted that "the support given to the show by the people of Grand Forks was splendidly illustrated by the large attendance of businessmen and other residents of the city at the banquet."

Hayes also commented on the remarkable growth of the All-American Turkey Show in four years. There were 150 entries in 1924 and 574 in 1928; 47 attended the banquet in 1924 and 387 in 1928. But, Hayes said, his ambition was to manage a turkey show at which there would be over six hundred exhibits from at least thirty states, and, in the hope of reaching that goal, he would again be manager in 1929. Dinner over and program concluded, the dining-room tables were removed and dancing ensued.

A feather-guessing contest piqued interest and provided excitement at the 1928 All-American Turkey Show. Claude Sipple, contest manager, offered a twenty-pound young Bronze tom, valued at $50, to the person whose guess was closest to the number of feathers on another twenty-pound young Bronze tom. More than two thousand contestants submitted their best guesses. At 3:00 p.m. on Wednesday, February 1, the young tom was killed and his feathers plucked and counted. He had 5,727 feathers. Mrs. Alvin Hunter of Gilby, North Dakota, was the winner. Her guess was 5,731.

What the writer for the *Grand Forks Herald* termed "a battle royal" provided the suspense and drama for the 1928 All-American Turkey Show.

What occurred was to be expected, when the purpose of the All-American Turkey Show was to bring turkey varieties as close to perfection as was humanly possible. In contention for the show's Grand Champion prize were two yearling toms, both owned and exhibited by women: a Bronze exhibited by Mrs. J. O. Beck of Orchard, Idaho, and a White Holland exhibited by Mrs. C. H. Folz of Drayton, North Dakota.

"So superlative was the quality and so evenly matched were the fine characteristics of the two yearling toms," wrote E. H. Cooley for the *Grand Forks Herald*, that the judges became deadlocked 2 to 2, "each judge standing his ground, refusing to recede in any particular from his decision." Unthinkable and unprecedented at any exposition noted for amiability, community spirit, and good sportsmanship, the All-American Turkey Show, for the first time, closed without a decision on the show's Grand Champion. At previous shows, more often than not, the judges were unanimous in their choice of Grand Champion.

If newspaper headlines, front page articles, and article headlines scream, the screams of those in the *Grand Forks Herald* in the closing days of the 1928 All-American Turkey Show were heard in every part of the turkey world. On Wednesday, February 1, 1928, the headline read, "Turkey Show Champs To Be Named Today." The turkey banquet was scheduled for 6:30 p.m. at the Hotel Dacotah, and it was "hoped that the grand championship winner" would be announced "at that time."

On Friday, February 3, the headline announced, "Judging Of All Turkey Classes Ended," noting that the grand champion had not yet been determined. The judges were certain to finalize the naming of the grand champion yet that evening. They could not concur, however, on which tom should win, and so Saturday's headline read, "Turkey Judge Recalled to Break Deadlock," explaining that "the four judges and Ed L. Hayes, manager of the show, have appealed to Judge M. C. Herner of Winnipeg to return and help settle the question." A second article noted, "White Holland And Bronze In Race For Honor: Each Yearling Tom Supported By Two All-American Judges."

Hotel Dacotah on North Third Street. Courtesy of Special Collections, Chester Fritz Library, University of North Dakota.

Because Judge Herner's return was delayed, "the two 'thousand dollar' turkeys" had been "transferred Saturday afternoon from the show quarters in the city auditorium to the window of the Herald and proved considerable of attraction to the late Saturday crowds." Just as the judges had failed to agree on which of the turkeys was the superior, so did the spectators disagree, so that "some very heated arguments" were heard. Spectators may have disagreed on the qualities of each turkey, but everyone agreed that, "the two birds, each weighing close to 40 pounds, would look well if their year's accumulation of meat were reposing on a banquet board."

The headlines for Tuesday, February 7, 1928, announced, "Bronze Tom Declared World's Best Turkey" and "Judge Herner Comes Back From Winnipeg To Cast Deciding Vote, Breaking The Deadlock Which Held

Long After Close of Show." The "momentous decision" had been "awaited for three days by turkey fans all over North Dakota." Judge Herner's word "was final."

The battle royal opened and was waged this way. The four judges were George W. Hackett, W. E. Stanfield, O. J. Weisner, and Oscar Grow, renowned judges all and each a specialist in his field. "After the most difficult week of judging ever faced by poultry judges," noted the *Grand Forks Herald*, during which each judge had placed winners in "world record classes" of one hundred or more turkeys of a variety, each judge "was positive that his selection for the highest honor was absolutely right."[24]

W. E. Stanfield, poultry expert and associate editor of the *Farmstead Stock and Home*, Minneapolis, Minnesota, judged White Hollands at the 1928 All-American Turkey Show. With so many entries in each of the classes, and with the best birds in each class being comparable in quality, he'd had to make fine distinctions when placing the winners in each class.

Stanfield's choice for the Grand Champion of the 1928 All-American Turkey Show was the White Holland tom. "This White Holland tom," he insisted, "is one of the most outstanding birds of merit that I have ever handled in my show experience covering over twenty years of judging in poultry shows on the American continent, including Madison Square Garden, Chicago, and West Coast expositions." Because the White Holland tom was a bird of ideal type, color, weight, and general proportions, Stanfield refused to yield.

Judge Oscar Grow of Iowa also judged the White Hollands at the 1928 All-American Turkey Show. He took "a firm stand for the White Holland by the side of Judge Stanfield." Hackett, recognized poultry expert and an acknowledged authority on Bronze, judged the Bronze at the 1928 All-American Turkey Show, and he favored the Bronze tom as Grand Champion.

"I cannot recede from my decision," he said, "in placing this Bronze tom before all others in the entire show." The difference of opinion on which tom was the better developed, Hackett said, arose because of a difference in conception and interpretation of the description in the *Stan-*

[24] *Grand Forks Herald*, February 5, 1928.

dard of Perfection. "This is a point," he insisted, "which can be settled only by the authority we have regarding the intent of the Standard and the application of the terms used therein." The Bronze tom, he insisted, had "most excellent quality," and he was "right in size and type" to represent the Bronze.

The Bronze tom was "admittedly a little slack in breast. He showed weariness from his long journey from Idaho and continuous cooping. However, when placed out on the floor, he very quickly developed the exceptional type I saw in him before placing him under the blue ribbon. He possesses most excellent color in every section, and when out upon the floor he immediately fills the eye."

For a time, it appeared that a compromise would be to choose a young Bourbon Red tom as Grand Champion, O. J. Weisner's choice. Weisner judged the Bourbon Reds at the 1928 All-American Turkey Show. The Bourbon Red young tom, also owned and exhibited by a woman, Gladys Honssinger of Lebanon, Missouri, had won the Chamberlain Feed Company's large cup. When the other judges would not accept the Bourbon Red young tom as Grand Champion—it was less than one year old and too immature—Weisner sided with Hackett on the Bronze. The result, a deadlock.

The fifth judge at the 1928 All-American Turkey Show, M. C. Herner, poultry specialist at the University of Manitoba in Winnipeg, had returned to Winnipeg on business and to judge another turkey show. "The nature of the call to the Manitoba turkey expert" to return to Grand Forks to break the deadlock, "and the manner of his coming," noted the *Grand Forks Herald*, added "an interesting chapter to the romance of the great new turkey industry as reflected in this 'world's greatest' turkey show, but of even greater note is the compliment it pays to the splendid quality that is so evenly found in the two turkeys, kings of their separate breeds."[25]

When Herner returned to Grand Forks, according to the paper on February 5, and rendered a decision that named the grand champion, the winner would receive "the highest title ever accorded a feathered champion in the world's history," and it would "make his owner the proudest woman in the turkey raising industry."

---

[25] *Grand Forks Herald*, February 4, 1928.

Expected to arrive in Grand Forks at noon on the Northern Pacific train on Sunday, February 5, Herner did not arrive until noon on February 7. He was met at the depot by Ed L. Hayes, show manager; Ray Andrews, Petersburg, North Dakota, superintendent of the All-American Turkey Show; and two representatives of the *Grand Forks Herald*. Herner was taken by automobile directly to the Herald building. No one was allowed to speak to him and no one was allowed in the room with him and the two young toms.

For two hours, Herner carefully studied the two turkeys—feather by feather, bone by bone, and portion by portion. While he did so, noted the writer for the *Grand Forks Herald*, "anxious spectators thronged the street" outside the Herald building. Among them were Mrs. C. H. Folz, owner of the White Holland tom, and Hackett, "who had placed the Bronze for grand championship honors. Each declared it the happiest day of their lives, to one the award of reserve champion meant a place close to the top in the turkey breeders' world, to the other it meant the biggest event in his life as a poultry judge."

After what anxious spectators must have considered an interminable wait, Herner announced his decision. The Bronze young tom was the 1928 All-American Turkey Show Grand Champion. The young toms were "two good birds," he said. "I don't wonder that the judges found it impossible to decide between them. These two birds are outstanding in their own breeds, as high class specimens as I have ever seen."

The Bronze, Herner declared,

> has the finer qualities, however. He is a bird of outstanding quality of bone. I never saw nicer wings, especially in the secondaries. He is exceptionally well marked in the tail coverts, better than any I saw in the adult class in the show. He has . . . magnificent bronzing. His condition is only fair because of a long cooping; during the 12 days he has no doubt lost a little weight. He is carrying a little less weight than he might for his size. This bird is just coming into his prime, he should be even better a year from now than he is today.

The World's Two Finest

At the 1928 All-American Turkey Show, judges were deadlocked on the choice of the show's Grand Champion. Shown above is the Bronze yearling tom exhibited by Mrs. J. O. Beck of Orchard, Idaho. Shown below is the White Holland yearling tom exhibited by Mrs. C. H. Folz, of Drayton, North Dakota. Image from February 5, 1928; used with permission of Grand Forks Herald.

The White Holland, Herner said, has "just passed the prime." The fault he found with the White Holland was that he was "a shade coarse . . . and fat." He was, however, "a remarkably fine bird," and he had "crowded the grand champion very closely for the supreme honors." The White Holland tom was awarded the reserve championship of the 1928 All-American Turkey Show, "an honor almost equally coveted by experts in the turkey industry."

After Herner's momentous decision, the Grand Champion was taken to a photographer's studio where photographs were taken, "destined to be used in the news reels of the movies." The judges, Herner and Hackett, "to whom the bird" owed the decision, "posed with him, as did also Miss Mabel Benson, recorder of the show, and circulation manager of the *Turkey Journal*."

The battle royal that (finally) concluded the 1928 All-American Turkey Show ended with the combatants retiring from the field, honor served and both sides proclaiming victory. The Bronze forces having captured the Grand Championship and the White Holland forces the reserve championship.

CHAPTER 10

Gentlemen Prefer Bronze (Turkeys)

(with apologies to Jane Russell and Marilyn Monroe)

AN UNQUALIFIED SUCCESS BY 1928, the All-American Turkey Show was acknowledged to be a legal entity when it was incorporated in November. Fittingly, the Certificate of Incorporation was dated November 19, near the time when the All-American's namesake graced Thanksgiving Day dinner tables. The single-page Certificate of Incorporation is as brief as it is interesting, interesting not only for its articles or clauses, but also for the individuals whose names are listed as directors or trustees.

The Certificate of Incorporation was headed, "ARTICLES OF INCORPORATION OF All-American TURKEY SHOW." Those undersigned had associated "themselves together for the purpose of forming a corporation under the laws of the State of North Dakota," and they had agreed to six brief clauses.[1] The third of the six clauses in the Cer-

[1] The first clause declared that the name of the corporation was, "All American Turkey Show." Parsing the words and noting their arrangement allows the show's name to be taken in a number of ways, all speculative perhaps, but interesting nevertheless. In the Certificate of Incorporation, the words "All" and "American" stand alone, no hyphen. References to the show, in the *Grand Forks Herald*, for example, often made "All" and "American" a compound word by inserting a hyphen, as if "All-American" was an adjective describing a turkey show. The show indeed became "All-American" when breeders from every turkey-producing state believed it incumbent on them to exhibit their birds at the show, and when the American Turkey Breeders' Association was organized. It may be that the phrase "All American" meant turkeys from the entire American continent, including Mexico and the Canadian prairie provinces. It may be a stretch to assume that the words "All American Turkey" were to serve as a reminder that turkeys were native only to America and that all turkey varieties had their origin or their development in America. The second clause in the Certificate of Incorporation merely reiterated what lecturers had emphasized since the inception of the All American Turkey Show. "The object and business" of the corporation was "to foster, promote and advocate the raising of turkeys, poultry and other live stock to the end that diversified farming be encouraged and developed." To further this "object and business," the corporation was to "conduct the exposition heretofore held at Grand Forks which has been called the 'All American Turkey Show' and such other turkey, poultry and agricultural shows as may be deemed advisable." For consistency, the hyphenated form is used throughout this text.

tificate of Incorporation stated the obvious. "The place of the principal business of this corporation shall be in the City of Grand Forks, State of North Dakota."

On the last day of the 1928 All-American Turkey Show, Friday, February 3, when several hundred people crowded into the balcony of the city auditorium to hear him speak, Dr. E. Billings, extension veterinary of the University of Minnesota Farm School, "lauded his listeners for the great enthusiasm shown at this, the greatest turkey show in the world."

He then "admonished" his listeners "to see that it did not get away from Grand Forks." Given the success and the reputation of the All-American Turkey Show by 1928 and the great enthusiasm with which it was supported, it was unlikely that Grand Forks business owners and residents would have allowed the show to be moved away from the city. Because Grand Forks was acknowledged to be the center of the largest turkey-producing area of the world, it was even more unlikely that any city would have had the temerity to challenge Grand Forks for the honor of holding the show and of reveling in its prestige.

George W. Hackett epitomized the All-American Turkey Show. Believing in the All-American with every fiber of his being, Hackett was convinced that this remarkable event would continue for many years, and he chided the naysayers who predicted that exhibitors, promotors, and townspeople would soon tire of them. The fourth clause in the Certificate of Incorporation suggests that those incorporating the show shared Hackett's conviction. "This corporation," the clause stated, "shall exist for the period of twenty years from the date thereof."

The Certificate of Incorporation sets the number of directors or trustees at eleven, identifies them, and notes where each lived. The individuals were either prominent businessmen or turkey breeders of note and most of them lived in or near Grand Forks, North Dakota, or East Grand Forks, Minnesota.[2]

---

[2] "The number of directors or trustees shall be eleven, and the names and residences of those to serve until their successors are elected and qualified" were as follows: Alfred Malmberg, Crookston, Minnesota; Ray Andrews, Petersburg, North Dakota; Victor Hartl, New Rockford, North Dakota; G. S. Mairs, Lisbon, North Dakota; Manville A. Johnson, Michigan, North Dakota; J. C. Sherlock, East Grand Forks, Minnesota; Julius F. Bacon, R. F. Bridgeman, I. Papermaster, D. F. McGowan, and W. W. Blain, all of Grand Forks, North Dakota.

The sixth clause of the Certificate of Incorporation stated that "the amount of the capital stock shall be nothing." The purpose of the All-American Turkey Show was "to foster, promote and advocate the raising of turkeys," not to issue stock, pay dividends, or enrich stockholders.[3]

J. C. Sherlock, I. Papermaster, and W. W. Blain were the three who "executed the foregoing instrument" before Harry P. Rice, Notary Public, North Dakota. Rice's Commission expired October 31, 1930. Those associated with the All-American Turkey Show since its genesis and familiar with the one held in 1928 could be forgiven for assuming that the All-American show could not get better.

But, it did, with the sixth show, held January 28–February 1, 1929. By every measure, except in the number of turkeys exhibited (512), the 1929 All-American Turkey Show surpassed any previous show. Neither superlatives nor Chamber-of-Commerce-like pronouncements do justice to descriptions. Nor did a blizzard, blocked roads, or frigid temperatures dampen enthusiasm or deter breeders—even those from a distance—from attending and exhibiting their turkeys.

For the entire month of January 1929, the temperature in Grand Forks remained below zero, and the thermometer hovered at -12°F during the week of the show. On February 1, the last day of the show, for the first time since January 22 the temperature rose above zero—to 5°F. The temperature at Minot, North Dakota, on January 25 was -34°F, the coldest point in the Northwest. The low that day in Grand Forks was -22°F and the high was -9°F. On Saturday, January 26, when "the blue bloods of American turkeydom" were arriving by rail, automobile, and George E. Lamb's familiar "hauler," it was -30°F in Grand Forks.

Turkeys, bred and raised in the Northwest, did not mind the cold, and their owners, accustomed to the weather, were inured to it. Colonel C. M. Martin, proprietor of the Martin-Laney Farms at Dallas, Texas, was asked how he liked the weather. When turkey people get together, he replied, "breaking the ice" was the expression used when referring to the starting of conversations. But, he said, it would require "a lot of breaking up" in Grand Forks, because there was "plenty of ice and it is thick."

---

[3] For commentary on each of the members, see Appendix E: Brief Biographies of the Directors and Trustees Listed on the All-American Turkey Show Articles of Incorporation.

"Like the sempiternal circus that each year is 'bigger and better,'" noted the *Grand Forks Herald*, "the sixth annual All-American Turkey Show would open its doors to the public at 9:00 a.m. on Monday." George W. Hackett, nationally known turkey authority, promised that the show would have "the largest assortment of superfine turkeys" that had ever been "assembled under one roof." Hackett admitted that his promise might have sounded like "the regular old ballyhoo," but he was "absolutely sincere" when he said that "never before in the history of the world" had "so many fine turkeys been assembled in one place as those that were exhibited" at the 1929 All-American Turkey Show.[4]

In its Sunday edition, the *Grand Forks Herald* quoted Hackett, the show manager. "There is no question," he said, that the show had "outdone all past events of the organization by a wide margin." Articles on the paper's front page that day read, "Officials Declare Turkey Show Was Best World Has Seen," "Hackett Declares Exposition Has Outdone All Former Shows: General Upgrade in Quality Birds Shows," and "Show Birds in Better Form."[5]

Judges, exhibitors, and spectators reportedly "were unanimous in their decision that the 1929 exposition was the best turkey show that the world had ever seen." Judge A. D. Walker of Memphis, Missouri, had attended previous All-American Turkey shows, but, he said, the 1929 show was "the best thing in the line of a turkey show" that he had ever seen, and "the class of the show was forty to fifty percent above what it had ever been before."

Colonel Martin, who had exhibited turkeys at all the eastern shows, was "particularly enthusiastic" about the All-American Turkey Show, because, of all the shows, it was "the best." G. C. Darrah, of Ray, Montana, another enthusiast, said that he would "bring turkeys to the All-American if they did not have a feather on them," and he would be satisfied if they did not win prizes because the All-American "was worth coming a thousand miles to see." Were the All-American Turkey Show held a great distance from Grand Forks, he was certain, city residents, "would pay money to go to it."

---

[4] *Grand Forks Herald*, Sunday edition, January 27, 1929.

[5] *Grand Forks Herald*, Sunday edition, February 3, 1929.

No "ballyhoo" this, the enthusiasts spoke true. "Aristocrats of the turkey world" from ten states and two Canadian provinces were exhibited at the 1929 All-American Turkey Show where they faced "the keenest competition ever experienced." Newspaper accounts observed "[t]he most experienced and progressive breeders" in the country were now exhibiting higher quality birds. Mrs. C. H. Folz of Drayton, North Dakota, had exhibited her White Holland turkeys at every show, and her young tom was runner-up Grand Champion in 1928. At the first All-American Turkey Show, she said, there were only a few birds and "almost any turkey" was "a show bird." Now, however, "to get any attention at all a turkey must be better than good."[6]

Judge O. J. Weissner, Fargo, North Dakota, had judged a "very strong class" of twenty-nine Bronze "good quality yearling hens." Competition was keen, because "all were well bronzed" and all were of "very good type and size." The first-prize bird, however, "had the amount and quality of bronze" and the "exceptional size and type" to make it "a very easy winner."

Quality was evident in every class judged at the 1929 All-American Turkey Show. In the White Holland young hen class, twelve of twenty-six entries placed. In the adult Bronze tom class of thirteen birds, five received awards. Of the seven entries in the Bourbon Red yearling hen class, five won prizes. All the Bourbon Red entries, the judges agreed, were "greatly improved" in quality over those shown in 1928. "This is proof of the good that the turkey shows" were accomplishing, they noted, and, "in the last analysis," that was "the aim of the turkey shows."

"Another noteworthy advance," evident at the 1929 All-American Turkey Show, was that breeders had shown "better knowledge" in selecting turkeys for exhibition and in the "more critical examination" of the turkeys they purchased. Among the busiest places in Grand Forks during the week of the show was the All-American Turkey Show's business office,

---

[6] Judge A. D. Walker, Memphis, Missouri, was impressed with the quality of the entries in the 1929 All-American. He judged a bronze cockrel class of twenty-five, an "outstanding class of toms, all very high quality." Of a class of 125 young bronze hens, the largest class he had ever judged and possibly the largest class ever shown in the country, 75 were "high class pullets" and they placed 1–25. "The competition was very close," he said, and he had rendered his decision "on some very minor defects." Walker had also judged a class of twenty-nine White Holland young toms, all "very high class birds," and he had placed down to fifteen places. The twenty-three young Bourbon Red hens in the class he judged were "very outstanding birds."

"hid[den] well in one of the stage dressing rooms." Mrs. Joe Ingram had a force of workers, busily checking entries and other clerical work connected with sales of prize turkeys.

Sales in 1929 were "one of the outstanding features of the show." Buyers and sellers were so numerous that the sales department "could not keep up with the pace of them." Ida Midgarden of Grafton, North Dakota, whose White Hollands had taken eleven prizes in five classes, sold three pens of hens and over $1,000 worth of eggs at $5 each. One man purchased forty of them. Colonel Martin from Dallas, Texas, sold more than enough turkeys to pay his expenses to the show.

In 1929, much more than before, the *Grand Forks Herald* devoted more column inches to explanations of how the turkeys were transported to the show, how they were received, and how they were cared for. The reason for this, at least in part, may have been because George Hackett was now show manager. "In gaily colored shipping coops," the paper noted, "the birds arrived on every express train" from "all points of the compass." Owners were assured their turkeys would have safe delivery to the City Auditorium by specially trained employees. These "trained employees," more than likely were employees of the Railway Express Agency (REA), the United Parcel Service and FedEx of its day.[7]

In 1917, President Woodrow Wilson commandeered the railroads to move federal troops, their supplies, and coal, and all existing express companies were consolidated in a single company, American Railway Express Agency. In 1929, the American Railway Express Agency's assets and operations were transferred to the REA, owned by eighty-six railroads in proportion to the express traffic on their lines. With its headquarters in New York, the REA served every part of the country with over 27,000 offices and representatives.

The REA's red-diamond trademark, green delivery trucks, baggage carts, and railroad equipment were a common sight, and people depended on the company for their shipping needs. The REA shipped virtually everything—ladders, auto parts, farm machinery, explosives, groceries, livestock, baby chicks—and the prize turkeys exhibited each year at the All-American Turkey Show.

---

[7] *Grand Forks Herald*, Sunday edition, January 27, 1929.

Once in the auditorium, decorated with American and Canadian flags and strings of pennants hanging from the ceiling, the turkeys were turned over to the show's assistant superintendent and then examined by a veterinarian who checked them for injuries and disease. The assistant superintendent then checked the long lists of entries, to determine to which cage the bird had been assigned on the show room floor.

"To the uninitiated observer," the *Grand Forks Herald* noted, there was "all the formality to this that is shown in a hotel when the clerk assigns a distinguished guest to a parlor suite, on the second floor, southern exposure." This done, someone held the turkey while a metal band was placed on its leg with "a pincer-like tool." The metal band had the number of the coop in which the turkey would remain throughout the show. A tag on the coop noted the bird's class and owner.

Once in the coops, the *Herald* continued, the turkeys had "nothing to do except to peck at the food" and "drink water," while awaiting "the crowd of mere humans to come and stare" at them. Members of the "show force" cared for the birds, providing feed and water and keeping the coops clean. A veterinarian examined the birds each day for disease. "Looking after distinguished personages," noted the *Herald*, was "a ticklish business," because the hens, toms, and young birds exhibited at the All-American Turkey Show were "aristocrats." The "aristocrats" in 1929 consisted of five varieties of turkeys: Bronze, White Holland, Narragansett, Bourbon Red, and Slate.

The educational programs were again an outstanding feature of the show, but exhibitors were more interested in them than ever before. According to the *Herald*, the educational sessions provided "a liberal education in turkey raising." Those attending the programs could learn from "successful specialists" how to make the turkey industry more profitable. Doctor A. F. Schlag spoke on "Disease Losses" for the Northwest in 1928, and Dr. W. Billings spoke on keeping turkeys in confinement, believed by many breeders to be the best way to raise them.

Billings also noted that because turkey numbers remained low, there was no danger of overproduction. In 1900, he said, 6.5 million turkeys were produced, but only 3.5 million in 1920. In 1929, perhaps 4 million would be produced, and there was a marked increase in the country's human population. So, he concluded, "we have a long way to go before we need to get excited over glutting the market." Turkeys were a profitable crop, he noted, and flock owners in North Dakota played "an important part in the growth of the industry."

The title of Judge W. E. Stanfield's address was "Gentlemen Prefer Bronze," and, he said, the All-American was "lending an influence" that was encouraging the turkey industry. Mrs. Clara Sutton of St. Paul, Minnesota, also spoke. Sutton was best known for her columns in the *Farmers' Wife*, a magazine popular with farm women.

Mrs. Sarah Reitz, editor of *Turkey World*, spoke on "My Contact and Experience With Turkey Raisers." The previous eight years, Reitz said, had been the "most wonderful" years of her life. She had devoted more time to turkeys than to anything else, and she "had breathed more prayers for the life of some little birds" than for herself. Eight years earlier, she had been "turned out on her brother's farm to die." On the road to the farm, she saw seven turkey chicks and asked for them. They were her first turkeys, and, since then, she had "lived entirely with the birds in the open." She had recently been called to Chicago to attend a show and she remained there—with *Turkey World*.

Civic and public support for the All-American Turkey Show may have been even more evident in 1929 than at previous shows. Judges, show officials, and many exhibitors were luncheon guests at both the Lions and Kiwanis Clubs. Representing the Grand Forks Commercial Club, C. W. Ross assured show officials of "the continued support and cooperation of the citizens of the city." Colonel Martin was particularly impressed with the All-American Turkey Show and how it was managed. There was always "teamwork" among turkey breeders, he said, "but the All-American . . . brought out teamwork from an entire community."

Because the All-American Turkey Show had begun quite modestly, each show had one or more "firsts." The 1929 show, however, outdid previous ones, not only in the number of firsts, but also in their significance. Ed L. Hayes, who aspired to manage a show at which six hundred turkeys were exhibited, had agreed to manage the 1929 All-American Turkey Show.

Unfortunately, he did not live to accomplish either goal, and George W. Hackett was named manager, a position he retained until 1940. Hackett acted as toastmaster at the 1929 turkey banquet and asked the nearly four hundred guests to stand in silent memory of the late Hayes. W. E. Stanfield of St. Paul, Minnesota, veteran turkey judge, read aloud a dedication of the 1929 All-American Turkey Show to Hayes.

Hackett had judged at each of the All-American Turkey Shows, but now, for the first time, he was show manager. The All-American Turkey Show "had a real manager in Mr. Hackett," Col. Martin observed, and the show was "bound to grow under his direction." It did, because Hackett, like his Creator, now had the opportunity to make the All-American Turkey Show in his own image.

Hackett was a stickler on the standards specified in the *Standard of Perfection*. At the 1929 All-American Turkey Show, judges used special scorecards of his design "in accordance with the various points that go to make up the perfect turkey." The "All-American Plan of judging," noted the *Grand Forks Herald*, "attracted attention" and "proved to be of inestimable value to the large number who congregated about the judge's table."[8] Those who believed that too much attention was paid to color were soon convinced that the judges also required "good bone, frame, and vigor." The new features were well received, noted the paper, and would be continued.

A Court of Honor, introduced at the 1929 All-American Turkey Show, was another new feature. The champion turkeys of each class of each variety were placed on the auditorium stage in one group, their cages flanking that of the show's Grand Champion. This way, spectators could see the best turkeys from nearly every turkey-producing state and Canadian province. Those who did not visit the All-American Turkey Show and view the Court of Honor missed a rare treat.

Colonel Martin exhibited at the 1929 All-American Turkey Show, and his farm's Bronze turkeys were described by the paper as being among the "outstanding features of the show." Martin's yearling Bronze tom,

[8] *Grand Forks Herald*, February 3, 1929.

the show's Grand Champion, was awarded a silver cup. Martin also received the Long Distance award, having brought his turkeys all the way from Dallas.

After the All-American Turkey Show closed, Martin wrote a letter to the *Grand Forks Herald* in which he expressed his thanks for the paper's coverage of the show. He also thanked show managers for the educational programs and entertainment, and he included a special thank you to the people of Grand Forks for making their guests feel welcomed during the week of the show.

Described by the writer for the *Grand Forks Herald* as probably the happiest person at the 1929 All-American Turkey Show, Borghild Brager, of Hoople, North Dakota, was a first-time exhibitor. At the age of thirteen years, she was the youngest exhibitor, and she was also the youngest exhibitor whose turkeys had taken prizes at an All-American Turkey Show. Miss Brager had finished grammar school at Christmas, and she planned to attend Park River Agricultural College in the spring.

Borghild Brager's parents raised Bronze turkeys, but, wanting to begin with a different variety, she had chosen to raise Slate turkeys, described by the Livestock Conservancy as a variety having a difficult color to perfect. From eggs she purchased from the O. B. Harmonson Farm in Justin, Texas, Brager raised five turkeys. The young tom and young hen she exhibited at the All-American took awards in their classes.

Slender, blond, bashful, looking younger than her thirteen years, and difficult to interview because she was unaccustomed to receiving honors, noted the writer for the *Grand Forks Herald*, "the smile that glows beneath her bright red hat shows the happiness she feels." Brager planned to continue raising Slate turkeys and exhibit them at future All-American Turkey Shows.

The 1929 All-American Turkey Show was also the first at which a movie actress made an appearance. Using the stage name "Hope," according to the writer for the *Grand Forks Herald*, the actress came to Grand Forks from New York, where she had appeared at Madison Square Garden. There, her stately carriage, stunning good looks, and fine figure had captured people's hearts and gained her many admirers. Hope, a Bourbon

Red young hen owned by Mrs. Gladys Honssinger of Lebanon, Missouri, was the pet of the Madison Square Garden show.

After she had placed first in her class, newsreel photographers "insisted on having Hope pose," which she did "most beautifully." Hope was "easily the pet of the show" in Grand Forks as well, when she performed at the All-American. Her performance included obeying her owner's commands to sit, roll over, follow her, and behave "as a proper young hen should."

By the time of the sixth annual All-American, the show had its regulars. One was George E. Lamb of Philip, South Dakota. Lamb transported his Bronze turkeys to Grand Forks, together with those he picked up along the way, in his "hauler," the trailer that he towed behind his coupe. Walter Burton, a judge, was "the prize storyteller of the show," and he had a story "on almost every prominent turkey breeder in the country." Burton was from Texas, "and his Lone Star drawl" did not "hurt the telling of the stories."

Max Morgan was another teller of turkey stories. One he told was about a fellow named Sam who asked a Texas town marshal for advice about a threatening, anonymous letter he had received, someone who had warned Sam to, "Leave my turkeys alone or you'll be killed." The marshal told Sam that he need only stay away from "that turkey roost." With that advice, Sam scratched his head and said, yes, "but I don't know whose roost that is."

The turkeys also had cutups among them. At the 1929 All-American Turkey Show, a Bronze young tom belonging to Alfred Malmberg of Crookston, Minnesota, developed "goat tendencies." He "showed these traits," according to the *Grand Forks Herald*, by consuming "the tag bearing his entry number and other data," pecking "longingly at a visitor's hat," and ending up attempting to masticate a steel pointer that he seized from the hand of a judge.[9]

The 1929 All-American Turkey Show closed on Friday evening, February 1. In its next day morning edition, the *Grand Forks Herald* noted that "exhibitors at the All-American Turkey Show, which has been under way here since Monday, packed up their turkeys and their prize awards

[9] *Grand Forks Herald*, January 30, 1929.

and departed to their homes Friday evening, after declaring that the 1929 All-American was the greatest ever put across."[10]

The Sunday editorial included the observation, "now that the show is over, the exaltation of the winners has subsided somewhat and the disappointment of the losers is not so keen as at the close of the judging, a fair estimate of the benefits of the show can be made."[11] The show had brought together turkey industry leaders from all over the country, and the advice and information they shared were "always beneficial." Breeders and laymen had been afforded the opportunity to compare turkeys and their varieties.

"Every step in the development of the turkey," the article continued, had been on display at the show, from the wild turkeys exhibited by Martin Magnus of Sterling, North Dakota, to the prize-winning "Magnificent Bronze," Bourbon Reds, and White Hollands shown by "nationally known authorities on turkeys." Breeders had to have been aware that "no greater group of turkey experts" had ever been assembled in the Northwest than those gathered in Grand Forks for the All-American. And yet, noted the editorial, "all these things are not the main advantage of the show."

"The main advantage" had to be measured in the "dollars and cents" that the All-American Turkey Show would bring to the Northwest. "Even a conservative estimate" of the amount would seem "absurdly large" to someone not in the turkey business. The best turkeys in the country had been exhibited at the show, and farmers had seen the turkeys that were "the very height of turkey perfection."

The imperfections of their own turkeys would have been glaring, had their birds been placed next to the prize-winning turkeys from all over the country. They would also have been made aware of the advantages of raising prize-winning turkeys when they saw that orders had "fairly poured into the hands" of those exhibiting them. Higher prices were paid for "superior birds."

"Better turkeys will come to North Dakota as a result of the show," the editorial predicted, because the market paid "a premium for birds fitted

[10] Men worked through Friday night, placing turkeys in shipping crates and sending them on their way to express offices, from where they would be sent to places ranging from Texas to Canada, and from Idaho to Missouri.

[11] *Grand Forks Herald*, Sunday edition, February 3, 1929.

in the North, and when the best birds of other sections are transplanted to the North, they will add to the premium now paid." "There," concluded the editorial, was "the main value of the All-American Show, the dollars and cents in additional payments for turkeys that will be forthcoming next year and for many years to come."

An article in the paper that day noted that "nothing is left of the 1929 All-American Turkey Show but the echoes of the praise of judges, exhibitors and spectators who saw the show and who were unanimous in their decision that the 1929 exposition was the best turkey show that the world has ever seen." Hackett, after managing the show for the first time, agreed. "There is no question," he said but that the show had "outdone all past events of the organization by a wide margin."

Perhaps the best measure of the 1929 show's success was that used by the *Grand Forks Herald* on February 2, 1929. "George W. Hackett, to whom is given credit for making the 1929 All-American the outstanding success in the realm of exclusive turkey shows," noted the writer, was re-appointed as manager, and he would be in charge of the 1930 show.

A fitting way to conclude the discussion of the early years of the All-American Turkey Show is to evaluate what was said and written about the seventh show, held January 27–31, 1930. Held three months after the stock market crash on October 29, 1929, in its features and numbers of visitors, exhibitors, and entries, the seventh All-American Turkey Show exhibited few effects of what Manager Hackett referred to as "the business depression."

Comments on the show, however, together with the show's features, serve as a portent of the All-American Turkey Show's future. The more perceptive, realizing the import of the comments and observing the show's features, might have framed an impertinent, if rhetorical, question: Were the purposes for which the All-American Turkey Show was inaugurated being served, and what purpose would be served by continuing to hold it?

Turkeys began arriving in Grand Forks on Saturday, January 25. The day before, a blinding snowstorm swept across much of the state, blocking roads and making travel difficult. All traffic on some roads was suspended, and many travelers resorted to using sleighs and bobsleds. Trains ran on schedule. On Sunday, February 2, for the first time since January 4, the mercury stayed above zero for twenty-four hours. On January 9 and 18, the highs were -18°F. The low on January 9 was -33°F.

Defying weather and the stock-market crash, exhibitors from across the country, many of them competing at the All-American for the first time, entered more than five hundred turkeys in the show, in "the greatest aggregation of show birds ever assembled in America." These breeders, noted the *Grand Forks Herald*, were those who were "stabilizing the industry throughout the country, making it at once a fascinating occupation, a profitable addition to the farm income, and a food source that consumers the nation over are greeting each year with finer appreciation."[12]

On January 24, the paper observed, the farmers of the Northwest were "just beginning to realize the value" of turkey raising as a "branch of agriculture," and "of the All-American as a factor in centralizing the interests of the turkey industry." No other "turkey organization in the United States" was "recognized as having the same importance to the industry," and this was "due largely to the educational programs held in connection with the shows held in the past six years." The 1930 show would emphasize "the marketing end of turkey raising" and would extend interest "beyond anything seen in the past."

Speaking at the banquet on the evening of January 29, M. C. Herner, one of the show's judges, noted that the All-American Turkey Show was "exercising a tremendous influence for good on the turkey industry." A "great inspiration for better turkey work" was "going forth annually" from the show. It was "significant," noted the writer for the *Grand Forks Herald*, that with the start of the All-American Turkey Show in 1924, no fewer than six exclusive turkey shows had sprung up across the United States, and "all look to the All-American as a leader in promotion of Standard bred turkeys and providing management in the conducting of such shows."

There was no gainsaying that the All-American Turkey Show had served many of the purposes for which it was intended, if not all of them. Features, new to the 1930 show, also spoke to its accomplishments. For the first time, two dressed turkeys were given away each day to the holders of lucky numbers on paid admission tickets. Hackett doted on feathers, their

---

[12] *Grand Forks Herald*, January 26, 1930.

pattern and color. To him, a dressed turkey was an insult, not only to the breeder who bred for color, but also to the turkey, the world's most noble bird. Encouraged by the All-American Turkey Show, consumers were eating more turkey, but they ate the flesh, not the feathers. Consumers' tastes were changing, and, to them, feather patterns and color were immaterial.

A headline in the *Herald* on January 28, the second day of the 1930 show, was, "Long List of Money Prizes and Trophies Will Give Winners of All-American Right to Gobble." What the winners gobbled about were the $1,500 in cash prizes and the more than thirty Special Awards. "Of special interest" were the awards giving the show "a thoroughly cosmopolitan character." Only those exhibitors who had come five hundred miles or more were eligible to receive one of them.

"To a large extent," winners of Special Awards now received a bronze plaque instead of a silver cup. Newly designed, the plaque bore "the form

Touting the All-American Turkey Show as the only original turkey show in the world. Image from January 28, 1930; used with permission of Grand Forks Herald.

of a turkey molded in relief and appropriately decorated." The donor's name and that of the award were placed on a plate that was attached to the plaque. The show's Grand Champion received the All-American Turkey Show bronze plaque.

What the *Grand Forks Herald* described as a "Long List of Money Prizes and Trophies" also indicated that the All-American Turkey Show was serving the purposes for which it was intended. Evident at the 1930 show was that the quality of the turkeys of each variety had been remarkably improved over what it was when the first show was held in 1924.

"The outstanding feature of the show" in 1930 was not the large number of entries, noted judges and officials, "but their remarkably high quality." "I never saw any real turkeys until I got here," was the comment of F. E. Cross, judge from the University of Minnesota. "There were seven Bourbon Reds judged today," commented Judge A. D. Waller, of Memphis, Missouri, "that made up the finest class I have ever seen in any show." Hackett, show manager and authority on the Bronze, said of the young toms in the Bronze division that they were "the finest shown anywhere. I'm satisfied that I've never seen so many birds of high quality anywhere as there are in this show."

Judges and show officials "laid stress on the fact that this was the best balanced show ever held" in Grand Forks, "the high quality being found in all breeds and classes." Judge Walter Burton, from Texas, waxed enthusiastic over the young tom class of Bronze. "I never saw a better class," he declared. "The folks that want to produce better turkeys can't do better than to talk to those who produced some of these leaders, and study their methods." The day after the show closed, February 1, the *Grand Forks Herald* noted that, "most of the high class turkeys which put this show in a class by itself for both numbers and quality of exhibits are on their way to their home roosts."

Because the best birds from the best flocks exhibited by the best breeders competed for the long list of awards and prizes, competition became keen and the work of the judges that much more challenging. To facilitate their work, judges in 1930 used Hackett's new scoring sheets, and, new to the show, a sliding scale of awards and ringside judging—all made necessary because of the large number of entries in each class and because so many of the entries were now of such high quality.[13]

What the *Grand Forks Herald* described as a "Long List of Money Prizes and Trophies" indicated another of the All-American Turkey Show's accomplishments. Lacking records, one may assume that prize lists grew ever longer as more donors—wishing to have their publication, their organization, their business, or their product associated with the world's first and greatest exclusive turkey show—donated ever more prizes.

Not only did more prizes make the work of the judges more challenging, it must also have indicated that more prizes were sometimes donated than could be awarded for qualities specified in the *Standard of Perfection*, such as excellence in conformation, bone, feather pattern, or color. Rather than offend a donor by declining to accept a prize, officials must have resorted to devising a category, so that the prize could be offered. The devising became more and more necessary the longer the All-American Turkey Shows continued.

Exhibitors, show officials, and others commended the *Grand Forks Herald* for its lengthy articles covering the All-American Turkey Shows, many of which appeared with E. H. Cooley's byline. The paper's Sunday edition on January 26, the day before the 1930 show opened, was no exception. It contained another of Cooley's lengthy and insightful pieces in which he reported on two prizewinning flocks of Bronze turkeys.

One flock was that of Mr. and Mrs. Miller Engh of McCanna, North Dakota, and the other was that of Mrs. Helen Baker, whose estate was on the banks of the Chester River on Maryland's Eastern Shore. Cooley's reporting will serve not only as a comment on some of the 1930 All-American Turkey Show's features, but also as a reflection on the first few years of the All-American Turkey Show and—to the more perceptive person who reads between the lines—as a portent of the All-American Turkey Show's future.

[13] With the sliding scale, when eighty or more birds of a variety were entered in one class, for example, twenty-five were placed: 1–5, $5 each; 6–10, $4 each; and so on down to the last group of 5, in which each received $1. With 4–8 in a class, 3 were placed: $4, $3, and $2.

Cooley had visited the Engh's farm at McCanna, and "much of our conversation," he wrote, "had to do with the All-American Turkey Show." Among the state's foremost breeders of Bronze turkeys, Mrs. Engh was a "consistent exhibitor" at the All-American. That she attended, she said, as much to learn as to win awards, was evident. In the Engh flock, Cooley wrote, "one may see a fair illustration" of what the All-American Turkey Show was doing for the turkey flocks of the country; the quality of turkeys had improved significantly since the show had become "an annual institution."

The improvement, Cooley said, was primarily because of the "increased knowledge of the industry which the educational side of the show inspires, and the opportunity afforded by these shows for the breeders to obtain the best strains in the country for the upbuilding of their flocks." Obtaining toms from "the best strains in the country," with which to "upbuild" their flock, was exactly what Mr. and Mrs. Engh had done. The results, so in evidence at the 1930 All-American Show, were so many turkeys of such high quality that they made the work of the judges that much more challenging.[14]

It is easy to see where this is heading, and Mr. and Mrs. Engh were not alone in "upbuilding their flocks" by purchasing toms from "the best strains in the country." Martin, R. H. Stapleton, and Mrs. Helen Baker also exhibited their Bronze at the 1930 All-American Turkey Show. A Martin-Laney Copper-Back Bronze was the show's Grand Champion in 1929.

One of Stapleton's Bronze frequently took first in the heaviest young tom division. Baker raised turkeys only for exhibition—1,500 of them in 1929—and sold them for $100 to $300 apiece. The best Bronze in each of the Engh pens, therefore, would be competing with the best Bronze from the very flocks from which they had purchased the toms with which to upbuild their flock at McCanna. Of such competition were turkey judges' headaches made.

---

[14] The Enghs had raised turkeys for more than a dozen years, 235 in 1929 and 370 in 1930. They sold their breeding toms for up to $35 each, and their dressed birds as No. 1 Fancy. Preferring to purchase toms rather than eggs, the Enghs attended the All-American Turkey Shows, where they could see toms with the quality and characteristics they wanted. They also purchased toms from the best Bronze breeders in the country.

The second of E. H. Cooley's prize-winning Bronze flocks was that of Mrs. Helen Baker, who raised turkeys on her historic estate on Maryland's Eastern Shore. Cooley larded his comments with material from an article published in the *Washington Post* on November 24, 1929, and sent to him by North Dakota Congressman Olger B. Burtness, a Grand Forks attorney.

The article, by Lucy Salamanca, was briefly titled, "Turkeys a Bargain at $300 Each." The subtitle, however, was lengthy, "But Not for Thanksgiving, Christmas or Other Table Use—A Unique Industry Flourishes Within a Few Miles of the National Capital Under the Guiding Genius of a Woman Who Violated All the Rules—Her Independence Has Made Her a World Authority on Turkey Culture."

Lucy Salamanca was Cooley's equal as a writer, and the opening paragraph of her article bears quoting. "Just what the Puritanical forbears of the 1929 Thanksgiving 'piece de resistance' would say," Salamanca began, could they look upon their effete descendants "strutting their stuff" before a movie camera and getting their gobbles recorded for all time to come in the talkies, is difficult to conjecture."

Not all of the turkeys on the Baker estate, Salamanca admitted, commanded the "munificent price of $300." It had, however, taken "a woman—and a city woman at that" to discover that such a price could be obtained "for the aristocrats of turkeydom." The woman was Mrs. Helen M. Baker. Newlyweds, Helen Baker and her husband, wanting to leave behind the congestion, noise, and "the turmoil of life" in New York, determined to find "a place where we could see the blue sky over our heads, feel the turf under our feet and hear the birds sing outside our windows." They found their ideal spot, "a beautiful old estate" of four hundred acres on the banks of Maryland's Chester River.

Their first years of farming, in an attempt to make the estate pay, Mrs. Baker confessed, were "best passed over." After a series of failures with crops, animals, and poultry, Helen Baker settled on raising turkeys, although even a bulletin issued by the Department of Agriculture advised that "turkey-raising was one of the most precarious of industries" and that turkeys were not only subject to "all sorts of diseases," they could also not be raised except "in small groups." Baker's first brood was twelve chicks, and she "unwittingly" killed "every one of them," when attempting to warm them on hot-water bottles. Her next broods drowned or died from exposure.

After five years of attempting to raise turkeys according to authoritative bulletins and after being "led astray" by "farm mothers" and "their red and black pepper, calomel, custards, cooked curds, linaments or tonics," Baker determined to "try the exact opposite" of everything she had read on raising turkeys and keep in mind "the habits of the wild turkey and not overlooking Mother Nature's share in the job." She devised a program by which she "might raise turkeys successfully," and, "with the courage of her convictions," followed the program "to the letter," starting with the best four unrelated pullets and one good cockrel that she could find. The results were "most gratifying," and they convinced her that she was "on the right track."

In 1927, Mrs. Baker raised 1,007 poults to maturity from forty hens and four toms, besides selling a hatch of day-old poults and four stock gobblers. In 1929, she raised over 1,500 turkeys. She accomplished this, Salamanca wrote, "by following entirely her own methods." Baker's "flock of beautiful pedigreed 'Bronze Beauties,' strutting and preening in the sunshine," testified "to the soundness of the methods she evolved."

Baker was recognized the world over for her successful methods of raising turkeys—in the United States, Canada, Ireland, Holland, Brazil, India, and Australia. In 1928, she was awarded the Turkey World's Women's Cup at the international turkey show in Chicago for her "outstanding service to the turkey industry." In 1929, at the June commencement exercises at the University of Maryland, she was awarded an honorary certificate of merit in agriculture in recognition of her contributions to the turkey industry.[15]

---

[15] Her book titled, *Turkeys: Common Sense Theories, Practical Management, Incubation and Brooding in Detail, Feeding Directions, Feeding Formulas,* published in 1928, was sold in every turkey-raising part of the world. Accepted as a standardized reference book among poultrymen, her book was published in a Second Revised Edition in 1933. In her book, Baker incorporated all her "hard-gained theories for successful turkey raising" so that a farm woman, "the humble wife of the farmer who works day in and day out on one of the most thankless jobs in the universe," could learn how to raise turkeys profitably for market. With the money, the farmwife could make her life easier and more pleasant by purchasing labor-saving devices and the "niceties of life" that would provide more comforts for herself and her children.

Baker was not content to rest on her laurels as a remarkably successful breeder of award-winning turkeys. Concluding her article, Lucy Salamanca noted that Baker was devoting "much time and effort" to what she believed was an "economically sound" marketing scheme, one that she had presented to market specialists in the United States Department of Agriculture. If implemented, the marketing scheme portended the future of the All-American Turkey Show.

Baker believed that turkeys should be marketed the year round, not just at Thanksgiving and Christmas, and at weights of thirty-five, forty, and forty-five pounds. Few families, she observed, could dispose of such a large turkey "without growing weary of the perforced sameness of their menu, which must for economy's sake, endure until the bird is consumed." There was "no reason in the world," she said, why people should not eat "delicious and nourishing turkey meat the year round," just as they ate steaks, chops, and other cuts of meat.

Beef steaks, lamb chops, and pork roasts were sold by the pound, Baker said, and she saw no reason why turkey could not be sold the same way. Butchers sold chicken portions throughout the year and there was no reason why they could not sell turkey portions the same way. "Who is there," she asked, "who would not welcome the more frequent appearance of this delectable dish on the dinner table," and why would the housewife not "welcome the opportunity for another meat"?

Baker said that she intended to devote her efforts to putting her "marketing scheme across." If it were, she believed, "the much-talked-of 'overproduction' possibility would vanish into thin air," and farmers and consumers alike would profit. With her observations, Helen Baker had, perhaps unwittingly, predicted that these consumers, desiring to eat this "delicious and nourishing turkey meat the year round," would not care from which variety of turkey the "delectable" meat had come nor what the bird's color and feather patterns were.

Baker had the All-American Turkey Show to thank for making possible the marketing scheme she advocated. The All-American Turkey Show had attracted national attention and had made American consumers aware not only of turkeys, but also that turkeys provided a "delicious and nourishing" meat, a "delectable" alternative to such standard fare as chicken, beef steaks, lamb chops, and pork roasts.

Flock owners, noting the qualities of prize-winning birds and attending educational sessions at the All-American Turkey Show, had learned how to improve the quality of their turkeys and how to market them more profitably. Farmers attending the All-American Turkey Show, where they heard the exhortations of those preaching the gospel of diversification, had seen the light and had been converted from the worship of King Wheat. Joining the diversification crusade, they were diversifying production and adding to their incomes by raising more turkeys.

The All-American Turkey Show had also made good the claims that the country's best turkeys were those produced in North Dakota and that Grand Forks, North Dakota, was the Turkey Capital of the World. From an uncertain beginning in a motor company's showroom in downtown Grand Forks in 1924, in little more than six years the All-American Turkey Show had, indeed, become "a really big shew."

CHAPTER 11

The Greatest (Turkey) Show on Earth

(with apologies to P. T. Barnum)

AT THE CLOSE OF the seventh All-American Turkey Show, on January 30, 1930, Judge Walter Burton of Texas "expressed his conviction" to a *Grand Forks Herald* reporter that the 1931 show was "bound to be better" than the one in 1930. "It can't help but improve," he said, "with the leadership of George W. Hackett." Judge Burton spoke true. The All-American Turkey Show owed its inception to Ed L. Hayes, "whose genius conceived of the idea of an exclusive turkey show." It was George Hackett, however, who formed the All-American Turkey Show, shaping it in his image and becoming the show's heart and soul by infusing it with his spirit.

The headlines and phrases used by the *Grand Forks Herald* in 1931 to announce the opening of the eighth All-American Turkey Show, more often than not without benefit of punctuation, were a measure of how well-grounded judge Burton was in his conviction. "TURKEY SHOW TO OPEN HERE MONDAY" was the paper's front-page headline, followed by "Thousands to see Birds at All-American"—"Many raisers arrive"—"Display to Be Greatest in Local History"[1]

The commentary continued: "Brilliantly plumaged birds, feathered aristocrats of the United States and Canada, Monday will draw thousands of persons to the city auditorium as the greatest All-American Turkey Show in history gets under way." "Kings of the various breeds, Mammoth Bronze, White Hollands, Narragansetts and Bourbon Reds will strut proudly in cages set beneath the green branches of evergreens."

Many of the "feathered aristocrats" were priced for sale at five hundred dollars each, conservatively estimated by manager Hackett to have

[1] *Grand Forks Herald*, January 25, 1931.

an aggregate value of twenty thousand dollars. Championship and Sweepstake prizes would be presented Wednesday morning at the conclusion of the judging. "Immediately the burnished bird upon which has been heaped the greatest honor in the power of judges to bestow—the Championship of the All-American—would be crowned."

The headlines continued: "Five Days of Events Crowded With Business and Entertainment" and "Education Stressed." "Supplementing what experts predict will be the most extraordinary collection of turkeys ever shown in the United States" would be an educational program "of unusual scope and practicality." Featured speakers were "specialists known throughout this part of the country for their work with turkeys." "The entertainment side of the five-day exposition will also be stressed." The Turkey banquet, "annually one of the most successful functions of the show," would be held Wednesday evening at 6:30 p.m.

The Turkey Hen Club, described in the paper as "one of the most colorful groups" in the exposition, would hold its fourth annual dinner on Thursday evening at 6:30 p.m. at the Hotel Ryan. *Turkeytorials*, the "official publication of the All-American," would be issued for the first time Sunday evening, January 25, 1931, under Monday's dateline with "last-minute news of the show" and "gossip about exhibitors, birds and other matters of note."

Beginning with the eighth in 1931 and continuing until the last in 1942, the All-American Turkey Show enjoyed its greatest successes. It also scored its most spectacular publicity coups, publicity coups that were unmatched by any other livestock show in the country.

The 1930s, however, were unkind to North Dakota. Historian D. Jerome Tweton and journalist Daniel F. Rylance explain in the Preface of *The Years of Despair* why they titled their work as they did. "The decade of the Thirties," they explain, "was a difficult time for most North Dakotans. Ravaged by wind, drought, grasshoppers, and the Depression, the state faced ruin, and its people felt deep pangs of despair. Some never recovered.

The authors introduced their treatment of what they called "years of despair" with the experience of the Alfred Berg farm family.

> A tall, lean, lantern-jawed Norwegian, Berg had come to North Dakota in 1909 and settled in Divide

Turkeytorials

Official Publication of the

ALL-AMERICAN TURKEY SHOW, GRAND FORKS, N. D. JAN. 26-30, 1931

CITY AUDITORIUM — SUNDAY, JANUARY 25, 1931

EDITORIAL

NOTHING SUCCEEDS LIKE SUCCESS

Again the All-American "goes over the top" and again have the raisers of high class turkeys set their stamp of approval on the plan and policies of the All-American, by their entries and their moral support. With a square deal to all and special favors to none—a welcome to the smallest entry the same as to the largest—a careful, systematic handling of all birds and their safe return to their owner guaranteed; prizes paid promptly and in full, all these together with annual home-coming social events, puts the All-American in a class by itself.

The success we have attained could not have been possible except through the splendid co-operation of both the exhibitors and the live business interests of Grand Forks to whom the Management wishes to express sincere thanks.

MARKET CATALOGUE

The Thursday Edition of Turkitorials will contain a complete list of exhibitors, entries and awards. On sale at show room Thursday and Friday for 10c. By mail, 15c.

The following list of exhibitors have entered from twelve birds up and have helped materially in putting over a full entry at this Eighth Annual Show:

Mrs. Martin Ellingson, 17; Theodore Bergstrom, 16; Mr. and Mrs. Rosengren, 15. These three are all exhibiting in the Narragansett Class.

Alfred Malmberg, 18; Mr. and Mrs. Al. C. Johnson, 15; George E. Lamm, 14; Mrs. D. C. McLeod, 14; Glen C. Bidleman, 13; and Ray Andrews, 12. These are all exhibitors of Bronze.

Among the interesting commercial exhibits at the show will be found the Duo-Breeder, exhibited by the Ringsrud Manufacturing Company, Elk Point, South Dakota. This breeder represents a new but practical idea in turkey breeding. See Mr. Ringsrud and his assistants who will be glad to demonstrate the workings o fthis breeder.

WELCOME NEW EXHIBITORS!

In looking over the list of exhibitors for the present show, we find the following new names of those who are from a distance and in most cases are well known to turkey breeders throughout the country:

Glen C. Bidleman, Kinsley, Kansas, a prominent breeder of Bronze Turkeys and a regular exhibitor at the Chicago International is with us this year for the first time.

T. C. Amos, also a Bronze breeder of Missouri, has made his first entry at the All-American. Mr. Amos was formerly Secretary of the American Bronze Turkey Club.

Edwin H. Burns, a Bronze breeder of Wisconsin, and familiarly known as "Turkey Burns," has also made an entry this year and it is hoped he will be present.

From far off sunny California comes an entry of Bronze from Mrs. R. G. Weidemier. This is the first entry to be received from California.

L. P. Rawlings, also of California, has made entry in the Narragansett Class.

From down in the Lone Star State comes an entry from A. J. Burks, a Narragansett breeder of note, and President of the International Narragansett Club. Sorry Mr. Burks cannot be present at the Show this year, but glad that we can see some of his good birds.

From Ohio we have received an entry from Mrs. Emma Snyder in the Bourbon Red class. This records the first entry we have ever received from the "Buckeye State."

Another entry of Bourbon Reds comes from Mr. LeRoy Kirby, a new exhibitor from Simms, Montana.

William M. Mikelsen of Montivedio, Minnesota, an enthusiastic breeder of White Hollands, has made his first entry with us. Mr. Mikelsen will attend the show.

Mr. and Mrs. Herbert Rosengren of the "Hawk-Eye Turkey Farm," Indianola, Iowa, comes in with the first entry that we have received from that State.

We also welcome a large number of new exhibitors from more local distances, whose entries even though small in many instances are greatly appreciated.

***Turkeytorials* was the All-American Turkey Show's first official publication.**
Courtesy Institute for Regional Studies, North Dakota State University, Fargo.

> County. His 160-acre farm provided well during the early years. By 1920 he had built a new barn, purchased new machinery, and rented additional acreage for some cattle.

Those were "good days," Berg recalled. He raised wheat and barley and owned a hundred head of cattle. The future "seemed bright."

By 1929, however, the Bergs were in trouble. Yields declined, as did prices, for wheat and cattle. The Bergs considered leaving the farm but could not. It was their life. Then drought struck, followed by grasshoppers. Crop failure followed crop failure. When the farm was foreclosed on, Berg moved his family to California with the hope of starting a new life. "The depression," noted the authors, "had finally broken him."

Lois Phillips Hudson aptly titled her autobiographical books on the 1930s, *The Bones of Plenty* and *Reapers of the Dust*. The titles conjure up images of nothing but bones remaining of something that was once sleek and fat, of an animal reduced by starvation to skin and bones, of bones bleached white by the heat and sun, of the dust that buried fence lines and machinery and blew into buildings and people's eyes and throats, and of there being nothing to harvest but dust.

Frederick Manfred, whose *The Golden Bowl* is favorably compared to John Steinbeck's *Grapes of Wrath*, wrote of Hudson's *Reapers of the Dust* that it recaptured the years and experiences of the 1930s. Farm families, after investing lifetimes of work, love, and pride in their land "saw that land blow up in their faces; they literally watched it drift miles high toward nowhere." "It was fortunate," Manfred noted, that Lois Phillips Hudson "was there to record it."

Authors Tweton and Theodore B. Jelliff, in *North Dakota: The Heritage of a People*, borrow the title, "Years of Despair" for their chapter on the Depression. The authors, one as knowledgeable on the state's history as the other, describe the 1930s as "bad days for everyone." They compared the Depression to "an angry tornado" that "ripped across the country" destroying everything in its path. "Drought, hot dry winds, and grasshoppers made life in North Dakota more unbearable," however, "than in most other places."

Tweton and Jelliff note that rain "was scarce" and the heat "was intense." Swarms of grasshoppers blocked out the sun and ate whatever crops the wind and heat had not already destroyed. Newspapers, the authors noted, carried "stories of desperation" and headed their front pages with phrases such as "HEAT, DUST STORMS DAMAGE CROPS AND MAKE LIFE DREARY." North Dakota, the authors concluded, "was the hardest hit of the forty-eight states, and its citizens suffered immensely."

In *North Dakota: A Bicentennial History*, Robert P. Wilkins and Wynona H. Wilkins titled their sixth chapter, "The Three Dreadful D's." The three "D's" were Drought, Depression, and Dust, and they were "dreadful" because the three words described North Dakota in the 1930s all too well. The phrase was coined by Protestant Episcopal Bishop Douglas H. Atwill, who had come to North Dakota "when the triple threat to well-being had not yet dissipated." The three "D" words and their effects, the Wilkenses noted, were "a vivid memory" for generations of North Dakotans long after the 1930s had ended.

The conditions responsible for the "dirty thirties" and the "dreadful D's" in North Dakota, the authors wrote, "first appeared as a cloud no bigger than a man's hand in 1929 when rainfall was below average" (an analogy drawn on I Kings 18:44). The years of the 1930s were characterized by "moisture deficits," and just how scanty rainfall was during these years was recalled by Episcopal priest Alexander Macbeth, a Scot, at Williston, North Dakota. Raised in India and well-traveled, Macbeth said he had seen more rain in the Sahara Desert in three months than he had seen in eighteen months in North Dakota in 1930–1931.

Grasshoppers so numerous they frustrated efforts at controlling them, compounded the dreadfulness of the "D" words. In Mott, in southwestern North Dakota, clouds of them were so large they darkened the sky in 1933, and streetlights were turned on in the middle of the day. Residents built large fires at street intersections in an attempt to kill the grasshoppers as they swarmed along the streets. At Killdeer, just west of what is now Lake Sakakawea, swarms of grasshoppers piled up four inches deep on the streets.

Robert and Wynona Wilkins concluded their chapter by stating that "with half its people on relief, insolvent and in need of even the necessaries of life, the state by its own efforts could not have kept its head above water." And, "there is," they believed, "little exaggeration in the proposition that federal money saved the state from bankruptcy, and not alone in the fiscal sense. Without that help there would have been a mass exodus which would have undercut the foundations of the social and political structure as it had developed over the preceding seventy years."

Reducing Bishop Atwill's "D's" by one, historian Elwyn B. Robinson used "Drought and Depression" to discuss the Thirties in his *History of North Dakota*. Introducing readers to his adopted state, Robinson noted that the state's climate was one of extremes; just how extreme was evident in 1936. "The year 1936," Robinson wrote, "was the coldest (-60°F at Parshall on February 15), the hottest (121°F at Steele on July 6), and the driest (8.8 inches) ever reported. Summer was a disaster." From July 5 to July 18, temperatures were 100°F almost every day, and eggs could be fried on city sidewalks. Robinson described prairie grass as "a drought-enduring flora," but, in 1936, no prairie grass grew west of the Red River Valley. Farmers and ranchers, short of feed, culled their herds and fed cottonseed cake, old straw, brush, and any available roughage to the animals they kept as breeding stock. The federal government bought some of the stock, paying four dollars per head for calves and twenty dollars for cows.

For North Dakota, according to Robinson, the 1930s were "a traumatic experience," and the state "suffered more than much of the rest of the nation." The effects of the drought and the Depression were compounded by North Dakota's status as a producer of raw materials, the price of which the state could not control when they were bought and sold. As a result, "thousands lost their farms; more than one-third of the population lived on relief; many people left the state."

Grasshoppers, appearing in Pembina and Adams counties in 1931, added "their ravages to the damage of drought." Before the 1930s ended, grasshoppers spread over much of the state, destroying crops and trees, cutting the twine on grain bundles, chewing holes in clothing, and damaging the siding on buildings and roughening shovel and pitchfork handles with their powerful mandibles. In (mostly futile) attempts to control the grasshoppers, farmers spread poisoned bait composed of a mixture of molasses, bran, arsenic, and water; gathered grasshoppers with machines resembling grain binders that killed them in troughs containing kerosene; and prevented them from reproducing by plowing under unhatched eggs. In the end, North Dakota expended more money—$3,600,000—on controlling grasshoppers than any other state in the Union.

The years of drought and grasshoppers, Robinson wrote, "were also depression years, and depression prices for meager crops brought double hardship." In 1932, the worst year, wheat sold for thirty-six cents per bushel, oats for nine cents, barley for fourteen cents, potatoes for twenty-three cents, and flaxseed for eighty-seven cents. Prices remained low for wheat throughout the 1930s, recovering only to fifty-three cents per bushel in 1938.

Not only were grain prices low, yields declined as well. From 1929 to 1941, North Dakota farmers had only two respectable wheat crops—totals of more than 100,000,000 bushels. In 1936, the worst year according to Robinson, the crop "was a pitiful 19,000,000 bushels." Farm incomes were correspondingly reduced, only $61,000,000, or about $760 per farm in 1932. In 1933, when it was the lowest, the per capita personal income in North Dakota was $145; that in the United States was $375. During the worst years, 1932–1937, the per capita personal income in North Dakota was 47 percent of the national average. "Plainly," Robinson believed, "the 1930s brought much greater hardship to North Dakota than they did to the nation as a whole."

Lack of income, Robinson concluded, "caused a multitude of misfortunes: a decline in land values, delinquent loans and foreclosures on mortgages, unpaid taxes, increasing tenancy, growing public ownership of land, large numbers of people on relief, and a great movement of people out of the state." By 1940, nearly 87,000 people had left "the stricken state." Forty-three of North Dakota's fifty-three counties lost population, those in the western part of the state losing the most.

Many "defeated people," in Robinson's words, left farms and small towns and moved to the state's largest cities. University of North Dakota Professor John M. Gillette, known as "the father of rural sociology," described the plight of the defeated after making a six-thousand-mile tour of the state in 1939. "Stoves," reported Gillette,

> are giving out; bedding is wearing out; curtains, carpets, and furniture becoming unusable; clothes have become shabby and indecent. . . . Along with these changes has come a decided loss of morale. Ambition . . . has been killed; and there is little hope of ever being anything but a WPAer. The children born into and being reared in this situation are decidedly underprivileged. They are cut off from association with middle class children, feel themselves to be outcasts and inferiors, have little or no recreational privileges, and come to absorb an atmosphere of defeatism and parasitism. . . . It is a most serious situation for the young men and women on farms who are just coming to maturity. Farming no longer has power to absorb them; and after remaining idle parasites on the farmstead for a time they float into towns and villages, marry, and join the WPA forces.

Dependence was among Robinson's six themes of North Dakota history, and "the depression emphasized North Dakota's dependence on outside resources in a new way." By the end of 1932, counties and private charities could no longer provide sufficient relief to those needing it. The state became dependent on such New Deal programs as the Reconstruction Finance Corporation, the Civil Works Administration, the Federal Emergency Relief Administration, the Civilian Conservation Corps, and the Works Progress Administration.

In short, Robinson wrote, "relief soon became the biggest business in the state." Between 1933 and 1940, the federal government spent some $266,000,000 in North Dakota. This "massive outpouring of federal funds," in Robinson's words, contributed much to the state's "survival and well-being."

Something more than a "massive outpouring of federal funds" contributed to North Dakota's "survival and well-being." That "something more" was what Robinson described as "the North Dakota character."

Robinson was invited to speak at the convocation commemorating the Seventy-Fifth Anniversary of the founding of the University of North

Dakota in 1883. In his address, delivered on November 6, 1958, Robinson named his "themes" of North Dakota history. These themes are as follows: remoteness, dependence, radicalism, a position of economic disadvantage, the Too-Much Mistake, and adjustment to the imperatives of a cool, sub-humid grassland.

Robinson revised his address and later published it in *North Dakota History: Journal of the Northern Plains* with the title, "The Themes of North Dakota History." In the article, Robinson explained how the themes were "tied to the most fundamental facts about the state" and why the "influence of these facts" could be seen "in every aspect of North Dakota history." The "six great themes," Robinson advised readers, "run through the state's history, virtually from the beginning to the present day." The themes, Robinson confessed, represented his "understanding of the North Dakota story—the story of a big spring wheat country." But, he admitted, acknowledging that he was a North Dakotan, they "are hard, disagreeable truths—hurting our self-esteem; yet, I believe, we must face them honestly, if we would use the lessons of the past to make a better future."[2]

Kindly refraining from leaving readers to ponder "disagreeable truths," Robinson hastened to remind them, that "one thing more remains to be said." The themes he had named, the "fundamental facts" with which they were associated, and, probably, the "winnowing process of pioneer settlement" had "placed a stamp upon the people, producing the North Dakota character." The "typical North Dakotan," Robinson believed, was "friendly and warm-hearted—ready to lend a helping hand."

With "a strong loyalty to the state," the typical North Dakotan, as described by Robinson, was "democratic, suspicious of the 'interests,' and something of a radical." He had "an independent courageous, stubborn, and aggressive spirit" and he "admired a fighter." Because the typical North Dakotan could endure hardship and suffering, he was often "pessimistic and cautious." He was "an energetic person, full of hustle," who took "pride in withstanding the rigors of North Dakota weather." Finally, the typical North Dakotan was "intelligent and alert."

Robinson admitted that perhaps no one was exactly the typical North Dakotan, but, he believed, "we all share many of his qualities and atti-

[2] Robinson, "The Themes of North Dakota History," *North Dakota History: Journal of the Northern Plains* (Winter1959).

tudes." Indeed, he wrote, "we can scarcely help ourselves, for they spring from the North Dakota experience and environment." And, he believed, "there is much evidence that these qualities and attitudes are fairly typical of North Dakotans."

Not only that, it was the "qualities and attitudes" of "the typical North Dakotan" that were so much in evidence during the 1930s, and it was "the North Dakota character" that enabled "courageous" and "stubborn" North Dakotans to endure "hardship and suffering" and withstand "the rigors of North Dakota weather" during the years of drought, dust, and the Depression.

With apologies to Elwyn B. Robinson for appropriating his words, "one thing more . . . needs to be said." Turkeys also "contributed much" to the "survival and well-being" of many North Dakota farm families during the 1930s, the turkeys that Ann Marie Low admitted were "a lot of darned hard work."

When a serious infestation of grain rust significantly reduced yields in the Northwood area, Mrs. William Eddie was thankful that turkeys did not rust. For Mrs. Roy Vasper, of Neche, turkeys were "life savers" during the years of crop failure. A writer for the *American Turkey Journal* noted that those farmers who raised turkeys were better able to provide for their families and stave off foreclosure than those farmers who relied only on crops for income.[3]

Those who raised turkeys during the 1930s may have been among those who learned how to improve their flocks and raise and market their birds more profitably by attending the education sessions that were features of the All-American Turkey Show. Having heard the diversification gospel, they had been converted and had joined the diversification crusade. Defying drought, dust, and Depression, they participated in the stellar successes of the All-American Turkey Show, and they reveled in its publicity coups.

The first seven All-American Turkey Shows, each one better than the previous one, were all of a pattern. Prize turkeys were checked in at the City Auditorium and readied for display and judging. The regulars renewed acquaintances and welcomed first-time exhibitors and visitors.

---

[3] *American Turkey Journal*, August 1941.

Storytellers told tales, much of it good-natured ribbing, on judges and breeders. Representatives of feed and poultry equipment companies explained why flock owners should use their products, and feature writers from leading farm and poultry journals circulated among exhibitors and visitors, gathering material for their columns. Education sessions were held, and there was the usual round of gatherings, luncheons, and dinners. All the while, the judges were deciding which of the hundreds of turkeys would be named champions in their classes.

The All-American Turkey Shows held in the 1930s continued the same pattern. Beginning with the eighth in 1931, however, and continuing throughout the 1930s, officials added one or more new or unusual features. The features made the shows, if not unique, at least all the more noteworthy and appealing to exhibitors and visitors.

The eighth All-American Turkey Show, reported as "the most successful on record," featured more than five hundred turkeys, "the highest quality birds" that agricultural experts and exhibitors had ever seen. The turkeys competed for $1,500 in cash prizes, plus trophies, medals, and ribbons. There were also the Sweepstakes Awards and the Grand Champion trophy. The number of Special Awards alone suggests that few exhibitors left the show without receiving at least one award of some kind.

At the close of the eighth show, manager Hackett and W. C. Blain, show secretary, predicted that even more awards would be pledged for the ninth All-American Turkey Show in 1932. With little attempt to conceal excitement, the headline at the end of January in the *Grand Forks Herald* was "Cando Turkey All-American Grand Champion," followed by "North Dakota Raiser Takes Banner Prize." A Mammoth Bronze yearling tom owned by Mrs. Henry Botz of Cando, North Dakota, was named the Grand Champion of the eighth All-American Turkey Show. Judges scored the bird as "the most perfect ever exhibited at the Grand Forks exposition."[4]

The award marked "the ascendancy of North Dakota to a place unique among states producing the nation's finest turkeys." The award also marked the return of the championship to North Dakota. In 1930, the

[4] *Grand Forks Herald*, January 29, 1931.

Grand Champion, also a Mammoth Bronze tom, was exhibited by Mrs. Howard Lathrop of Delta, Colorado. The Cando turkey was also chosen as champion in the Bronze division, and it also received the Sweepstakes Award as "the most perfectly shaped and colored Bronze" in the show. Mrs. Gladys Honssinger of Lebanon, Missouri, retained her position of "unchallenged supremacy among Bourbon Red breeders in the United States." Her birds took three championships: young tom, best four young toms, and best four young hens.

Glen C. Bidleman, Kingsley, Kansas, received the Master Breeder Award for his best all-around display of turkeys. Bidleman, "a prominent breeder of Bronze turkeys and a regular exhibitor at the Chicago International," exhibited at the All-American Turkey Show for the first time in 1931. He had exhibited at turkey shows all over the country, he said, and the All-American Turkey Show compared favorably with the best. His highest praise, however, was that the All-American Turkey Show was "the best managed, best arranged and most satisfactorily cooped" show he had ever attended.

Manville A. Johnson, of Michigan, North Dakota, offered advice to flock owners in his "Observations of a Country Buyer on Methods of Handling, Dressing, and Marketing." The title of Johnson's presentation explained why he was a part of the education program described as unusual in scope and practicality.[5] Manville A. Johnson was the Johnson of the prestigious Johnson Stores.[6] Johnson titled a chapter in his *Fifty Years*

---

[5] The education program of 1931 included a report on "Experiment Station Results of the Last Year" by Professor O. A. Stevens, poultry expert at the North Dakota Agricultural College. B. O. Norby, manager of the Land O' Lakes poultry plant at Thief River Falls, Minnesota, demonstrated the grading and packing of turkeys for shipment, and Frank Moore, the new poultry extension agent at North Dakota Agricultural College, reported on "Turkey Production in Idaho and the Mountain States." Mrs. Claude E. Wright, prominent turkey breeder from Atkins, Minnesota, spoke on trap nesting and breeding. In her remarks, Clara Sutton of *The Farmer* of St. Paul, Minnesota, described the turkeys exhibited at the last World's Poultry Congress.

[6] Manville A. Johnson was born and raised in northern Minnesota, land of timber and lumberjacks. Eighteen years of age, with a graduation certificate in bookkeeping from the International Correspondence Schools and carrying his entire wardrobe in a canvas-covered cardboard suitcase, he arrived in Michigan, North Dakota, on April 13, 1907. Manville A. Johnson got his start in business working as a clerk in the Michigan Mercantile Co., owned by his uncle Marcus Johnson, eventually becoming a store manager. When M. A. Johnson and stockholders bought out the owners of the Michigan Mercantile Co., they formed the Johnson Stores Co., a company that established a chain of Johnson Stores, one in each of seventeen small towns in North Dakota. Each Johnson store was "a shopping center on the prairie," and, if the store did not stock an item, customers did not need it.

OUTSTANDING AT THE ALL-AMERICAN

Above, left to right: Mrs. Howard Lathrop, Delta, Colorado, owner of the 1930 Grand Champion, a Bronze yearling tom, shown second. Mrs. W. J. Janda, St. Hilaire, Minnesota, secretary of the All-American Bronze Turkey Club. Below, left to right: Mrs. E. B. Aheren, Lisbon, North Dakota; Mrs. Andre Johnson, McVille, North Dakota, wearing her coat made of feathers of forty Bronze turkeys raised on her farm; Mrs. August Swenson, Gilby, North Dakota, the Hen Club's "official caller." Image from February 2, 1930; used with permission of Grand Forks Herald.

of Country Storekeeping, "Farm Produce Saves the Day." "I am doubtful," he began, "that our company would have survived the depression and drought years of the '30s had it not been for our trade in farm produce." The trade in farm produce included cream, butter, eggs, and poultry.

"With our chain store organization," Johnson wrote, the company could combine the farm produce from all the stores, resulting "in more economical handling—such as live poultry shipped in carload lots direct to eastern markets, carloads of dressed turkeys that commanded premiums on the best markets, cream churned right in our own town . . . rather than shipping cream out of the state." These services "provided sources of extra income for our farm customers during a period when their crops were meager incomes at best."

"Especially interesting," Johnson observed, "perhaps because it presented some of the worst problems, was the marketing of dressed turkeys." Before the day of commercial dressing plants, "all the dressed turkeys that went to market were dressed right on the farm," Johnson explained. But, "few people understood what was needed to grow a turkey that would be an attractive market bird and [what seemed still more difficult] how to prepare it for market." "Our Northern turkeys," Johnson had learned from experience,

> when properly raised and dressed, were superior to those from warmer climates and would bring a premium price. Here then, during the depression when almost everything a farmer had to sell was at bottom price, the North Dakota farmer who could put a No. 1 quality turkey on the market would have a very profitable crop. For us, this fact gave opportunity to aid the income of our farm customers—if we could interest the farmer (or, more likely his wife) in producing a flock of market turkeys.

Johnson set about stimulating this interest and educating farmers and their wives on the methods of handling, dressing, and marketing turkeys.

Johnson became an expert on dressing turkeys by studying procedures at a plant in Geneseo, Illinois, where he observed "turkeys were dry-picked, professionally." At the plant, he learned that "the most important step in the process was proper bleeding of the bird—stabbing the brain in such a way that feathers would loosen for easy picking." He passed these lessons on to his farm customers. "Many of our customers," Johnson wrote,

> began doing a fine job in producing and dressing turkeys so that we were able to build up a special market for our turkeys in the East. Though our output was not large, when compared with the national crop, we were able to get our pack recognized in New York, Boston and Philadelphia turkey markets so that we obtained attractive premium prices. To foster this turkey industry in our area, we conducted various educational campaigns, held meetings, worked with the farm women's turkey clubs, and conducted drives out among turkey growers.

"Producers responded to the extent," he commented, "that in one Christmas shopping season we shipped 13 carloads of dressed turkeys."

The All-American Turkey Show's annual banquet in 1931 was a turkey dinner in fact as well as in name. Roast turkey was the *piéce de résistance,* and dessert was individual portions of ice cream, molded in the shape of turkeys. "Hundreds of guests" proclaimed the dinner "the outstanding function in All-American history."

Entertainment included music and addresses by such notaries as Miss Clara Sutton of *The Farmer* of St. Paul, Minnesota. Dr. W. A. Billings of University Farm, St. Paul, Minnesota, also spoke and pointed out that turkeys were selling for two to three times the amount received for other farm products, that "nothing but a brilliant future" lay ahead for the industry, and that there was little likelihood of overproduction. Dinner was followed by the "annual turkey show hop."

In 1931, for the first time, the All-American Turkey Show had its "official publication." *Turkeytorials*, a handsome sheet dated January 26–30,

1931, contained "last-minute news of the show" and "gossip about exhibitors, birds and other matters of note." The first editorial was headed: "NOTHING SUCCEEDS LIKE SUCCESS."[7]

The first issue of *Turkeytorials* also included a measure of the All-American Turkey Show's expanding reach: a listing of first-time exhibitors and the states represented at the show for the first time. In most cases, the first-time exhibitors were "well known to turkey breeders throughout the country."[8]

J. C. Sherlock, known as a booster supreme and All-American Turkey Show director-turned statistician, provided figures documenting the importance of the All-American Turkey Show. Greater Grand Forks poultry buyers had purchased over one million pounds of turkeys in the Grand Forks district in the six weeks preceding the show. That was $270,000 in cash that had been brought into the trade area. In addition, more than thirty thousand pounds of turkeys had been purchased by Grand Forks poultry businesses, plus a "conservative estimate" of twenty thousand to thirty thousand pounds of breeding stock had been sold. Much of this cash income, Sherlock believed, "was attributed to the influence" of the All-American Turkey Show.

---

[7] "Again," the writer noted, "the All-American 'goes over the top' and again have the raisers of high class turkeys set their stamp of approval on the plan and policies of the All-American by their entries and their moral support. With a square deal to all and special favors to none—a welcome to the smallest entry the same as to the largest—a careful, systematic handling of all birds and their safe return to their owner guaranteed; prizes paid promptly and in full, all these together with annual home-coming social events puts the All-American in a class by itself."

[8] As found in *Turkeytorials*, T. C. Amos, a Bronze breeder from Missouri and former Secretary of the American Bronze Turkey Club, "made his first entry at the All-American." Another Bronze breeder, Edwin H. Burns of Wisconsin, familiarly known as "Turkey Burns," also made his first entry. "From far off sunny California," Mrs. R. G. Weidemier brought an exhibit of four Bronze turkeys, "the first entry to be received from California." Before moving to California, Weidemier was "a prominent turkey raiser" in the vicinity of Park River, North Dakota. L. P. Rawlings, also from California, exhibited his Narragansetts for the first time at the All-American Turkey Show. "From down in the Lone Star State," came an entry of A. J. Burks's Narragansetts. Burks was "a Narragansett breeder of note, and President of the International Narragansett Club." The first ever entry from the "Buckeye State" were the Ohio Bourbon Reds of Mrs. Emma Snyder. Another entry of Bourbon Reds were those of another first-time exhibitor, LeRoy Kirby, Simms, Montana. "An enthusiastic breeder of White Hollands," William M. Mikelsen, Montevideo, Minnesota, also "made his first entry" at the All-American. Iowa was represented for the first time at the All-American Turkey show by Mr. and Mrs. Herbert Rosengren and their Hawk-Eye Turkey Farm Narragansetts. The show also welcomed "a large number of new exhibitors from more local distances." Their entries, "though small in many instances," were "greatly appreciated."

"Gossip" included the report that Judge Walter Burton, one of the best storytellers in turkey circles, was prevented from attending the show by an injury received in an automobile accident. Max Morgan of Fairmont Creamery Farms at Fargo, the second best, would attend, however.

"The great indoor sport" at the All-American Turkey Show was "holding turkeys on the judging table." The show's old-timers, such as H. G. Link, agent for the *Turkey World*, a trade publication; O. K. Ose, Thief River Falls, Minnesota; Mrs. S. Birk of Maxbass, North Dakota; and other experienced "holders," seemed to have no trouble with the birds, but "many a novice" learned "just how much the slap of a turkey wing hurt."[9]

The honor of driving the longest distance to the show—more than six hundred miles—went to Mr. and Mrs. Herbert Rosengren of Indianola, Iowa. They had transported their Narragansett turkeys in the rear seat of their automobile. George E. Lamm of Philip, South Dakota, beat out the Rosengrens on "a bird-mile basis." He hauled his eleven turkeys six hundred miles in a trailer behind his coupe.

Copies of the final edition of the 1931 *Turkeytorials* were mailed to exhibitors a few days after the All-American Turkey Show closed. The issue included a catalog of all entries and lists of all prizes and awards. A Court of Honor was another innovation in 1931. Placed at the front of the City Auditorium and an attention-getter for exhibitors and visitors alike, the Court of Honor was a ring of coops, a coop for the champion of each turkey variety. The Grand Champion's coop was placed at the center of the ring with a spotlight playing on "the ruler of all Turkeydom." Flanking the Grand Champion's coop were those of the lesser champions. A. J. Bentley, the city electrician in charge of lighting, hoping that a Bronze turkey was chosen as Grand Champion, had rigged up bronze-colored lights "to play on the 'champion of champions' at the show."

[9] William Huggins was in charge of arranging coops, and his wife, Ann, was the show's official ticket seller. Mr. and Mrs. C. M. Rusche of Grand Forks operated the lunch counter.

CHAPTER 12

Dressed Turkeys: Naked to Our Judges

(with apologies to Charles W. Ferguson, ***Naked to Mine Enemies***, 1958)

IF NOT SOMETHING UNIQUE in a turkey show, at least something unusual, the "Turkey hen nest" was among the first sections of the 1931 All-American Turkey Show to be constructed and readied for use. Decorated with evergreens, in keeping with the appointments in the rest of the auditorium, the Turkey hen nest contained "everything needed for a real turkey nest," according to Mrs. W. J. Janda of Mahnomen, Minnesota, Turkey Hen Club president.

Turkey hens preferred to nest in secluded areas, screened by brush or tall grass, where they felt safe. In the Turkey Hen Club's turkey hen nest, to which no male was admitted, women exhibitors could rest, screened from the rest of the auditorium, insulated from crowds and distractions. In addition to having "everything needed for a real turkey hen nest," the nest had roses for the women exhibitors from California, palms for those from the South, pines for those from the North, and bright lights for "the fanciers" from the East.

Acknowledging changing tastes and acceding to exhibitors' requests (and waiving the requirement that exhibits had to have feathers in order to be entered into competition and judging), All-American Turkey Show officials in 1931 accepted more than fifty entries of dressed turkeys. Allowing exhibits of dressed turkeys presented a problem, however. The dressed birds had to be chilled, and they had to be on display for judges and visitors, not kept in a cold storage locker. Those in charge of arrangements improvised a huge, refrigerated area for the dressed turkey display by enclosing the entire north side of the auditorium balcony and opening the balcony windows. With January temperatures in Grand Forks predictably 0°F and below, the dressed exhibits could be kept chilled with natural refrigeration, which was not only economical, but also reliable.[1]

Exhibitors and show officials were unanimous in declaring that the 1931 All-American Turkey Show had been "the most successful on record from the show standpoint." Manager Hackett, and few would be in a better position to judge than he, believed that the 1931 All-American show was not only the most successful on record, but also the best. Given the 1931 show's added features, the many "firsts," and the striking innovations, these statements were neither ritualized comments nor empty boasts.

There were other affirmations of the 1931 show's success. Hackett was reappointed and would manage the 1932 All-American Turkey Show. C. Dyke Page would serve as president. J. R. Carley, president of the First National Bank, notified the All-American Turkey Show's board of directors that the bank would double its support for the 1932 show. The bank would also provide a $25 cash prize and a special trophy for the winner of the Master Breeder Award. The cash prize and the trophy were expected to increase interest in "this division of the show."

The 1931 All-American Turkey Show may have been the most successful and the best on record, but neither claim held true after the 1932 show. In large, bold letters, the headline on the January 25, 1932, edition of the *Grand Forks Herald* was, "TURKEY EXPOSITION OPENS," followed by, "Hundreds of regal bluebloods of American Turkeydom strutted proudly in doubledecked cages at the city auditorium show room as the greatest All-American Turkey Show opened in Grand Forks."[2]

Manager Hackett described the 1932 show as "unquestionably the largest and the finest in the nine-year history of the event." Exhibitors had entered 400 live turkeys and more than 440 dressed birds, making the dressed division the outstanding feature of the show. "The quality of live and dressed entries," entered by eighty-seven exhibitors, Hackett declared, exceeded "anything exhibited at the All-American to date."

---

[1] Sealed bids, taken on the dressed birds, were opened at 1:30 Friday afternoon, the last day of the show. Dressed birds remaining unsold were purchased, often at better than market prices, by one of the city's produce houses.

[2] *Grand Forks Herald*, January 25, 1932.

Indisputably, in Hackett's estimation, the All-American Turkey Show had become "the show window of the turkey industry." It represented the industry "as the National Dairy Show does the dairy industry of America and as the American Royal does the beef cattle division of the country."

A "more dazzling Court of Honor for the strutting prize birds" was another distinguishing feature of the 1932 All-American Turkey Show. The Court of Honor, placed at the front center of the auditorium, was made more dazzling by being provided with more space and better lighting. Also outstanding was a brand new Turkey Hen Nest for women visitors and exhibitors, "all in a newly decorated setting of more beauty than ever before." The nest was in a large room at the front of the auditorium.

Two "turkey bluebloods," White Holland hens owned by Henry W. Domes, of McCoy, Oregon, provided another first for the All-American Turkey Show: the pair produced the first turkey eggs ever laid at a Grand Forks show.

Booths with displays of poultry feed and products were another feature of the All-American Turkey Shows. Purina Mills; Hubbard Milling Company; Northrup, King & Company; and the Commercial Feeds Division of the North Dakota State Mill & Elevator—all advertised balanced rations for every stage of a turkey's life, from egg to market. The Grand Forks Seed Company; Newday Products of Fargo, North Dakota; and others offered poultry equipment and medicines. The Northern Packing Company and Red River Valley Produce, both of Grand Forks, had attractive displays of poultry, eggs, and other products.[3]

Publication of the *American Turkey Journal* was among the All-American Turkey Show's major publicity coups that other livestock shows would have been hard put to match. Copies of the first issue were distributed at the 1932 show. Hackett edited the *American Turkey Journal* from its beginning in February 1932 until it was sold to *Turkey World* ten years later, in February 1942. The heart and soul of the All-American Tur-

---

[3] The "outstanding" dressed turkeys commanded high prices when auctioned off on the last day of the show. The Belmont Café paid $1 per pound for the Grand Champion bird, a Bronze. For the best pair of young hens, twelve pounds and over, the Poppler Piano Company paid 50 cents per pound, as did the First National Bank, for the best pair of old hens. The J. C. Penney store paid 50 cents per pound for the best pair of young toms, twelve pounds and over. For the best six young hens, the Hubbard Milling Company paid 26 cents per pound. Land O'Lakes Creamery paid higher than prevailing market prices for the dressed exhibits not sold to bidders.

key Show, Hackett also invested himself in the *American Turkey Journal*. The journal's every feature bore his imprint, from his thoughtful editorials at the beginning of every issue to the Club News and ads at the back.[4]

Hackett's editorials were as eagerly read as they were awaited, and in his first one he introduced the *American Turkey Journal* and explained its purpose. "This friends," he began, "is the formal introduction of the AMERICAN TURKEY JOURNAL, dedicated to AMERICA'S GREATEST BIRD, and devoted to the SINGLE PURPOSE of serving a great industry and the promotion of the same." Professing an unlimited faith in the future of the turkey industry, Hackett pledged the journal's "EVERY EFFORT" to its advancement.

The *American Turkey Journal* was a handsome publication of thirty-four to forty or more pages. Color was reserved for the journal's cover, always a picture of something associated with the All-American Turkey Show, usually a turkey and, most often, a champion in its class or a show's Grand Champion. The sole purpose of the *American Turkey Journal* was the promotion of the turkey industry, not to return a profit.

The *American Turkey Journal*, he promised, offered its "uncompromising support" for Standard turkeys and it would allow "no compromise for any but the BEST OF MARKET TYPE." The *Journal* would endeavor to maintain standards and to improve the breeding that had transformed "the original diminutive wild turkey" into the current varieties of "marvelous turkeys."

February issues often carried detailed accounts of the previous All-American Turkey Show, including lists of exhibitors and prizes, and reading the accounts served almost as well as attending the show, observing the prizewinning turkeys in their coops, and mingling with turkey folk on the show room floor. The cover of the February 1934 issue carries a picture of the Grand Champion of the eleventh Annual All-American Turkey Show, held January 22–27, 1934. The turkey was bred and owned by George E. Lamm, Philip, South Dakota. Shown with its ribbons, the

[4] The *American Turkey Journal*, a monthly, was published by the Page Printing Company at 105 South Third Street in Grand Forks. C. Dyke Page, owner of the Page Printing Company, also served as president of the All-American Turkey Show. Subscription rates were modest: 50 cents per year in the United States, 75 cents in Canada, and $1.00 in foreign countries. Single issues were 10 cents.Reminding subscribers and readers that Grand Forks, North Dakota, was the Turkey Capital of the World, every issue of the *American Turkey Journal* included a notation at the bottom of its first page: "Published Monthly by the PAGE PRINTING CO. at 105 South 3rd Street, Grand Forks, N.D."

Grand Champion was almost dwarfed by its trophy. Hackett's account of the 1934 All-American Turkey Show began on page three, under a large picture of the Court of Honor and the heading, "11th Annual All-American A Big Success."[5]

Several companies had eye-catching, full-page ads, replete with testimonials, lauding their turkey rations and touting the results flock owners could expect when using them. Purina Mills, 812 Checkerboard Square, St. Louis, Missouri, included a reproduction of the company's well-known trademark—two opened feed sacks bearing the checkerboard pattern—and a photograph of dressed turkeys on a judging table with the words, "They Brought the Money!" The ad went on to say, "Feed your poults Purina Turkey Startena for the first six weeks. Then switch to Purina Turkey Growing and Fattening Chow. You'll have birds big in frame, good in finish, straight-breasted and weighing extra pounds on market day."

Northrup, King & Company, Minneapolis, Minnesota (Dependable Since 1884), made its point with a photograph of a large flock of Bronze turkeys on the company's experiment farm at Litchfield, Minnesota. "At the Litchfield Turkey Farm, season 1933," noted the caption, "Sterling Turkey Starter raised a higher percentage of poults with less Feed consumption—and at the same time produced birds with excellent frames, feathering, health, weight and vitality. WHY? Because over 60% of this feed is easily and quickly assimilated. Other large farms have had similar results."[6]

The Hubbard Milling Company, Mankato, Minnesota, placed its ad on the issue's inside back cover, an enviable location. The company's slogan, "Feed the HUBBARD SUNSHINE Way," was emphasized with the notation: "NEW INFORMATION!! The New HUBBARD SUNSHINE Feeding Manual—a complete directory for feeding livestock and poultry. A special chapter on Raising turkeys. Ask for your Free copy today."

Among the testimonials was that of Mrs. Homer Price of Newark, Ohio, "prominent White Holland breeder." Mrs. Price had "excellent re-

---

[5] For more details on the February 1934, see Appendix F: *The American Turkey Journal* in Brief.

[6] Sterling Turkey Starter was a product of Northrup, King & Company, and "Sold By Dealers AT REASONABLE PRICES."

sults in feeding SUNSHINE Starting Mash, and after a visit to the mills in Mankato" she could confidently "recommend HUBBARD'S SUNSHINE Feeds to turkey raisers." She planned to use "HUBBARD'S SUNSHINE CONCENTRATE for a laying mash" in 1934 and would also "feed the starting and growing mashes."

As evidence of Mrs. Price's excellent results, she had exhibited her White Holland turkeys at the World's Fair Poultry Show in Chicago, the New York Poultry Show in New York, and the Pennsylvania Farm Show in Harrisburg, Pennsylvania. She had entered her White Hollands in eighteen classes and had taken First Prize in seventeen of them.

The Northern Packing Company, located on the extreme northern edge of Grand Forks, also advertised in the *American Turkey Journal.* Reminding flock owners that no grain had "enough protein," the company advised them that "Meat Meal must be added to feed turkeys right." "Meat Meal," the ad explained, was "highly concentrated animal protein in its most digestible form. By mixing Meat Meal with the grain you have on the farm, you get a feeding mash with *all essential* food elements at the lowest possible price." Those with questions were invited to "WRITE FOR DETAILS."[7]

Each issue of the *American Turkey Journal* included what might be termed an Advice or Opinion section. In the February 1934 issue, Hackett, under the heading, "Quality vs Big Production," again expressed his belief that raising turkeys in small flocks was the best way to assure quality. "By QUALITY, as we here use it," he wrote, "we do not necessarily mean turkeys that will win at the shows or the kind that would satisfy our leading turkey breeders, but rather, the market quality of the great bulk

---

[7] Lists of exhibitors, prizes, and awards, always included in the first issue following an All-American Turkey Show, filled eight pages in the February 1934 issue of the *American Turkey Journal.* Awards included Silver Cup Trophies, Silver Trays, Gold Medals, Special Design Plaques, one fine grade "A" Bronze Tom, a setting of ten eggs from high-class mating, five hundred pounds of Sterling Turkey Starter, one hundred pounds of Northern Packing Company Meat Meal, Beautiful Club Ribbons, and $2.00 in cash. Some categories appear to have been devised in order to avoid having to offend a donor by refusing the offer of an award. There were awards for the exhibitor entering six or more birds and winning the least points, the exhibitor entering six or more birds from the greatest distance, the lady exhibitor in attendance at the show from the greatest distance, the exhibitor making the largest entry and winning the least points, and the exhibitor having the second lowest scoring display of any variety.

of turkeys that are to go on the market next fall and winter." Hackett concluded that it was more profitable to raise smaller flocks in prime condition than to glut the market with inferior flocks that would bring "ruinous" prices.

Another perennial issue, also addressed in the February 1934 issue of the *American Turkey Journal*, was early egg production, because turkey hens, unlike chicken hens, do not produce eggs throughout the year. Under the heading, "Prominent South Dakota Breeder Discusses Early Egg Production," Al C. Johnson, Bath, South Dakota, offered advice on how to get turkey hens to begin laying eggs earlier in the season. "Last season," Johnson wrote,

> we started to gather eggs the 12th of February and hatched the first poults April 9th. Likewise do we expect the same results this season as the same care is again given. In fact this is our third year that we use the brooder houses for our pens.
>
> We mate our pens right after the All-American show and should the weather become severe as it usually does in February, here in the Dakotas, we build a fire in the stove. We use the regular brooder stove without the canopy. We think using the brooder house has many advantages with very little extra work, that of keeping it clean. We use about 6 inches of straw on the floor for litter which should be changed every week or 10 days. . . .
>
> Another advantage is, we move the brooder houses out on the green, when the weather permits, put up a fence and let the birds out, but they will continue using the brooder house for laying quarters and the eggs will not be chilled.
>
> Don't forget the cod liver oil in your laying mash. This is what we feed our hens: a mixture consisting of, 40 lbs. coarse ground corn, 20 lbs. wheat, 10 lbs. heavy oats, 20 lbs. meat scraps, 5 lbs. dried buttermilk, 4 lbs. bone meal, 1 lb. fine sifted

> salt. To this add 1 quart cod liver oil. This mash and whole cleaned oats we keep before them all the time. So are grit and oyster shells available; milk is very good and of course always fresh water. We also feed alfalfa in a rack or set some corn bundles in the corner for them to peck on.

With this method, Johnson said, he had good results, and "hatchability" was "very good."[8]

AUNTIE SUE'S HOME TALK was a folksy feature of the *American Turkey Journal.* The column began with one of Edgar A. Guest's homey poems titled "OLD YEARS AND NEW."

Old years and new years, all blended
into one,
The best of what there is to be, the
best of what is gone—
Let's bury all the failures in the dim
and dusty past
And keep the smiles of friendship
and laughter to the last.

Old years and new years, with all
their pain and strife,
Are but the bricks and steel and
stone with which we fashion
life;
So put the sin and shame away, and
keep the fine and true,
And on the glory of the past let's
build the better new.[9]

What followed was a light-hearted review of the previous All-American Turkey Show titled "Glimpses of Social Life at the All-American." "Auntie Sue" concluded her column with the recipe for Heavenly Hash, submitted by Mrs. A. D. Walker, Memphis, Missouri:

---

[8] *American Turkey Journal*, February 1934.

[9] Ibid.

> 1 cup golden drip syrup, 1 cup cream, and if milk is used add small piece of butter. Boil to a soft ball stage, remove from the fire and beat until creamy. A cup of any kind of nuts or shredded coconut is added before pouring in the buttered dish.

The last several pages of each issue of the *American Turkey Journal* were given over to Turkey Club news and to Classified Advertisements.

The 1932 All-American Turkey Show had a number of "dazzling," "distinguishing," and "outstanding" features and events, but the most significant was the distribution of the first issue of the *American Turkey Journal*. The All-American Turkey Show was its own best promoter, but the *American Turkey Journal* enhanced the show's reputation and extended that reputation across the entirety of North America and beyond.

The tenth Annual All-American Turkey Show opened on Monday morning, January 23, 1933, to run for a week. Because the show was often the first newsworthy event of the new year, promoters and the *Grand Forks Herald* capitalized on it. Almost every issue of the paper for a week preceding the show's opening carried articles publicizing the show.

By the morning of January 22, it was expected that the show would have the largest display ever of dressed turkeys, and over four hundred live birds—"turkey aristocrats" all—had been entered from seven states: North Dakota, South Dakota, Minnesota, Wisconsin, Colorado, Idaho, and Oregon. These "bluebloods of the turkey realm" would compete for prizes, and "some of the classiest birds in Turkeydom" would strut and preen in their coops in the Court of Honor when judging had been completed and prizes awarded.

The Grand Forks City Auditorium was decorated with small evergreen trees and bunting, and colored lights hung from the ceiling. Tiers of coops covered the main floor and stage. The Turkey Hens Nest and the lunch stand were at the front of the main floor. The entire north side of the balcony had been converted into a large refrigerated room for the dressed exhibits, and commercial displays of poultry feed and equipment were on the south and west sides of the balcony.

Other measures were taken besides publicity to encourage attendance. Both the Great Northern and the Northern Pacific railroads offered reduced passenger fares for those attending the show. The 1933 All-American Turkey Show provided evidence, were more needed, that the purposes of the show were being realized. Flock owners were improving the quality of their turkeys and more farmers were raising more of them. The "general quality" of the more than four hundred live turkeys entered in 1933, noted the *Grand Forks Herald*, was "the highest of the 10-year history of the show."

George E. Lamm of Philip, South Dakota, drove all night to arrive at the show by opening time. Forsaking his "hauler," the famous trailer pulled behind his coupe, Lamm used a truck to haul forty turkeys—his and those of other exhibitors—to the show. His efforts were rewarded. He was awarded the Master Breeder gold medal for "the general excellence of his large string of Bronze turkeys in the competition."[10]

"A feathered blueblood" owned by Wallace Jerome of Barron, Wisconsin, survived "some of the most spirited competition in the 10-year history of the show" and was crowned Grand Champion. Wallace Jerome was twenty-four years of age in 1933 when he exhibited for the first time at the All-American Turkey Show. He would make his mark at the 1934 show with his Bronze turkeys.

Judges at the 1933 All-American Turkey Show were reported to be "loud in their praises of the quality of the entries." F. E. Cross of Minneapolis, Minnesota, said that "the classes in general were 'away ahead'" of those in any show he had ever judged, including the eight years he had judged at the International Exposition in Chicago. P. M. Pierce of Denver, Colorado, said that "the quality of the classes as a whole" in 1933 was "unrivaled in turkey show annals."

Exhibitors were learning what good turkeys were and how to produce them as a result of the All-American shows. A. J. Burke, a new judge at the 1933 show, had heard about the All-American Turkey Show when judging at shows in Texas, Canada, and Mexico. Given "the class of birds"

---

[10] The Master Breeders' gold medal originated with the All-American Turkey Show, and it was awarded to the exhibitor who entered a pen of six birds of the same variety, to include a young tom, young hen, yearling tom, yearling hen, adult tom, and adult hen.

he judged in Grand Forks in 1933, he proclaimed the All-American Turkey Show had "more than lived up to its fame."[11]

The Hubbard Milling Company of Mankato, Minnesota, for the first time at an All-American Turkey Show, outfitted a small motion picture theater on the auditorium's stage. Lloyd S. Larson showed films he had taken of the 1932 show-officers, exhibitors, prizewinners, and "other interesting highlights." He also showed an educational film produced by the company on the proper feeding and care of turkeys. In honor of Turkey Week, and to encourage people to attend the All-American Turkey Show, the Paramount Theater showed a sound film featuring turkeys. The film was furnished by the Hubbard Milling Company.[12]

Three hundred guests attended the Turkey Banquet, always the most popular event of an All-American Turkey Show. Professor John E. Howard of the University of North Dakota's music department was in charge of the entertainment. The program included community singing, a saxophone solo by Orville Blackstad, a violin solo by Professor Howard, numbers by the Lions Club quartet, and feats of magic by Eddie McCoch.

The auction of dressed exhibits on Friday afternoon had become one of the most popular features of All-American Turkey Shows. Bidding was spirited, but good natured. The Grand Champion of the 1933 show, exhibited by Mr. and Mrs. John O. Allen of Radium, Minnesota, was a twenty-two-pound Bronze tom. It was purchased by the Belmont Café for eighty cents per pound, for a total of $17.60.[13]

The All-American Turkey Show in 1933 once again asserted its exclusiveness by scoring another major publicity coup that no other livestock show could match. An article in the *Herald* announced that the show's

---

[11] Eliminating any bird had been difficult for him and for the other judges. Another indication of the All-American Turkey Show's effect on the quality of turkeys was the large number of high quality live turkey entries sold by exhibitors to other breeders who wanted to improve the quality of their flocks and of their entries in future shows.

[12] An exhibit of wild pheasants also created much interest at the 1933 All-American Turkey Show. The exhibit included two Ringnecks, two Lady Amherst, two Golden, two Silver, and two Melanistic mutant pheasants. All the birds were from the F. E. Murphy Farms at Battle Lake, Minnesota.

[13] Its mate, also exhibited by the Allens, sold for forty cents per pound. The Reserve Champion, a fourteen-pound young hen exhibited by Miss Emma Ulrope of St. Thomas, North Dakota, brought thirty-five cents per pound. The dressed birds not in the prize class and remaining unsold were purchased by the Red River Valley Produce Company of Grand Forks for an average of three cents per pound above market price.

sweepstakes winners would be awarded Cable vases, "specially designed trophy vases made from North Dakota clay at the University here."[14]

Awarding Cable vases as prizes at an event with the reputation that the All-American Turkey Show enjoyed demonstrated the skill with which show officials drew attention, not only to the All-American Turkey Show and turkeys, but also to North Dakota and its resources. Wheat made North Dakota known as "the land of the No. 1 hard." The All-American Turkey Show made North Dakota known for its turkeys. Cable pottery made North Dakota known for its pottery clays.

Cable pottery had its inception with Earle J. Babcock, a graduate of the University of Minnesota in chemistry who joined the University of North Dakota's natural sciences faculty in 1889 at the age of twenty-four. Described by Louis G. Geiger in his history of the University of North Dakota as "fresh out of college and scarcely more than a boy, the redheaded and enthusiastic" Babcock "more than compensated for his lack of technical training by his energy and imagination."[15]

Having done the groundwork for ceramics with his field work and testing, Babcock convinced university president Frank L. McVey that the School of Mines needed a full-time instructor and researcher in ceramics. The first instructor, Margaret Kelly Cable, was hired in 1911.

Cable, a native of Ohio, began her training in pottery by taking a two-year course of study at the Handicraft Guild of Minneapolis, Minnesota, where bookbinding, metal smithing, jewelry making, leather craft, and pottery were the major decorative arts taught. At the Handicraft Guild, Cable noted, she worked with "simple folk" who "loved beauty" for the sake of beauty.

They "wrought with their hands and hearts as well as their brain" and they "put into each creation a bit of themselves." Apparently, it was here that Cable came to realize that "the work of the potter, especially that

[14] *Grand Forks Herald*, January 26, 1933.

[15] Intent on developing North Dakota's natural resources, Babcock made his chemistry laboratory a testing room for lignite and clay, and in 1891 and 1892 he published two pioneering pamphlets, one on the state's coals and the other on its clays. Although he did not have formal training in geology, Babcock was appointed state geologist, and, in 1897, he was appointed director of the university's School of Mines.

of the art potter," combined "in an unusual way industry, science, art and culture." It stimulated "the love of labor, the quest of knowledge, the study of art and the joy of culture."[16]

All clay, for Cable, was a metaphor for life, and her "impressions in clay," according to Miller, visual arts professor at University of North Dakota, were "not only insightful, but also inspirational."[17] Cable wrote of North Dakota clay in her poem, "The Song of the Pot."

> I am a lucky little lump of North Dakota clay,
> My heart is filled with gladness
> And I sing a song all day.
> For a potter found me worthy of his very finest ware,
> And fashioned me upon his wheel
> With tender loving care.
> With the magic of his fingers gave me form and life
> and soul.
> Transformed me from a shapeless clod
> Into a flower bowl.
> And as I hold within my arms a prairie rose bouquet
> I bless the hand that made me
> All from North Dakota clay.

To identify the pieces made from North Dakota clays at the university, Cable developed a seal, and, by 1913, according to Miller, "all pieces of any significance—pottery, figurines, tiles, medallions and, frequently, commercial production samples—bore the University of North Dakota seal." Of this distinctive seal, Cable wrote, "as long as a fragment of the seal remains the color of the printing will be legible as it was the day it came from the kiln, and will still proclaim to the world the worth and value of North Dakota clay." This seal was on the bottoms of the specially designed

---

[16] Donald Miller, "Margaret Kelly Cable: Thoughtful Impressions in Clay," *North Dakota History: Journal of the Northern Plains* (Spring/Summer 1996). Cable modified the philosophy of the Arts and Craft Movement of the late-nineteenth and early-twentieth centuries that the human touch had been destroyed by the Industrial Revolution and that it should be brought back into the making of useful objects. She believed that "industry could play an important and meaningful role in the development of the decorative arts." She put her research on western North Dakota's high-quality clays to use, not only in the development of such utilitarian products as drain tile, sewer tile, facing brick, and porcelain insulators, but also in the production of pottery.

[17] Miller described Cable "as the heart and soul of UND Pottery" in his article, "Margaret Kelly Cable."

trophy vases awarded to Sweepstake winners at the 1933 All-American Turkey Show.[18] Coveted as prizes at the All-American Turkey Show in 1933, and again in 1934, today Cable turkey vases are prized by collectors for their distinctiveness and rarity.[19]

Margaret Kelly Cable throwing in the Ceramics Department. All-American Turkey Show pot in front of the wheel on the left. From *The Cable Years: The Golden Anniversary of Margaret Kelly Cable's Retirement* (Grand Forks: University of North Dakota, 1999, Second Edition). Used with permission of Donald Miller.

---

[18] In a letter to Margaret Cable dated December 22, 1933, C. Dyke Page, president of the All-American Turkey Show, ordered five large pottery vases and twelve small pottery vases for "the coming All-American Show, the lettering we spoke of on the bottom of each piece can be taken care of after they are finished." On the original of the letter, in the University of North Dakota's Elwyn B. Robinson Department of Special Collections, Cable jotted, "check for 102.00 received for the 17 pots at 6.00 per-Feb. 1933."

[19] In June 2009, one of the vases was offered at auction by Rago Arts in Lambertville, New Jersey. Estimated to have a value of $4,000 to $6,000, the vase sold for $14,400.

CHAPTER 13

All in the (Turkey) Family

(with apologies to Archie and Edith Bunker)

WHEN THE ELEVENTH ALL-American Turkey Show opened on January 22, 1934, the Great Depression was in its fifth year, and North Dakota residents were in its grip. Historian Elwyn Robinson described them as "a stricken people." With only 13.5 inches of precipitation, 1933 had been the fourth-driest year on record, but 1934, with only 9.5 inches of rain, set a record of being the driest year yet.

"The state," Robinson wrote, "was drying up." And it was blowing away. Dust storms occasionally made travel by plane or automobile hazardous, and drifting soil covered fences, farm machinery, and ditches. The dust killed livestock and made life miserable for humans.

By April 1934, three months after the eleventh All-American Turkey Show closed, more than one-fifth of the state's residents were on relief. The percentage rose to 37 percent a year later, and relief, Robinson wrote, had become "the biggest business in the state."

Defying the Depression and the conditions in a drought-stricken state, fifty-six exhibitors "from the far corners of the United States entered hundreds of classy birds" that promised to make the eleventh All-American Turkey Show "one of the most successful on record." Seven states were represented—North Dakota, South Dakota, Minnesota, Wisconsin, Colorado, Idaho, and Oregon—and five turkey varieties—Bronze, Narragansett, Bourbon Red, White Holland, and Black.

"We Met Them at the Show" was a folksy feature of the *American Turkey Journal* that readers looked forward to reading as eagerly as they looked forward to reading news of their neighbors in their weekly newspapers. Included in the February issue, the feature provided news of the

All-American Turkey Show held the month before. Entries in that winter issue of the *American Turkey Journal* were typical.[1]

"Miss Clara Sutter, representing the *Farmer, Farm-Stock & Home* (St. Paul), was present most of the week and did we enjoy her." Miss Sutter spoke at the education sessions, over the radio, at "the get-together meetings," and at both banquets. "It would hardly be an All-American without Miss Sutter present."

Mrs. C. H. Folz, one of the two exhibitors who had exhibited at each of the eleven annual shows, "made her usual good entry of White Hollands and was present during the show." She and her husband, Alfred, were honored at the show in 1933 "for their long and loyal support."

"Wearing a smile that never rubs off and always busy with her fine turkeys on exhibit," Mrs. W. J. Janda, Mahnomen, Minnesota, "was present and everywhere." Mrs. Janda was secretary of the All-American Bronze Club. Mr. and Mrs. W. E. Hoynes of Mapes, North Dakota, among the exhibitors "who can always be depended on for an entry, and whose presence we can also count on," remained at the show all week "and enjoyed every function." They also won "some of the coveted prizes" that were "hotly competed for."

"Mrs. Mack Burnett doesn't talk much unless you get her alone. But how she can work and in any spot or place." She was present at the show all week. Mrs. William Eddie, Northwood, North Dakota, was "another enthusiastic Narragansett breeder who broke into the high awards" when her yearling tom was named champion in the Narragansett class. Mrs. Eddie was the Narragansett Club secretary and the newly-elected president of the All-American Turkey Hen Club. She was "doing good work" for the All-American Turkey Show. Mr. and Mrs. Martin Ellingson of Evansville, Minnesota, exhibited "a fine string of Narragansetts" and won "the lion's share" of awards in the live bird exhibit. They also "copped off the Grand Champion in the dressed bird division," which was "a boon" to Narragansetts.

Mrs. Bertha Tarr, a "well-known Bronze breeder" from Milton, North Dakota, intended to have a large entry at the show, "but the snow was too deep for auto travel to town and the distance too far from team

[1] *American Turkey Journal*, February 1934.

Miss Clara Sutter of *The Farmer*, St. Paul Minnesota, January 24, 1935. Image used with permission of Grand Forks Herald.

and sled." George E. Lamm "and his 'All-American Express' again made the 600 mile trip from Philip, S. Dakota, with a bigger cargo than ever before." Driving by way of Watertown and Aberdeen, Lamm picked up turkeys for Miss Grace Baxter of Hazel, for Mrs. R. Egge of Bristol, and for the A. C. Johnsons at Bath. Lamm's "cargo" also included a number of dressed turkeys for the show's dressed bird exhibit.

A fellow would have to be intrepid and determined to haul turkeys six hundred miles over bad roads in the dead of winter in a truck of that era. Lamm apparently was that sort of fellow, and his determination was rewarded. One of his Bronze turkeys was named the Grand Champion of the 1934 All-American Turkey Show, and Lamm received a Silver Cup Trophy provided by the Red River National Bank and Trust Company of Grand Forks.

In his *American Turkey Journal* editorial, George W. Hackett acknowledged that it was not his "to decree" whether or not the All-American Turkey Show was "doing the job it was established to do" and whether or not it was doing it "in the best and most satisfactory way." That would have to be determined, he wrote, "from comments and expressions from visitors and exhibitors."

However, Hacket believed no one was in a better position than he "to judge the pulse" of the show, to assure that "the machinery which must be in operation during show week" was functioning, and that all departments were working "smoothly and expeditiously," and that "steps" had been taken to make the show "a little better than the previous one."[2]

The "rearrangement" of the coops "made the floor work easier" and "gave a better view of the show room and added to the capacity of cooping space." The "greatest improvement of all," he believed, was "the arranging of the 'Court of Honor' on the auditorium stage." The new arrangement "gave an effect, the equal of which" was "seldom seen" at a turkey show, and it "elicited more favorable comment" than any other change made in the seven years he had been manager.

The new arrangement, with the "fine silver cup trophies" and the "specially designed trays" all displayed "in artistic arrangement" in the center of the Court of Honor, "had to be seen to be appreciated." Hackett assumed that the "whole setting had much to do with the cheerful and friendly attitude that pervaded the show room.

Although Hackett believed that it was "always an injustice to good specimens to put them on display denuded of part of their plumage," he was, nevertheless, "especially proud" of the dressed exhibits in 1934. Thomas W. Heitz, agriculture expert from the Department of Agriculture, Washington, D.C., judged the exhibits. Their quality, he said, was the best he had ever seen.

The Grand Champion dressed turkey, a fifteen-pound young Narragansett hen exhibited by Mrs. Martin Ellingson of Evansville, Minnesota, sold for $2.00 per pound, the highest price ever paid in the history of the All-American Turkey Show. Russell Whaley of New York City, a prominent turkey buyer in the Northwest, was the buyer, and, at his request, the dressed bird was given to the Grand Forks Home for the Aged.[3]

---

[2] *American Turkey Journal*, February 1934.

[3] The Red River Produce Company purchased the dressed birds remaining unsold at the end of the show, paying 19¼ cents per pound for young toms, 17 cents per pound for old toms, and 16 cents per pound for old hens.

To describe the show's educational program in 1934 as unusually good would be an understatement. "Judging from the intense interest shown and the great number of questions asked," according to a feature in the *American Turkey Journal*, never before at an All-American Turkey Show had there been a demonstration that "was more eagerly followed through" than the Turkey Health and Disease Clinic conducted by Martin B. Potratz and George C. Dibble of the Salsbury Laboratories, Charles City, Iowa.[4]

In the clinic portion of his presentation, Potratz posted two ailing turkeys, taken from a flock that had sustained heavy losses from disease. The posting revealed that cholera was the cause of the ailment, and, therefore, the cause of the heavy losses in the flock. Potratz concluded his presentation by explaining how flock owners could detect cholera in their turkeys and why a program of "strict sanitation" could help to ward it off in their flocks.[5]

That each All-American Turkey Show was better than the preceding one was, for Hackett, a matter of pride, and it was each show's "firsts" that helped to make each show better than the preceding one.

At the 1934 All-American Turkey Show, the Hubbard Milling Company went other poultry feed companies one better. According to the *American Turkey Journal*, the Hubbard Milling Company

> was on hand again this year with bells on, as usual. Their booth was arranged in the form of an intimate little movie theatre seating about 30 persons and under the capable direction of the indefatiguable Lloyd S. Larson, the movie impressario, a continuous performance of their motion picture features went on. Scenes from the All-American Show of 1932 and 1933 were shown as well as highly

---

[4] Dibble drew on his field work experience for Salsbury Laboratories when speaking on the diseases and ailments affecting turkeys and how to treat them. Potratz reviewed the origin of domesticated turkeys and explained that, in the process of domestication, turkeys had lost much of the "strong constitutional vigor" of wild turkeys and were, therefore, susceptible to diseases.

[5] The presentations of Dibble and Portratz were so favorably received by the large number of people filling the large lecture room in the auditorium's basement that they requested that a similar clinic be held at the show in 1935.

> instructive scenes of noted flocks of Northwestern turkeys, illustrating methods and improvements.[6]

Lloyd S. Larson, who had "become an All-American 'regular'" at the All-American Turkey Show in 1934, "again became the demanding movie director for a time and took scenes of all features of the current show," which would be made available to turkey shows and club meetings free of charge.[7]

The Footlight Parade was the most striking new feature of the All-American Turkey Show in 1934. Utilizing terms and routines common to the entertainment world to heighten interest and generate publicity, the show in 1934 featured Turkey Queens parading "turkey aristocrats" around the Court of Honor in a Footlight Parade. The women controlled their charges with reins made of colorful ribbons, while Larson shot the proceedings with his movie camera.

The T. R. Miller Milling Company, with the genial and agreeable T. I. Foster of Chicago, Illinois, in charge, provided another first for the 1934 show. The firm's "new light-weight shipping containers for both live and dressed birds" had made the company's name "well-known to turkey growers far and wide," and its booth "was a center of attraction" during the show.

Foster, who had "entered into the spirit of show week and was in the middle of everything that was going on," published a scandal sheet, containing "highlights on all show events." His special edition exposing "a newly-discovered Nudist colony" was a sensation. The nudist colony turned out to be the show's dressed turkey exhibits, but the exposé endeared him to the All-American crowd.

---

[6] *American Turkey Journal,* February 1934.

[7] Another new feature in 1934, "which met with marked approval," was KFJM, the University of North Dakota's radio station, reporting on judging and other events "right from the show room each afternoon." The Hubbard Milling Company purchased time on KFJM and the broadcasts were made from the show's Hubbard Sunshine Theátre. KFJM also broadcast many of the show's other features during the week, including some of the education sessions. By 1930, station KFJM, the only one in the area, had expanded from a noon-hour broadcast and an hour every Wednesday evening to about twenty hours per week, plus special broadcasts of music contests and sports events. Almost entirely self-supporting, the station charged $3.00 per hour for airtime. By the time the station began broadcasting from the All-American Turkey Show showroom, operating hours had been increased to sixty hours per week and the station had implemented a schedule of commercial fees.

The Court of Honor at the All-American Turkey Show in 1934, with the Grand Champion in center stage, surrounded by twelve other champions in their cages. *American Turkey Journal*, February 1934.

Another first in 1934 was for an exhibitor, not for the All-American Turkey Show. Wallace Jerome of Barron, Wisconsin, made his mark at the show, a mark still seen in the meat aisles of supermarkets. Wallace Jerome was born in Spooner, Wisconsin, March 22, 1909, and he spent his early years on a farm. Farm work prevented him from attending school regularly, but he graduated from Barron High School in Barron, Wisconsin, in 1928.

In 1930, Jerome enrolled in the Wisconsin Agricultural Short Course in Madison, and he started to work as a poultry and egg inspector for the Wisconsin Department of Agriculture. He also became an accredited poultry judge. In 1935, Jerome attended the University of Minnesota, but, wanting a degree from his home state, he enrolled in the University of

Turkey Queens and prize turkeys participating in the Footlight Parade, staged first at the All-American Turkey Show in 1934. *American Turkey Journal*, February 1934.

Wisconsin, Madison, and he received his degree in Poultry Husbandry in 1941.[8]

Jerome got his start in turkeys with fourteen eggs and two broody chicken hens. The hatch, fortunately, included one tom, and Jerome had the beginnings of his variety of Super Bronze turkeys, "Triple-bred for Exhibition, Egg Production, and Market Type." Between his fourteenth birthday and the time he completed high school in 1928, Jerome had hatched a flock of two hundred turkeys. A year later, he was named Wisconsin's Outstanding 4-H Poultry Club Member.

---

[8] Fortunately for researchers, Wallace Jerome's life, his business enterprises, and his awards are well documented. His daughter, Mary Ella Jerome, also of Barron, Wisconsin, has her father's records, and she graciously made them available.

Wallace Jerome of Barron, Wisconsin, exhibited his Standard Bronze turkeys for the first time at the All-American Turkey Show in 1933. He is shown with his Bronze Tom that was chosen as the show's Grand Champion. Photo courtesy Mary Ella Jerome, Barron, Wisconsin.

When judged at exhibitions, turkeys lose points or are disqualified if any tail feathers are damaged or missing. Jerome's method of transporting exhibition turkeys protected the feathers, and in 1933 he exhibited his Bronze turkeys for the first time at the All-American Turkey Show. One of his yearling toms was named the show's Grand Champion.[9]

Jerome was an innovator in the turkey industry. When large operations, producing hundreds of turkeys per year, were considered inefficient, unprofitable, and impossible, Jerome saw them as the method of the future. His strategy was to be involved in every aspect of the turkey business from conception to consumption; from the breeding, to the hatching, to the raising, to the processing, and to the marketing. Being independent, he was able to survive when others failed during the years of the Great Depression.

[9] At the 1934 All-American Turkey Show, Wallace Jerome exhibited fourteen turkeys. Twelve of them placed, and Jerome received five Sweepstakes trophies, including those for Best Bronze Display, Best Breeders' Display, Best Wisconsin Display, and Best All-American Bronze Turkey Club Display. His trophies included a Cable Turkey Vase, a Silver Trophy, and $10 cash. Jerome also received the Master Breeders' Award with Gold Medal, the All-American Turkey Show's highest and most coveted award. In 1934, Wallace Jerome was twenty-five years old.

Jerome discerned that consumers' tastes were changing. Some consumers wanted smaller turkeys, some wanted only turkey parts, and many wanted turkey meat available throughout the year, not just at Thanksgiving and Christmas. Responding to what consumers wanted in grocery store meat aisles, Jerome changed his breeding program and began crossing the Broad Breasted Bronze variety with the smaller White Holland turkey.

After improving turkey-raising systems to handle large numbers of turkeys, Jerome started a Home Farm in 1941 to produce New York dressed turkeys, whole birds that were plucked and bled. In 1950, he purchased an abandoned pea cannery in Barron, Wisconsin, that later became the location of Jerome Foods. After visiting a plant in Omaha, Nebraska, to learn how others were processing turkeys, he designed and developed his state-of-the-art, oven-ready processing plant, outfitting it to prepare "oven-ready turkeys."[10]

John and Amanda Allen were among the best-known Bronze turkey breeders in the Red River Valley. Often exhibiting as many as twenty-one birds at an All-American Turkey Show, they were consistently among the top award winners, including that of Grand Champion in the dressed division in 1933 and 1934. At the 1934 show, more than a dozen of their birds placed, and their awards included those for the Best Bronze Display, Champion Bronze Display, and the All-American Bronze Club Display. Amanda Allen was active in the Hen Club, and John Allen was vice-president of the All-American Turkey Show association. As had Jerome, the Allens started their turkey operation with two settings of turkey eggs.

---

[10] Wallace Jerome named the operation Badger Turkey Industries. Three years later, Jerome bought out Peter Fox Sons, a rival, and he converted its facility into another hatchery. Badger Turkey Industries became Jerome Foods, Inc., in 1964. With Jerome's innovative spirit and under his leadership, Jerome Foods, Inc., became one of the top ten turkey processors in the country, with processing plants in Barron, Wisconsin, and in Altura and Faribault, Minnesota. The company had hatcheries, growing facilities, and feed mills in Wisconsin and Minnesota. Fascinated with processing turkeys even further, Jerome introduced turkey steaks, tenderloins, ground turkey, turkey ham, turkey sausages, and specialty items such as GobbleStix—all marketed nationally under the Turkey Store trademark, introduced in 1984, and all helping to make turkey meat an everyday staple rather than the centerpiece on holiday dinner tables. Jerome Foods, Inc., stopped selling whole turkeys in 1996, and its name was changed to The Turkey Store Company in 1998. By 2001, the Turkey Store Company had become the sixth largest turkey business in the United States with 2,500 employees and annual sales of $309 million.

In 1935, Hackett visited "This Famous Bronze-raising Family" for an article in the *American Turkey Journal.* The Allens' farm of 320 acres was located near the village of Radium, Minnesota, on the eastern edge of the Red River Valley and one hundred miles south of the Canadian border. The area, "ideal for turkey raising," was broken prairie with sufficient brush and timber to furnish cover and protection for their flock. Much of the land was given over to raising grain, but turkeys, Mrs. Allen said, had been a life saver during the years of the Depression and drought.[11]

Amanda started raising turkeys in 1912, while single and teaching school. The woman with whom she boarded gave her two settings of turkey eggs of "common mixed breeding" that produced twenty poults. Two years later, she raised twenty-seven turkeys and sold them for $27.00. "Being of a progressive nature and not satisfied with . . . ordinary stock," she decided to raise better quality turkeys. In 1917, now Mrs. John Allen, she bought two purebred Bronze hens and a tom, "the best she could find." An owl took one hen and dogs killed most of the poults.

Undaunted, in 1921, Allen paid what she considered "quite an exorbitant price" for a tom from a noted eastern Bronze breeder. The results justified the purchase price. Continuing to improve the quality of her breeding stock, she began exhibiting turkeys in 1930. Husband John began taking part in rearing and exhibiting birds from "their fine stock," and year after year the Allens' turkeys had "made enviable winnings in the hottest of competition."

The Allens raised an average of 300 turkeys per year, but, in 1935, they had 550 "very fine, rapidly developing young birds" that promised "to give plenty of competition in both live and dressed divisions of the best shows." Following what they believed was a safe and sane plan, the Allens were particular when selecting their breeders, convinced that better-bred turkeys produced the best birds for exhibition and for market.

John and Amanda Allen had three children—Evelyn, Lloyd, and Fern—and turkeys still run deep in the Allen family. When his parents retired, Lloyd took over the farm and the turkey operation. Having grown

---

[11] *American Turkey Journal,* August 1935.

up with turkeys, Evelyn and Fern vowed never to marry turkey farmers. Both, however, did.[12]

Turkey breeders such as Jerome and the Allens provided the answer to Hackett's question, raised in his *American Turkey Journal* editorial, about whether the All-American Turkey Show was "doing the job it was established to do."[13] Not only did the All-American Turkey Show make Grand Forks, North Dakota, the Turkey Capital of the World, it also changed Americans' eating habits by encouraging breeders to raise more turkeys, to develop turkey products, and to make turkey available to consumers throughout the year, not just during the holiday season.

The twelfth Annual All-American Turkey Show, January 21–26, 1935, was marked not so much by "firsts" as it was by the conditions under which it was held. The Great Depression was in its sixth year, the New Deal in its second. The winter's cold and snow were as trying as the heat and dust of the previous summer.

Within days after taking office, President Franklin D. Roosevelt, on March 12, 1933, went on the radio to address his fellow citizens in the first of his fireside chats. Described in *The American Heritage History of the 20s and 30s* as "overtly informal talks," the fireside chats opened with the words, "my friends," a greeting that Roosevelt had used since his earliest days in politics. The new president delivered his first fireside chat in "a relaxed intimate voice" that made listeners "feel that they were closer to their president than ever before in history." Reassured by their president's voice and optimism, the "forgotten" people of America believed that they were finally being "remembered."

Three days before President Roosevelt chatted with his fellow citizens for the first time, Congress convened in special session for the Hundred Days, the three-month period during which it enacted many of the New Deal measures with which Roosevelt intended to address the causes and effects of the Depression.

---

[12] Fern Allen married Dale Peterson in 1948. Dale had grown up on a North Dakota farm, and he graduated from North Dakota State University in 1937 with a degree in poultry science. After graduation, Peterson formed a turkey operation with Howard Holden near Northfield, Minnesota. Howard Holden married Evelyn Allen. Holden Farms still raises turkeys, although the farm's main business is hog production.

[13] *American Turkey Journal*, February 1934.

In June 1933, Congress enacted the National Industrial Recovery Act (NIRA), designed to assist the country's economic recovery. Included in the provisions of the NIRA was the requirement that industries had to draft industry-wide codes of fair competition that fixed prices and wages and established production quotas. By executive order, President Roosevelt created the National Recovery Administration (NRA), an act he considered the most important of the early New Deal. Administered by Hugh S. Johnson, the NRA was intended to eliminate wasteful competition, encourage better regulated pricing and marketing policies, and provide higher wages and shorter hours.

During the NRA's first months, administrators drew up codes dealing with hours of work, rates of pay, and levels of prices for industries, more than five hundred codes in all. By the summer of 1933, the NRA's Blue Eagle, a symbol used to show compliance with the NIRA, was appearing all over America, in the windows of shops, restaurants, and industries; on the backs of framed pictures and on the sweatbands of men's hats; and on plates attached to truck license plate brackets.

With the establishment of the NRA and the drafting of its codes, the question in the minds of many turkey producers was the article title in the *American Turkey Journal*, "Shall Turkeys Go Under a Code?"[14] Flock owners across the country could share stories of "deadly cut-throat competition;" of unfair pricing; and of poor quality eggs, poults, and birds being offered for sale. "Cutting prices and underselling," Wallace Jerome noted in an article published a few months later in *Turkey World*, seemed "to be a universal custom in many industries which came into effect during the depression."[15]

Mrs. C. E. Brown, Littleton, Colorado, cited examples of "small turkey growers within a radius of one hundred to two hundred miles of Denver" who had "sacrificed" their turkeys in the fall of 1933 "for almost any old price they were offered," because markets were glutted with too many turkeys, many of them of poor quality, those from commercial turkey businesses. Brown had this advice for flock owners who were "planning on doubling or increasing their flocks" in an effort to remain in busi-

---

[14] *American Turkey Journal*, February 1934.

[15] *Turkey World*, May 1934.

ness. "Take warning," she said, "of the evil of over-production such as has caused the ruin of the wheat, cotton, tobacco and livestock industries and on which too our government is now spending billions of dollars to try to rehabilitate."

Certainly, Brown wrote, "restriction and a readjustment of the entire industry" was "bound to come," unless flock owners showed "a more cooperative spirit" and limited the numbers of turkeys raised "within a reasonable and measurable distance of consumptive requirements." Unfortunately, she believed, this would require "the compulsion of a code similar to those being put into effect by the government in other lines of agriculture."

The question of whether turkeys should go under a code was addressed in meetings held during the 1934 All-American Turkey Show. Prominent turkey breeders from all parts of the country were in favor of seeking protection for small producers under a turkey code. Commercial hatcheries were setting prices for turkey eggs that left the small breeder no chance for profit, and large numbers of these eggs were small, having been produced by immature hens. Jerome noted in his article titled, "Turkeys and the Code," that the quality of turkey poults was "greatly dependent upon the size of the eggs." From small eggs came small, weak chicks.

Jerome introduced his article by asking, "Are the turkey breeders and hatcheries going to receive the same benefits that the chicken breeders and hatcheries" were receiving "under the new National Breeder and Hatching Code?" The answer was no, because—noted those meeting during the 1934 All-American Turkey Show—when the hatchery code provisions were formulated, none were made for turkey breeders because there were no representatives from the industry. The managing agent stated, however, that provisions could be made for turkey breeders if "a representative group" drew up the provisions.

At the 1934 All-American Turkey Show, a committee was formed to formulate provisions to be presented to the National Commercial and Breeder Hatchery Code Committee, requesting that turkey breeders be included in the National Commercial and Breeder Hatchery Code.[16] A

[16] Ray Andrews, Petersburg, North Dakota; Wallace Jerome, Barron, Wisconsin; George W. Hackett, Wayzata, Minnesota; A. J. Burke, Midlothian, Texas; and Mrs. William Eddie, Northwood, North Dakota—some of the best known and most prominent individuals associated with the country's turkey industry—were among those named to the committee.

code, Jerome believed, would "do a great deal of good for the breeder, the breeder hatchery, the commercial grower and the industry in general."

First, because the code would eliminate much of the "cheap, cut price competition" that was "doing more harm than good to turkey growers." Second, because the code would specify a minimum weight for turkey eggs, which would "do much to increase the quality" of chicks, because the quality of turkey poults was dependent on the size of the eggs. Third, because the code would increase the price that commercial hatcheries paid producers for their eggs, thereby also raising the quality of the eggs and that of poults.

Turkeys, Wallace was certain, would be included in a code in 1935. His optimism was misplaced. In 1935 the question of whether turkeys should "go under a Code" became moot. For all that President Roosevelt expected of it, the NIRA did not last long enough to fully implement its policies. In "the sick chicken case," *Schechter Poultry Corp. v. United States*, the U.S. Supreme Court in May 1935 held in a lengthy and unanimous opinion that the NIRA, a central piece of President Roosevelt's New Deal program, was unconstitutional.

The Schechter Poultry Corporation purchased live poultry in New York City, butchered the birds, and sold the meat to butchers and retailers in Brooklyn. The company was convicted of breaking the provisions of the Live Poultry Code. Among the sixty charges, later reduced to eighteen, was that the Schechter Poultry Corporation had sold chickens unfit for human consumption. The company appealed the conviction, eventually to the U.S. Supreme Court.[17]

What Elwyn B. Robinson termed "the rigors" of North Dakota's "cold winters" provided the second of the conditions under which the twelfth All-American Turkey Show was held. On the first day of the show,

---

[17] The Supreme Court held that the Live Poultry Code violated the Constitution's Separation of Powers clause, because it delegated legislative power to the president, and NIRA administrators enforced the Code's provisions with no congressional direction. The Supreme Court also held that the Code's provisions did not constitute a regulation of interstate commerce. The Schechter Poultry Corporation's business was intrastate, not interstate. The decision also seemed to indicate that Supreme Court justices did not accept President Roosevelt's argument that radical measures were required to deal with the causes and effects of the Depression.

Monday, January 21, 1935, the temperature in Grand Forks was -30°F.[18] "Bitterly cold weather in this vicinity and throughout North Dakota and Minnesota," reported the *Grand Forks Herald* two days later, "interfered with normal traffic on the highways and in some areas, according to the Associated Press, forced schools to close. Because the extreme cold prevented truck and auto travel, city streets and area roads were "practically devoid of traffic."[19] Trains were running one to two hours late because of the cold, and passenger buses were also running late. Northwest planes were grounded on Saturday and Sunday, January 19 and 20, because of the extreme cold.

"Snow Blocked Roads Prevent Some Arrivals" was the article title in the *Grand Forks Herald*.[20] Some exhibitors were unable to reach Grand Forks until the show's second day. George E. Lamm of Philip, South Dakota, and his All-American Express arrived late, having been delayed by blocked roads and severe weather. Thomas W. Heitz, of the U.S. Department of Agriculture, was scheduled to judge the dressed exhibits. He was snowbound on a train in southern Minnesota for thirty-six hours.

On January 21, travel was hazardous, and many highways in the Grand Forks and Devils Lake areas were blocked. Snowplows dispatched to open the roads became stuck or broke down, leaving drivers and trucks stranded. Given the likelihood that blizzards would block the roads, many exhibitors—including Mrs. W. J. Janda of Mahnomen, Minnesota, one hundred miles to the southeast of Grand Forks—wisely elected to send their exhibits to Grand Forks by train and not attend the show. Travel conditions and the bitter cold also dampened sales of both breeding stock and dressed turkeys. "Blocked roads in the country," reported the *Herald*, "handicapped sales in the local field, as many growers from nearby points were unable to get into the show."[21]

Defying "both intense cold and snow blocked roads," reported the *Grand Forks Herald*, "turkey fanciers of the Northwest" were on hand

---

[18] Two days later, January 23, the temperature in Grand Forks at 7:00 a.m. was -35°F, having risen from a low of -40°F, a new record low for the season and the lowest temperature since December 28, 1933, when it had been -37°F. Every reporting station in the state recorded temperatures of -30°F or lower on January 23. Williston, on the far western edge of the state, was also cold. The temperature there was -40°F.

[19] *Grand Forks Herald*, January 23, 1935.

[20] *Grand Forks Herald*, January 21, 1935.

[21] *Grand Forks Herald*, January 25, 1935.

when the twelfth annual All-American Turkey Show opened on time.[22] The "more than 400 of the finest turkeys in the country. . . gobbled and strutted in coops at the auditorium when the doors were thrown open to the public at 10:00 A.M." Live birds would be judged, starting at 1:30 p.m., and those arriving late, up to Tuesday morning, would be cooped, be allowed to compete, and be judged with their classes.

The twelfth All-American Turkey Show had been well advertised, months in advance of its opening. "THE GREATEST OF ALL TURKEY SHOWS will be greater than ever when the doors open for the 12th annual show on Monday, January 21st, 1935," was the announcement in an earlier issue of the *American Turkey Journal.*[23]

The December ad in *Turkey World*, consisting of more phrases than complete sentences, was even more effusive: The All-American Turkey Show, "First and Foremost of All Turkey Shows," was scheduled for January 21–26, 1935. "All-American Features" included the following: "completely efficient management"; "the best judges in the country"; "fair and impartial judging in every respect"; "systematic, orderly, comfortable cooping (caging)"; "the most careful care of all birds"; "an outstanding educational program"; "the most elaborate trophies and awards of any show"; "a week of social enjoyment."[24]

"The internationally famous All-American," continued the ad, in 1935 would "mark 12 years of continuous progress and development in the staging of exclusive shows." Starting in 1924, "in a small storeroom," the All-American Turkey Show had "grown in size and prestige," and it was recognized "as the final forum in the selection of the country's best turkeys."

A win at the All-American Turkey Show "against the stiff competition offered there" was touted as "proof positive of the highest quality in every class." Exhibitors were not required to accompany their birds to the show. During the previous five years, many of the show's most coveted awards, including that of Grand Champion, had been won by birds unaccompanied by their owners.

---

[22] *Grand Forks Herald*, January 21, 1935.

[23] *American Turkey Journal*, October 1934.

[24] *Turkey World*, December 1934.

DAILY PROGRAM

ALL-AMERICAN TURKEY SHOW

City Auditorium, Grand Forks, N. Dak.

JANUARY 21 - 26, 1935

"A Week Both Educational and Social"

MONDAY, JANUARY 21st

10:00 A. M.—Doors open to the public. Free Admission.
12:00 Noon—Judges luncheon with Manager.
1:30 P. M.—Judging of live birds begins.
Registration of visitors at Information Booth.
6:00 P. M.—Entries close on dressed birds.

TUESDAY, JANUARY 22nd

9:00 A. M.—Judging resumed on main floor.
10:00 A. M.—Judging of dressed birds begins.
No afternoon program as judging occupies attention of all present.
6:30 P. M.—Grand "get-to-gether" meeting and luncheon at Belmont Cafe. Entertainment program will be furnished by the Grand Forks Business and Professional Women's Club. Introductions and informal social meeting to follow.

WEDNESDAY, JANUARY 23rd

9:00 A. M.—Judging resumed in both divisions.
1:30 P. M.—Educational Program opens at Auditorium.
"The Incubation of Turkey Eggs" by Glen F. Richards, of Jamesway Incubator Company.
2:30 P. M.—"Turkey Diseases, Their Treatment" followed by clinic on diseased turkeys. Talent from Dr. Salsbury's Laboratories, Charles City, Iowa.
6:30 P. M.—Hen Club Banquet and Program, Ryan Hotel.

THURSDAY, JANUARY 24th

9:00 A. M.—Announcing Class Champions, Sweepstakes Winners and placing winning birds in the "Court of Honor."
1:30 P. M.—"Woman's Work in Turkey Production," By Miss Clara Sutter of The Farmer, St. Paul, Minnesota.
2:15 P. M.—"Winter Care and Feed for the Breeders" by Prof. M. C. Herner, College of Agriculture, Winnipeg, Canada.
3:00 P. M.—"Advantages of Federal Grading of Turkeys" by Thos. W. Heitz, U. S. Department of Agriculture, Washington, D.C.
6:30 P. M.—Big Homecoming Banquet, entertainment and ball, Garden Room Hotel Dacotah. All exhibitors, visitors and other supporters of the Show are cordially invited. Reservations should be made early to insure plate at the banquet.

FRIDAY, JANUARY 25th

10:00 A. M.—Final posting of awards.
1:30 P. M.—"The Advantages of Trap Nesting As We Have Found Them," by, Mrs. Claude Wright, Aitkin, Minnesota.
2:30 P. M.—"Economic Turkey Production," by Prof. O. A. Barton, North Dakota College of Agriculture.
3:00 P. M.—"Turkey Production Outlook for 1935" by Dr. W. A. Billings, University Farm, St. Paul, Minnesota.

The Daily Program for the 1935 All-American Turkey Show. *American Turkey Journal*, January 1935.

For the benefit of visitors and first-time exhibitors, Hackett described the All-American Turkey Show's major awards in the *American Turkey Journal.* "Special Awards," he explained, served a number of purposes. They were intended to compensate those who had exercised "great care and untiring effort in placing their 'show string' on exhibit in a high-class manner, thus adding greatly to the attractiveness of the show, increasing the general interest, and encouraging greater determination on the part of the other exhibitors to duplicate such success at a later time."[25]

Those donating the trophies for the Special Awards often stipulated how the award was to be listed and for what it was to be awarded. The stipulation had to be honored, because businesses donated the trophies for advertising purposes and they wanted the best and most extensive exposure possible.[26]

Breed Champions had their own awards, an award for the champion bird of each turkey variety exhibited at the show. There were five in 1934: Bronze, Narragansett, White Holland, Bourbon Red, and Black. Breeders, Hackett cautioned, could not advertise Breed Champions as Grand Champions of their variety. Each show had only one Grand Champion. The Master Breeder Award, the highest and most coveted award offered at any turkey show, had been inaugurated at the All-American Turkey Show.[27]

Braving the weather and putting aside the country's economic woes for a time, exhibitors from a dozen states, including Minnesota, South Dakota, North Dakota, Wisconsin, Ohio, Iowa, Colorado, Texas, Missouri, and Kansas had entered more than six hundred live and dressed turkeys. The Jerome Twins—owners of Bronze birds—attracted Hackett's attention as he noted that Wallace Jerome, who had made his mark at the shows in 1933 and 1934, towered at almost six feet tall but was

[25] *American Turkey Journal,* January 1935.

[26] "Sweepstakes Awards," Hackett noted, could have no limitations placed on them if they were to be Sweepstakes Awards, and there could be as many Sweepstakes Awards as there were donors. There were twenty donors and Sweepstakes Award winners at the 1934 All-American Turkey Show.

[27] A Master Breeder Award entry consisted of at least one bird in each of the single classes, that is, one adult, yearling, and young tom, and one adult, yearling, and young hen. At least four of the six birds had to have been bred and raised by the exhibitor. This requirement was as necessary as it was important, because the award was to demonstrate the exhibitor's ability to mate and manage his breeders so as to produce high quality birds and thus "prove title to the highest honor known to turkeydom." Birds competing for the award had to have been entered singly and have competed in the open classes. The first-place winner in each variety could compete for the Master Breeder Award.

JEROME'S

SUPER BRONZE

Again Prove Their Superior, Triple-Bred Qualities by Winning

MASTER BREEDER'S AWARD

1934 All-American, Grand Forks

Also First All-American Bronze Club Display, Breeders Display, Sweepstakes Display. Placed 12 out of 14 birds shown.

JEROME'S STOCK IS TRIPLE-BRED FOR Exhibition, Egg Production and Market Type Trapnested and Pedigreed

WALLACE JEROME, Barron, Wisc.

Wallace Jerome of Barron, Wisconsin, made his mark at the 1934 All-American Turkey Show by receiving the Master Breeder's Award and Gold Medal. *American Turkey Journal*, February 1934.

known as the "little" twin. His brother, Willis, could gaze over his little brother's head.

Cooped on the auditorium's main floor were, as Hackett described, "425 of the finest turkeys in America, representing the pick of the turkey world from all parts of the United States." Included in the 425 was a display of Black turkeys, a variety developed in England. The commercial booths, also on the main floor, were arranged along the auditorium's north end.[28]

[28] Included were those of *Capper's Farmer* of Topeka, Kansas; Dr. Salsbury Laboratories from Charles City, Iowa; Purina Chow; Railway Express, with a miniature refrigerator car; Northrup King & Co.; F. D. Miller Mills; Conkey's poultry feed; and the General Box Co. from Kansas City, Kansas.

The famous Court of Honor was on the auditorium stage, with the show's Grand Champion in the center, flanked by the champions of their varieties. As the champions were selected, they were transferred to the honor row. The Turkey Hen Nest was also located on the stage in 1935. Mrs. William Eddie of Northwood, North Dakota, outgoing president of the All-American Turkey Hen Club, and Mrs. John Allen of Radium, Minnesota, president-elect of the Hen Club, were in charge of the nest.

The Twelfth Annual All-American Turkey Show's Educational Program opened at 1:30 p.m. on Wednesday, January 23. The session topics were not only timely, they also reflected the desire of turkey breeders to learn how to improve the quality and profitability of their flocks.[29]

The dressed birds, displayed in the cooler on the auditorium's balcony, were judged by Thomas W. Heitz of the U.S. Department of Agriculture and Professor O. A. Stevens of North Dakota's College of Agriculture. Stevens was reported to be enthusiastic about the dressed bird exhibit. "In spite of the shortage of feed in North Dakota," he said, "the birds from this section this year" were "in the best condition they have ever been."[30]

About two hundred exhibitors and visitors attended the annual turkey dinner, sponsored by the Grand Forks Chamber of Commerce. W. P. Davies extended the Chamber's greetings, and Thomas W. Heitz was among the speakers. Music was under the direction of John E. Howard, of the University of North Dakota's music department. C. Dyke Page, All-American Turkey Show president, presided at the dinner, and George

---

[29] Topics included the following: "The Incubation of Turkey Eggs," by Glen F. Richards, of Jamesway Incubator Company; "Turkey Diseases, Their Treatment," followed by a clinic of diseased turkeys, conducted by Dr. Eyvert Erikson of Dr. Salsbury's Laboratories, Charles City, Iowa; "Woman's Work in Turkey Production," by Miss Clara Sutter of *The Farmer*, St. Paul, Minnesota; "Winter Care and Feed for the Breeders," by Professor M. C. Herner, College of Agriculture, Winnipeg, Canada; "Advantages of Federal Grading of Turkeys," by Thomas W. Heitz, U.S. Department of Agriculture, Washington, D.C.; "The Advantages of Trap Nesting As We Have Found Them," by Mrs. Claude Wright, Aitkin, Minnesota; "Economic Turkey Production," by Professor O. A. Stevens, North Dakota College of Agriculture; and "Turkey Production Outlook for 1935," by Dr. W. A. Billings, University Farm, St. Paul, Minnesota.

[30] The Grand Champion, a Narragansett young tom weighing 23 ¼ pounds, was exhibited by Mrs. Martin Ellingson, of Evansville, Minnesota. At the Friday afternoon auction, the Grand Champion was purchased by the Belmont Café for $1.00 per pound, a total of $23.75, a princely sum in 1935. The Railway Express Co. paid 35 cents per pound for an 11 ½ pound bird, and the Hubbard Milling Co. of Mankato, Minnesota, purchased twelve dressed birds, including the First-place Black young hen from Oakdale Turkey Farms at Kensington, Minnesota, and the champion Narragansett young hen of Mrs. Martin Ellington. Dressed birds not sold at auction were purchased by a Grand Forks produce business for 25 cents per pound, 3 to 4 cents over market price.

Hackett presented the Grand Champion and Master Breeder trophies. The Show's Grand Champion, a Bronze tom, was exhibited by Mrs. Wilhelmina Grant, of Glyndon, Minnesota. As it was later stated, "The royal purple of turkeydom was placed on the youthful shoulders of Glen Beidleman of Kingsbury, Kansas, who was awarded the Master Breeder's prize of the twelfth All-American Show."

An article in the *Grand Forks Herald* the following day, announced that exhibition turkeys would be released to exhibitors Saturday night. The twelfth All-American Turkey Show would then "be history."[31]

Apart from a remarkably successful All-American Turkey Show, held January 20–25, the year 1936 in North Dakota had little to recommend it. "As 1936 arrived," Ann Marie Low wrote in her diary, "I seemed to have great hopes for it." Her hopes were dashed.

Ann Marie Low and her family lived on a stock farm in Stony Brook country, near the towns of Kensel, Pingree, and Edmunds, thirty miles northwest of Jamestown, North Dakota. For them, the years of crop failure began on Sunday, July 1, 1928, when a devastating hailstorm destroyed their crops. The family lost the farm when the government, during the Depression, took it and others in the neighborhood and established the Arrowwood National Wildlife Refuge. Young men in the Civilian Conservation Corps were employed in developing the Refuge. Ann Marie Low left an account of the year 1936 in her diary, as represented in the following excerpts.

> *January 3, 1936, Friday*
> I went for a ride on Roany and thought hopeful thoughts for the new year. The last six years have been tough, what with the Big Depression, my inability to get a good paying job, illnesses, a thousand petty discouragements, crop failures, drought, dust storms, poor cattle market, and now this game refuge thing costing us our home and Dad's work of a lifetime. Somehow we've made it so far. Surely in 1936 things will break for the better.

[31] *Grand Forks Herald*, January 25, 1935.

February 13, 1936, Thursday
This is the worst winter I ever saw. There is one blizzard after another, and for six weeks the warmest day has been 10 below zero. From there it has ranged down to 40 below. A number of days we have not held school because the buses can't travel.

February 19, 1936, Wednesday
The Strawberry roan gets out for a few minutes before sundown every day. He bounces around some, stands on his hind legs, claws the cold air with his front feet, does some fancy bucking and quite a lot of equine acrobatics and then is glad to get back in his warm stall.

The snowbanks right on main street here are so high we can look into second-story windows as we walk down the street. We have not been able to hold school this week, but plan to tomorrow.

July 11, 1936, Saturday
This is the eighth day of terrible heat.

Yesterday it was 110°F with a hot wind blowing. Today is the same. I'm writing this lying on the living room floor, dripping sweat and watching the dirt drift in the windows and across the floor. I've dusted this whole house twice today and won't do it again.

July 16, Thursday, 1936
It has stayed scorching hot.

Today it cooled down to 104°, but with a hot wind. Arrowwood Lake has developed a stench detectable for miles. A trench of seedlings right on the edge of the water in June is now 25 feet from it.

July 29, 1936, Wednesday
That rapidly drying [Arrowwood Lake] is the only body of water left in this area. Hundreds of little ducks, hatched on the prairie, perished before they could reach water.

> *August 1, 1936, Saturday*
> July has gone, and still no rain. This is the worst summer yet. The fields are nothing but grasshoppers and dried-up Russian thistle. The hills are burned to nothing but rocks and dry ground. The meadows have no grass except in former slough holes, and that has to be raked and stacked as soon as cut, or it blows away in these hot winds. There is one dust storm after another. It is the most disheartening situation I have seen yet. Livestock and humans are really suffering. I don't know how we keep going.
>
> The dirt quit blowing today, so I cleaned the house. What a mess! The same old business of scrubbing floors in all nine rooms, washing all the woodwork and windows, washing the bedding, curtains, and towels, taking all the rugs and sofa pillows out to beat the dust out of them, cleaning closets and cupboards, dusting all the books and furniture, washing the mirrors and every dish and cooking utensil. Cleaning up after dust storms has gone on year after year now. I'm getting awfully tired of it. The dust will probably blow again tomorrow.

Drought, dust, and Depression blight all they touch. In late August, before leaving for a teaching position at Sentinel Butte in western North Dakota, Ann Marie Low wrote in her diary of her last days in Stony Brook country.

> Sunday evening I went for a moonlight horseback ride—possibly my last in the Stony Brook country. I sat for a long time on a hill southeast of the lake looking over the area I once loved so much. Its beauty is gone. CCC roads penetrate coulees where only cattle, Roany, and I used to go. The river is spoiled with their dams. The camp and its activities are everywhere. Our old neighbors are all gone and their houses torn down. Now the landscape is blotted with that ugly and expensive Refuge Headquarters and even more ugly sprawling set of buildings housing the CCCs. This isn't home any more.

Detached and writing as a historian, Elwyn B. Robinson described the year 1936 in North Dakota as "the coldest (-60°F at Parshall on February 15), the hottest (121°F at Steele on July 6), and the driest (8.8 inches) ever reported. Summer was a disaster." Between July 5 and July 18, the temperature reached 100°F almost every day. Not even drought-resistant prairie grass grew west of the Red River Valley. The federal government bought cattle, and ranchers—attempting to save the best animals in their foundation herds—fed their cattle whatever feed they could find.

Given the conditions, many North Dakotans could not pay their taxes, and by the end of 1936, Robinson wrote, three-fourths of the taxes in the state's southwestern counties were delinquent. By October, after the heat and drought of the summer, nearly 61,000 people "were employed on W.P.A. projects, emergency conservation work, and the projects of other agencies." In the last half of the year, "about 330,000 persons (half the population) were on relief."

Those attending the Lucky 13th All-American Turkey Show, January 20–25, 1936, could agree with Robinson and Ann Marie Low that the winter of 1936 was one of the extremes and the worst winter they had ever seen. On Saturday and Sunday, when exhibits were arriving at the show, the temperature was -29°F.[32] The morning temperature for five days in a row was -30°F or below, and the forecast was for the freezing temperatures to continue for two weeks. All roads in the Grand Forks district were blocked, including Highways 2 and 81, and rural traffic was stalled. Driving in Grand Forks was deemed a hazard.

In spite of the severe weather, the *Grand Forks Herald* reported that attendance at the All-American Turkey Show was high. More than fifty exhibitors had entered 460 "turkeys of proudest lineage and distinction of individuality," the "royalty of American farm fowl." The dressed division had 396 entries, and the frigid temperatures during the week of the show provided the necessary refrigeration for the exhibits.[33]

---

[32] At 9:00 a.m. on January 20, opening day, the temperature was -30°F. On Wednesday morning, the temperature was -39°F, the lowest temperature since 1916, when the temperature had been -40°F. The high temperature on Wednesday, January 22, was -31°F.

[33] *Grand Forks Herald,* January 22, 1936. All six of the varieties allowed to compete for awards were represented: Bronze, Narragansett, White Holland, Bourbon Red, Black, and Slate. The live birds were entered by exhibitors from North Dakota, South Dakota, Minnesota, Montana, Missouri, Kansas, Idaho, Oregon, Texas, New York, Ohio, and a number of Canadian Provinces, including British Columbia.

In the December 1935 *American Turkey Journal*, Hackett announced the dates for the 1936 show and reminded readers of how far "the First Exclusive Turkey Show in the World" had come since its "primitive" beginning in 1924. Its methods and policies had been widely adopted nationally and internationally by other shows at which turkeys competed.

The All-American Turkey Show, "the climax of glory for turkeydom," where "the regal turkeys step into their own," enjoyed an "enviable record" among agricultural shows. This "Turkey Classic" was on a sound financial footing and had a modest-but-adequate reserve fund, "which the distressing days of the depression failed to make a dent in."[34]

In his comments, Hackett acknowledged that the All-American Turkey Show could not survive without the "whole-hearted cooperation and support" it received from individuals and businesses. This support was given year after year, "willingly and with a minimum of solicitation," because supporters appreciated what the show was doing for area businesses and for the turkey industry. Many business owners and prominent breeders also served as directors on the board of "this worthy enterprise."

Hackett promised that the 1936 All-American Turkey Show would be another "rousing turkey exposition and a great gathering of turkey growers, all bent on knowing new things about successful growing, and having a grand good time" while they were doing it. The show in 1936 would have "the most elaborate" awards—cash, trophies, and merchandise—"ever offered in the history of the show."

He extended a special invitation to turkey breeders who had not attended the show before. Were they to attend, he assured them, they would "find out more there in a day's visit" than they could glean in "many years' experience," and they would enjoy themselves "in the bargain with the finest folks" they had ever met.

William H. Kircher, field editor of *The Farmer* in St. Paul, Minnesota, believed that "the continued success" of the All-American Turkey Show, when many shows had failed, was because the one in Grand Forks "met the demands of the industry." The combination of live and dressed birds, in many cases produced by the same breeder, gave prospective buyers "an exceptional opportunity" to determine the quality of breeding lines from both the fancier's and market value viewpoints.

---

[34] *American Turkey Journal*, December 1935.

Judge Harry Lamon of New York commended the All-American Turkey Show for the large number and excellent quality of exhibits in both the live and dressed departments, making the All-American Turkey Show "the leading show in the United States."

H. W. Yoder, staff member for *Turkey World*, attended the 1936 All-American Turkey Show and reported on it in a later journal issue. Yoder had attended dozens of the largest turkey shows in the country, he wrote, but the 1936 All-American Turkey Show "was far superior in both quality and attendance and interest" to any show he had ever attended.[35]

Yoder went on to describe how the representatives of "feathered aristocracy," all of them "monarchs" of the various varieties, and the "royalty of the show," were judged by those "wise in turkey lore" who knew "just how every feather should lie and how its color should shade." The judges acknowledged that they would work with "the finest exhibit of turkeys in the history of Grand Forks and its leading national institution," that competition would be "extremely keen," and that their task of judging and placing the entries would not be an easy one. With so many fine birds in each class, the judges indeed had to be wise and skilled when determining points and placing the entries.[36]

Grand Forks County Agent William R. Page arranged the education sessions for the All-American Turkey Shows. The presentations arranged for the thirteenth All-American Turkey Show indicate the level of sophistication the show and its exhibitors had attained since the show's hesitant beginning in 1924. The presentations, all made by authorities in their fields, included those on "Getting Better Results from the Breeding Flock," "Progress Made Through U.S. Grading," "Transportation for the Turkey Industry," "Turkeys as a Cash Crop," and "Specialty Clubs and Their Service Value to the Industry."[37]

[35] *Turkey World*, February 1936.

[36] Of the 460 live turkeys entered in the 1936 show, 196 were Bronze, entered by 43 breeders. In his article, Yoder gave the final scoring: adult toms: 15 shown, 12 awards; yearling toms: 28 shown, 18 awards; young toms: 67 shown, 25 awards; adult hens: 11 shown, 10 awards; yearling hens: 17 shown, 14 awards; young hens: 58 shown, 25 awards; old pens: 5 shown, 5 awards; and young pens: 20 shown, 15 awards. For the Bronze division alone, the 1936 show had a total of 124 awards, 104 of them for individual birds, 20 for pens.

[37] William R. Page's Annual Reports are in the Institute for Regional Studies Archives located at the North Dakota State University in Fargo.

All-American Hen Club members once again furnished and maintained a Turkey Nest, an enclosed room with attractive appointments and comfortable chairs, for women attending the show. The Club's objective was described as "purely recreational," and many women considered the week of the show "a vacation period."[38]

For Hen Club women in 1936, the week's highlight was the Annual Hen Club Dinner at the Hotel Ryan. They were also guests of the Grand Forks Business and Professional Women's Club at a tea on Thursday afternoon from 3:00 to 5:00 p.m. in the Davis Hall Women's League Rooms at the University of North Dakota. Miss Tilda Natwick of the university's Home Economics faculty arranged the tea that was served by senior Home Economics students as a Special Project.

The *Herald* reminded Grand Forks residents that the show's judges called the All-American Turkey Show the city's "only nation-wide institution," and that they should, therefore, attend it. Their opportunity was provided by a feature new to the show in 1936, a Grand Forks Day. The next evening, Wednesday, January 22, judges, marketing experts, and well-known breeders explained the fine points of the live and dressed exhibits for the benefit of laymen and for the purpose of adding interest to the show and encouraging attendance.[39]

Also new in 1936 were the silver and blue color scheme in the City Auditorium—but with the usual small evergreens on the tops of the coops—and a daily fifteen-minute radio broadcast over Grand Forks and Fargo stations. During the broadcasts each noon, sponsored by the Hubbard Milling Company, breeders, judges, and visitors offered brief comments on the show. Hackett made arrangements to extend the National Broadcasting System broadcasts to a full hour in 1937.

To encourage flock owners to participate in the All-American Turkey Show and compete in their classes, Judge Frank Cross, Minneapolis, Minnesota, provided a new prize to be awarded at the 1936 show. The

---

[38] Hen Club members were mostly "Big Business" women, "with little consideration" for the feminist feminist cause, conducting their business under their husband's names. At the 1934 All-American Turkey Show, only 15 percent of the turkeys were entered in women's names, although more than 80 percent of the birds were owned, raised, and cared for by women.

[39] *Grand Forks Herald*, January 21, 1936.

Some of the afternoon crowd at the 1934 All-American Turkey Show. Note the evergreens used to decorate the City Auditorium at all of the shows and the dressed turkey division in the balcony-cum-refrigerated area upper right. *American Turkey Journal*, February 1934.

award was two prizes of $25 cash for the man and woman exhibitors who had given the greatest loyalty and service to the show and who had competed in the shows for five consecutive years.[40]

Also in 1936, the North Dakota State Mill and Elevator, a part of what Elwyn B. Robinson described as the Nonpartisan League's "system of state socialism," introduced a new line of Dakota Maid feed, especially

[40] Those receiving the prizes were elected by exhibitors, and each exhibitor was entitled to one vote for each $1.00 paid in entry fees. The prizes were awarded to Mrs. W. J. Janda of Mahnomen, Minnesota, and to Alfred Malmberg of Crookston, Minnesota.

formulated for turkeys from hatching to marketing: Turkey Starter, Turkey Grower, Laying Mash, and Turkey Finisher. Of all the poultry and livestock feeds in the country, Dakota Maid feeds were the only ones produced in a state-owned mill and elevator and in Grand Forks, the Turkey Capital of the World.

The North Dakota mill advertised its new line of commercial turkey feed as "Sure Fire," being produced from tested formulas with ingredients backed by research and produced in the most modern feed-grinding mills in the country. The mill maintained a large turkey flock on which to test its feeds, and it became obligatory for those attending All-American Turkey Shows to tour the plant, see the mill's flock, listen to officials explain the advantages of using Dakota Maid turkey feed, and enjoy a dinner featuring turkey from the mill's flock.[41]

"ND Turkey Wins Grand Championship," was the headline in the *Grand Forks Herald*, on January 22, 1936, The Grand Champion was an adult Bronze tom exhibited by Mr. and Mrs. Frank Ralston of Crystal, North Dakota. The adult tom was also the Bronze class champion. The Master Breeder Gold Medal, the highest award, was presented to Mr. and Mrs. Al C. Johnson of Bath, South Dakota.[42]

The dressed division Grand Champion was a twenty-two pound Bronze young tom. The young tom was also the Bronze class champion. Described in *Turkey World* as a "superb bird, perfect, or as nearly perfect as a turkey could be," the Bronze tom was exhibited by Robert B. Mitchell of Sunnyside, California. With this bird— and an unexpected connection to a famous boxer — the Lucky 13th All-American Turkey Show scored a publicity coup that was not matched by any turkey or livestock show in America.

In 1935, boxer Jack Dempsey opened Jack Dempsey's Restaurant on Eighth Avenue and 50th Street in New York, directly across from the third Madison Square Garden, built in 1925. As much an American institution as its proprietor, the restaurant became known for its cheesecake, shipped

---

[41] Dakota Maid turkey feed was the first in the Northwest to contain Wheat Germ Oil, vital for stabilizing vitamin E and necessary for assuring "reproductivity, high fertility, and hatchability." Dakota Maid turkey feed was also among the first to contain manganese sulphate, important for helping to prevent Perosis (slipped tendon or leg deformity) in turkeys, caused by the lack of this mineral.

[42] *Grand Forks Herald*, January 23, 1936.

to customers all over the United States and Europe. World Heavyweight Champion from 1919–1926, Jack Dempsey was known as "the Manassa Mauler" for his aggressive fighting style and exceptionally powerful punches.[43]

The 1936 All-American Turkey Show dressed Grand Champion was purchased for $25.00 by Russell Whaley of New York City, acting as agent for the Jack Dempsey Restaurant. The "nearly perfect" bird, noted an article in the February 1936 issue of *Turkey World*, was "displayed in the window of the now famous Dempsey Restaurant in New York City." The publicity would "do much for the coming All-American Turkey Shows" and would "certainly create a lot of interest throughout the country by the publicity" it would "receive through the press."

The *American Turkey Journal* acknowledged the publicity possibilities and added that the bird "showed eastern turkey breeders what could be produced in the West." The Grand Champion's "carcass," according to the article, was in "the best of condition" and being exhibited "to the best advantage" with a placard bearing the words, "A fine stroke of advertising for the All-American and for Northwest turkeys."[44]

Made wiser by what the year 1936 had brought, and embittered by what the federal government had done to her beloved Stony Brook country, Ann Marie Low confided in her diary on January 2, 1937, Saturday: "A new year has come. I'm writing this in bed in the Butte Hotel. And I am remembering that a year ago I was welcoming the year 1936 with high hopes. After all that happened in 1936, I won't be silly enough to welcome 1937 with high hopes for better times. However, I can say again, as I said a year ago, we've made it so far."

With the advantage of hindsight, Elwyn Robinson noted that 1937 did bring slightly "better times" and that many of the state's residents had indeed made it so far. With federal funds, they had won the battle against grasshoppers, and, although precipitation in 1937 remained below nor-

---

[43] Among the most popular boxers in history and a cultural icon of the 1920s, Dempsey was such a draw that many of his matches set records for attendance and gate receipts, including the first million-dollar gate.

[44] *American Turkey Journal*, March 1936.

mal, the "disastrous droughts of 1934 and 1936 were not repeated." Crop yields, however, were still "scanty," prices remained low, and 1937 was still a Depression year.[45]

"Plows Clear Highways in Record Cold" was a headline in the *Grand Forks Herald* on Friday, January 22, 1937. "Principal highways in the Grand Forks district," noted the article, "were opened Thursday night and Friday as snowplows worked in the lowest temperatures in the winter." A blizzard on Wednesday, January 20, the first of the new year, had blocked highways in North Dakota and Minnesota.[46]

Even when opened, all highways were "slow," according to James Kennedy, Grand Forks division engineer, and highway departments advised against travel. Neither buses nor trains could operate, and a Rock Island passenger train had become stuck in a half-mile-long snowdrift on Thursday morning. Some schools were closed because of the blizzard and the cold that followed on its heels.[47]

Although "attendance was hampered by cold and snow," according to the *Herald*, hundreds "of the finest specimens of that grand American bird called the turkey" were on hand Monday morning "as the doors of the city auditorium opened to the public for the fourteenth annual show." In the live division, "more than 400 of the best turkeys in America gazed from coops," and five hundred birds, the largest number of dressed birds in the history of the All-American Turkey Show, were on display in the refrigerated section of the balcony.[48]

Five varieties of turkeys, "the proudest birds in the nation and Canada," were entered in the live division: Bronze, 150; Narragansett, 100; White Holland, 75; Bourbon Red, 45; and Black, 30. The *Herald* noted

---

[45] Noteworthy for its many "firsts" and for its stellar publicity coups, the fourteenth All-American Turkey Show, January 18–23, 1937, was among the year's few bright spots. The weather in Grand Forks during the week of the show was among the "firsts."

[46] *Grand Forks Herald*, January 22, 1937.

[47] The temperature was -8°F when the All-American Turkey Show opened on the morning of Monday, January 18, but, by closing time at 10:00 p.m., it was -18°F. On the second morning of the show, the temperature was -31°F, the season's coldest, and it was only -16°F at noon. Thursday night's low temperature in Grand Forks was -35°F, and it was -34°F when the show opened Friday morning. At noon, the thermometer read -16°F.

[48] *Grand Forks Herald*, January 20, 1937.

that "attracting considerable attention was the display of Black turkeys from Texas," where the variety was "becoming popular." Seven of the Black turkeys were entered by Fay Leatherwood, Oakland, Oregon, exhibiting at the All-American Turkey Show for the first time.

Exhibitors expressed pleasure with the number of entries other than Bronze, as the entries showed an increased interest in the other varieties. A coop of "bright plumaged partridges and pheasants," including golden pheasants, was admired as "one of the most beautiful displays of the show." The display, provided by Femco Farms, Battle Lake, Minnesota, was located in the auditorium balcony.

Among the better-known exhibitors with birds at the fourteenth All-American Turkey Show were A. P. Hudson, Tangent, Oregon, and Henry W. Domes, Richreall, Oregon. Mrs. Gladys Honssinger, Pleasant Valley, Missouri, with 25 Bourbon Red turkeys, was returning to the All-American Turkey Show after a long illness. Besides her award-winning Bourbon Reds in 1929, she was especially remembered for two of her entries—Sensation and Hope. Possessing admirable dispositions, the birds obeyed Mrs. Honssinger's beck and call, strutted when instructed to, and gobbled replies when asked questions.

In 1929, Sensation and Hope had performed not only at the All-American Turkey Show, but also at service club meetings. For many years, Mrs. Honssinger had entered her turkeys in shows in Grand Forks, Chicago, New York, St. Louis, and Texas. In each show, she emerged with winners in the Bourbon Red class. "Scores of other celebrities of Turkeydom" as reported in the *Herald,* entered their birds in the fourteenth All-American Turkey Show, including Robert D. Mitchell of Sunnyside, Washington, and Mrs. Ward Cockeram of Oakland, Oregon, "another well-known turkey fancier of Oregon."

Judging tables had been set up by the time the show opened Monday morning and, shortly after noon, judges Harry M. Lamon of New York, Frank E. Cross of Minneapolis, and H. P. Griffen of Salt Lake City began to score each entry. The judges would, as described by the *Herald*, "scan each bird for a discolored feather, a misplaced quill or an improper conformation." Given the number of entries in the live classes, the judges' tasks had to have been challenging, as the results of the first day's judging illustrated.[49]

"Judging was close in nearly all cases," observed Hackett, "with very few entries disqualified." Words well spoken. The judges must have been particularly discerning when scanning feather color and shading and configuration when they determined to give one young Bronze tom, for example, nineteenth place, but they gave another twentieth place.

In addition to blue ribbons, the fourteenth All-American Turkey Show awarded cash prizes of $1,000 and multiple awards and trophies, so many that most exhibitors must have received at least one.[50] In keeping with manager Hackett's intention of making each All-American Turkey Show better than the previous one, a new prize was offered in 1937. The prize was a farm radio, to be awarded to any exhibitor from previous shows who got the largest number of new exhibitors to enter birds in the live division that year.

The All-American Turkey Show also offered a new competition that year, known as the Special Pack competition. An entry, according to judging guidelines, consisted of a single dressed turkey "attractively wrapped in cellophane" and "resting" in a colored box "of appropriate size." Exhibitors were allowed freedom of choice as to the size of the dressed bird, the style of wrapping, and the kind of box.[51]

---

[49] Bronze: adult hens, 12 awards; young toms, 20 awards; yearling hens, 12 awards; adult toms, 5 awards; yearling toms, 12 awards. Narragansett: young toms, 5 awards; young hens, 12 awards; adult toms, 5 awards. White Holland: yearling hens, 5 awards; adult hens, 5 awards; young toms, 13 awards; yearling toms, 5 awards; adult toms, 3 awards. Bourbon Red: adult toms, 9 awards; adult hens, 2 awards; yearling toms, 3 awards; yearling hens, 4 awards. Black: adult hens, 2 awards; yearling hens, 4 awards; young hens, 4 awards; young toms, 5 awards; adult toms, 3 awards; yearling toms, 3 awards.

[50] Among the Special Prizes were those for the new exhibitor entering four or more Bronze; the J. J. Quam prize for a new exhibitor from North Dakota entering at least four birds and one for a new exhibitor showing six birds and receiving the most points; the *American Turkey Journal* award for the lady coming the greatest distance to show Bronze turkeys; the Dr. Salsbury Laboratory prize for the exhibitor with fifteen or more entries and scoring the fewest points; the Armour and Co. trophy for the second-highest-rated display of White Holland turkeys; the Senator Capper award for the best display from the Central Western States; the North Dakota Bankers' Association prize for the best young pen from North Dakota; the Kem Temple trophy for the best four young toms; and the Merchants' Association of Grand Forks and East Grand Forks prize for the best four young hens.

[51] George W. Hackett appointed a special committee consisting of Miss Clara Sutter of *The Farmer*, J. J. Fennesy of the Railway Express Agency, and Thomas W. Heitz of Washington, D.C., to judge the dozen entries. Of the four prizes offered, Mrs. John O. Allen of Radium, Minnesota, won first; and Mrs. William Eddie, of Northwood, North Dakota, won fourth.

The *Herald* noted that "Royalty of American Turkeydom Friday gazed proudly from the court of honor of the All-American Turkey Show."[52] On the auditorium's stage, in the center coop, "basking in the spot light, strutted the adult White Holland tom of Henry W. Domes of Richreall, Oregon, second in his breed to win the grand championship in the 14 year history of the show." The "pure white bird" occupied a coop "decked in royal purple."

Flanking the Grand Champion on either side, were "the lesser royalty, champions of the various breeds." In addition to "his purple grand championship ribbon," rosettes proclaimed the Grand Champion "the champion White Holland, a member of the best display of that breed, and the property of the winner of the Sweepstakes breeder award." Henry W. Domes was also awarded the Master Breeder Medal.

To allow exhibitors to observe the judging, the Education Sessions were scheduled for Wednesday afternoon, the third day of the show. Grand Forks County Agent William R. Page, anticipating exhibitors' interests, arranged sessions on a number of timely topics, including, "The Result of Producing and Attempting to Market Unfinished or Poor Quality Turkeys," "Giving Credit Where Due in Turkey Raising," "Improvement of Turkeys Through A.P.A. Inspecting and Banding," "The Proposed National Turkey Improvement Plan," and "Incubation for the Small Breeder."

Turkey Hen Club members maintained a sitting room, equipped with comfortable chairs, on the auditorium's stage. Dubbed the Turkey Hen Nest, the room was "arranged in accordance with well-known peculiarities of a turkey hen. Hedged in with evergreens for seclusion," the room allowed "a view of the entire auditorium." In the "least conspicuous corner among the evergreen boughs" was "a real nest with a clutch of turkey eggs." The eggs were "the product" of the White Holland hens shown by Henry W. Domes of Richreall, Oregon.

Turkey Hen Club members were described as businesswomen conducting "one of the important industries of the Northwest." They raised 80 percent of the birds exhibited at the All-American Turkey Show, and they constituted most of the audience members at the Education Sessions.

[52] *Grand Forks Herald,* January 22, 1937.

They were the ones who asked the questions and those who shared their experiences. So that all could become better acquainted, these businesswomen were guests of the Grand Forks Business and Professional Women's Club at an afternoon tea served by Home Economics students at the "practice house" on the University of North Dakota campus.

At the 1937 All-American Turkey Show, Hen club members broke with a tradition they had set when they formed the club nine years earlier. For the first time, they invited men to attend their annual dinner, held Wednesday evening at the Hotel Dacotah.

The publicity coups scored by the 1937 All-American Turkey Show remained unmatched by any other livestock or poultry show in the country. As reported by Harry Yoder in *Turkey World*, the Grand Champion dressed turkey, a 34½-pound Bronze tom, exhibited by Robert D. Mitchell, of Sunnyside, Washington, was shipped by Railway Express to President Franklin D. Roosevelt. The Reserve Champion, a Narragansett young tom, was sent to Henry A. Wallace, Secretary of Agriculture.[53]

The entire dressed bird exhibit, weighing 8,500 pounds, was purchased by the Great Atlantic and Pacific Tea Co. and placed on exhibit in Washington, D.C. The birds were purchased at two cents per pound under the New York price for prime turkeys, with premiums paid for prize-winning birds. The company paid $1.00 above market price for first-prize winners in all classes and fifty cents above market price for second-prize winners in all classes. Howard C. Pierce, who negotiated the purchase, said that when the dressed birds were placed on exhibit in the national capital, "all award ribbons" would "be attached." His company had purchased the birds, Pierce said, "as an inducement for Northwest buyers to produce still better turkeys."

Hackett noted that the sale of dressed birds to the Great Atlantic and Pacific Tea Co. brought exhibitors from five to ten cents per pound above the local market price. He was also pleased with the sale because "the wide publicity attracted by displaying the entire dressed exhibit in Washington, D.C." would "do much for the show" by drawing attention to what the All-American Turkey Show was accomplishing for the turkey industry.

---

[53] *Turkey World*, February 1937.

For Hackett, who believed it an insult to display America's regal bird without its plumage, and for exhibitors who were attempting to bring their birds as close as was humanly possible to the standards specified in the *Standard of Perfection*, the publicity coups must have been bittersweet. The five hundred dressed birds at the 1937 All-American Turkey Show outnumbered the four hundred live birds by a wide margin, and the coups had been scored with dressed birds, one of them Mitchell's 34½-pound heavy Bronze tom.[54]

The Grand Champion in the dressed division in the 1937 All-American Turkey Show, the bird presented to President Franklin D. Roosevelt, was the 34½-pound Bronze tom. The Grand Champion in the dressed division of the 1936 All-American Turkey Show—the bird placed on display in the famous Jack Dempsey Restaurant—was a 22-pound Bronze tom.

According to A. W. Brant, author of "A Brief History of the Turkey," in the early part of the twentieth century, Jesse Throssel, a turkey farmer in England, received a contract from Lord Rothschild, the financier. The contract called for Throssel to supply Lord Rothschild with the large turkeys he wanted to give as gifts at Christmas, and Throssel was to be paid a premium for those turkeys weighing over twenty pounds. He scoured the country for large turkeys to serve as breeders and developed the Sheffield Bronze, a large, meaty bird. At maturity, toms weighed up to forty pounds, hens up to twenty pounds.

When Lord Rothschild died and Throssel lost his contract, he emigrated to British Columbia in 1926 with his stock. In about 1930, Throssel took birds so large and meaty they had difficulty mating naturally to

---

[54] Regarding the new interest in dressed birds, note that all turkey varieties are descended from the American wild turkey, and varieties differ from one another primarily in plumage. A turkey variety can be produced from a sport, a chance genetic mutation. Once thought to be the product of a cross, the Slate variety, according to *The Livestock Conservancy*, was, however, produced from a legitimate mutation. In "A Brief History of the Turkey," A. W. Brant wrote that many varieties of turkeys had been developed over the bird's five hundred years of domestication. Until about the turn of the twentieth century, turkeys had been bred for show, and the focus was on plumage and other characteristics, with little attention paid to meat production. Brant noted that a "change," however, "was in the wind." In 1930, the Northwestern Turkey Show at Oakland, Oregon, added a new feature—a dressed turkey division. Subsequently, dressed turkeys became "a popular competition" in many shows. For display, only the turkey's feathers and blood were removed, and its head was neatly wrapped in paper. Judging was based on meatiness, conformation, and presentation. The *Standard of Perfection* also specified the weights for birds of each variety. In competition, if two birds were equal in all respects, judges were to award the prize to the lighter bird.

the Portland, Oregon, International Livestock Show, where they were a sensation, especially those entered in the dressed division. Oregon and Washington turkey breeders, who had been breeding for improved meat production, asked to purchase toms from Throssel's flock for breeding.

Making crosses with the Bronze variety, as it was the most popular at the time, breeders developed a new turkey variety, the Broad Breasted Bronze, which became the commercial turkey of choice throughout the 1940s. The Broad Breasted Bronze breeders formed a club to promote the new variety, and they advertised their breeding birds in the *American Turkey Journal*. They were allowed to display their turkeys at the All-American Turkey Show, but they were not allowed to enter the birds for judging, because the Broad Breasted Bronze variety was not approved for listing in the *Standard of Perfection*.

Breeders' preferences and consumers' tastes were changing. Breeders wanted to produce meatier turkeys, and consumers wanted more white meat. For both breeders and consumers, feather color, pattern, and shading were immaterial. Wallace Jerome may have stated it best, if cynically. He changed his breeding program and began producing meat-type birds, he said, because he could not make a living raising turkeys for their feathers. For Hackett and breeders to resist the trend was as futile as Dame Partington's attempt to keep back the Atlantic Ocean's encroaching waves with her mop.

Hackett and exhibitors were justified in taking pride in the publicity coups scored by the 1937 All-American Turkey Show. The coups did not, however, bode well for the All-American Turkey Show's future.

CHAPTER 14

Turkeys Take to the Air(Waves) on a Bea-u-u-tiful Day in Grand Forks

(with apologies to NBC's Everett Mitchell)

STORM CLOUDS WERE GATHERING as the fifteenth annual All-American Turkey Show opened on Monday morning, January 17, 1938. The clouds did not, however, presage the storms that frequently greeted exhibitors, those that produced numbing cold and blizzards that blocked roads, stalled trains and buses, and halted vehicle traffic. These were war clouds.

Seldom during the week of the All-American Turkey Shows was the show driven from the front page of the *Grand Forks Herald*. On January 22, 1938, however, the headline was "Long War Ahead, Jap Leaders Predict." Front page articles were headed, "Ask for More Troops, Funds in Tokyo Meet"; "Will Link China, Japan, Manchoukuo in Economic Union"; "Hit at Chiang Rule"; "Foes Must Pay Indemnity."

In what Elwyn Robinson described as this "time of great anxiety" for the United States, the fifteenth Annual All-American Turkey Show was held January 17–22, 1938. On Friday, January 14, William Huggins, floor manager, had begun the task of preparing the coops to receive entries, and crews had begun decorating the City Auditorium. Manager Hackett believed competition would be keen.

By the time the fifteenth All-American Turkey Show was held, the shows were of a pattern, so much so that those associated with the show—exhibitors, judges, officials, workers, and visitors—should have been familiar with which events would be scheduled and where and when they would be held. Each All-American Turkey Show, however, had one or more features that, if not making it better than previous shows, made it different, if not distinctive. Each feature of the fifteenth All-American Turkey Show gave evidence that not only was interest in this first and largest exclusive turkey show not waning, but also that its appeal might even be becoming greater. There were more first-time exhibitors in 1938 than there had been in any recent year.[1]

Seasoned exhibitors from both coasts and points in between entered their birds in the 1938 All-American Turkey Show, continuing to believe that it was necessary to pit their turkeys against the best in the country. Mrs. Mary E. Driscoll of Forest Hills, New York—who operated a turkey farm at Henning, Minnesota—entered ten Bronze. Mrs. Driscoll "was one of the top winners, placing first, second and third in the Bronze competition at the New York show in 1937."[2]

Breeders who had won nearly every prize and award offered by the All-American Turkey Show had entries in the 1938 show. Mr. and Mrs. Al C. Johnson, "prominent exhibitors at previous shows," had twenty-eight turkeys in the Bronze class. The Johnsons received the Master Breeder Award at the 1937 All-American Turkey Show. Mr. and Mrs. John O. Allen, Radium, Minnesota, also entered birds from their outstanding flock of Bronze turkeys.

More than 850 birds were entered from a dozen states, including Texas, Indiana, Kansas, Missouri, Washington, Oregon, Montana, Nebraska, Minnesota, Wisconsin, South Dakota, and North Dakota. In the live division, four hundred "brilliantly plumaged turkeys of proud lineage and individual distinction strutted and gobbled in the pens. . . . Monday morning as the fifteenth annual exhibition opened its doors to the public." Described as "exceptional birds," the turkeys "preened their feathers and waited their turns on the judges' tables." The newspaper noted that admission was free, and people were invited "to see one of the finest exhibits of high class turkeys ever assembled in the entire world."

Hundreds of entries in the dressed bird division arrived over the weekend prior to the show's opening, many of them from exhibitors as far away as Oregon and Washington. The number of Special Pack—cellophane-wrapped turkeys ready for the oven—entries in 1938 was

[1] First time exhibitors from North Dakota included Mrs. William Finkey, Eckman; Mrs. Frank Zimmerman, Anamoose; Herbert Brokeman, Sanborn; and Oscar Braaten, Kindred. Those from South Dakota included Howard Tanner, Gettysburg; H. C. Stevens, Clark; and Frank A. Olson, Sutton.

[2] Driscoll said that "from the standpoint of quality," the show in Grand Forks was "ahead of the New York exhibit."

twice the number entered in 1937, and exhibitors in this division competed for merchandise awards and for $25 in cash prizes. Special Pack cartons were mailed to exhibitors a week before the show opened.

Also evidencing its appeal, the fifteenth All-American Turkey Show continued to secure the country's most prominent turkey experts to serve as judges. In 1938, Thomas W. Heitz, from the U.S. Department of Agriculture, Washington, D.C., judged the entries in the dressed division. Frank E. Moore, poultry husbandman from the North Dakota Agricultural College in Fargo; Erle Smiley, from Seward, Nebraska; and Walter Burton, of Arlington, Texas, judged the entries in the live division. Judge Burton admitted to the significant contrast between the temperatures in Arlington, Texas, and Grand Forks, North Dakota, but, he said, by donning his "red flannels" he got through his stay in Grand Forks quite well.

Scattered throughout the main floor and the balcony were the commercial booths displaying the latest trends for raising and handling turkeys. During the week of the show, motion pictures were taken that would be added to the corpus of available views of previous shows. Indicating that flock owners were still interested in learning how to improve the quality of their turkeys and how best to market them, the Education Sessions arranged by Grand Forks County Agent William R. Page were well attended. "Had full meeting place for P.M. program with turkey presentation," was his comment on one session that "went over well." He admitted, however, that he had a "Hell of a time to get some of Geo. Hackett's extemporaneous talks started" and that he was "a little peeved at Hackett's oversights here and there."

The Education Sessions at the 1938 All-American Turkey Show, as in previous shows, were presented by authorities in their field. Mrs. Claud Wright, for example, described in the program as "one of the prominent exhibitors" at the All-American Turkey Show and "a consistent winner," spoke on "Turkey Improvement by Trapnesting."[3]

---

[3] Other examples include Dr. I. M. Roderick, North Dakota Agricultural College, presented "Turkey and Poultry Diseases and Their Control." Judge Thomas W. Heitz addressed "U.S. Grading Spells Progress in Turkey Markets," and Judge Frank E. Moore spoke on "Highlights of Scientific Progress in Turkey Breeding." Moore also presented the results of tests conducted in 1937 on three lots of turkeys that demonstrated the importance of proper type and maturing ability in turkey strains. The better-bred turkeys from the best strains had produced a higher percentage of prime and choice birds when marketed. Judges Moore and Heitz demonstrated how to grade, pack, and market the oven-ready turkey, and Misses Ruth M. Dawson and Amy Erickson of the North Dakota Agricultural College demonstrated the proper way to roast and serve them.

The Turkey Hen Club also continued to appeal to women exhibitors, and the *Grand Forks Herald* reported that many of the club's fifty members had reported at "the Turkey hen nest" and that "those who came early" had "a long list of recruits."[4]

At their annual meeting—the only one of the year—held in the auditorium on Thursday, January 20, club members elected Mrs. George Kirk, Niagara, North Dakota, president, to succeed Mrs. Roy Vosper, Neche, North Dakota, who was elected vice president. Mrs. Ole Nelson, Kensington, Minnesota, was elected secretary-treasurer. Club members named a committee to "co-operate with the sponsoring agencies of the show" to arrange the annual picnic to be held in the city's Riverside Park in July.

That the All-American Turkey Show was still of interest was also evident in *North Dakota: A Guide to the Northern Prairie State*, published in 1938 as part of the American Guide Series. According to the Preface, the publication was "something new in this part of the country. For the first time North Dakotans and their guests have a concise but comprehensive survey of the State, which tells them what should be seen, and why, and how. Our aim has been to produce a book not only to be used in touring the State, but to be enjoyed by fireside travelers and all who would deepen their understanding of North Dakota."[5] The first of the Annual Events that should be seen in Grand Forks, according to the *Guide*, was the "All-American Turkey Show, City Auditorium, usually last week in January."

---

[4] *Grand Forks Herald,* January 18, 1938. Among the "recruits" and those who had recently joined were Mrs. A. Swanson, Gilby, North Dakota, who had not missed an All-American Turkey Show in thirteen years; Miss Grace Baxter, Hazel, South Dakota; Mrs. W. T. Hall, Denton, Texas; Miss Sadie Caldwell, Broughton, Kansas; and Mrs. Mary E. Driscoll, Forest Hills, New York.

[5] The American Guide Series was the best-known of the hundreds of publications produced by Federal Writers' Project workers under the New Deal's Works Progress Administration (WPA). Guides were published for each of the forty-eight states, the Alaska Territory, Puerto Rico, and Washington, D.C. Not only did the Federal Writers' Project provide work-relief for unemployed writers, librarians, clerks, researchers, editors, historians, and others, it also created what was described as a "self-portrait of America." Generally uniform in format, each Guide included detailed histories of the state or territory; descriptions of every city and town; automobile travel routes; photographs; maps; and entries on resources, culture, and geography.

Some of the titles are prosaic, such as *Montana: A State Guide Book; South Dakota Guide*; *Tennessee: A Guide to the State*; and *Wyoming: A Guide to Its History, Highways, and People*. Some titles are particularly imaginative, such as *Missouri: A Guide to the 'Show Me' State*; and *Oregon: End of the Trail*. Other titles are descriptive: *Colorado: A Guide to the Highest State*; *Iowa: A Guide to the Hawkeye State*; *Kansas: A Guide to the Sunflower State*; *Kentucky: A Guide to the Bluegrass State*; and *Nebraska: A Guide to the Cornhusker State*. With the subtitle, *A Guide to the Northern Prairie State*, North Dakota stands alone, the only state identified as being in the north and on the Great Plains.

That interest in the All-American Turkey Show had not waned and that its appeal was still far-reaching was made evident in a telegram, sent to the *Grand Forks Herald,* which stated that *Life*, the national magazine, wanted to visit the 1938 All-American Turkey Show and attend some of its activities.[6]

All-American Turkey Show officials and sponsors welcomed *Life* magazine's announcement, as this was "the first real national recognition given the show by any magazine outside the turkey field." Realizing "the vast amount of valuable publicity that would result," show officials made arrangements for photographs.

The All-American Turkey Show's top awards were still appealing to exhibitors, and they continued to compete for them. Mr. and Mrs. Al C. Johnson had captured practically every honor but one at the past shows. They received that honor in 1938 when their yearling Bronze tom was named the show's Grand Champion. Mr. and Mrs. John O. Allen, Radium, Minnesota, had the Grand Champion in the dressed division, a seventeen-pound young Bronze hen.

"Texas Woman Takes Master Breeder Prize," ran the *Grand Forks Herald* headline on January 20, 1938, Mrs. W. T. Hall, Denton, Texas, had "swept the field with her display of Black turkeys." Her Black turkeys were the best birds in six classes, and one of her young Black hens was named the show's champion hen.

Friday, the last day of the show, was designated sales day. The manager's office united exhibitors with prospective buyers. The birds in the dressed division were auctioned, and the Belmont Café paid $1.00 per pound for the Grand Champion dressed bird. The remainder of the dressed bird division was purchased by the Red River Produce Co. of Grand Forks. The company paid up to twenty-five cents per pound over market for some of the choice birds.

---

[6] *Grand Forks Herald,* January 18, 1938. *Life* was a weekly, general-interest magazine, known for the quality of its photography. The first all-photographic American news magazine, *Life* dominated the market from 1936–1972, at one point selling more than 13.5 million copies per week. The magazine's role in the history of photojournalism is considered its most important contribution to publishing, and among its best-known photographs, taken by Alfred Eisenstaedt, is that of a sailor kissing a nurse in New York's Times Square on August 14, 1945. The date became known as V-J Day, the day Japan surrendered to the United States in World War II.

All exhibits were released to owners for removal by 3:00 p.m. on Saturday, January 22. When the show's gates closed at 5:00 p.m., the fifteenth Annual All-American Turkey Show was officially over.

C. Dyke Page, president of the All-American Turkey Show, and Hackett, manager, "were jubilant at the success of the 1938 show." Both believed it "to be one of the most successful shows both in the point of view of attendance and quality of exhibits in the 15-year history of the organization." Both men expressed their appreciation for "the cooperation the show had received from exhibitors, officials, visitors and Greater Grand Forks Business Men."

The sixteenth annual All-American Turkey Show was held January 16–20, 1939. On the morning of its opening, the temperature was -28°F, warming only to -9°F by early afternoon. Fortunately, winds were calm. The All-American Turkey Show's offices had been moved from City Hall to the City Auditorium the previous Sunday morning.

Phrases on the *Grand Forks Herald's* front pages that week gave promise that the 1939 All-American Turkey Show would be the most memorable in its history. "As the first of the world's exclusive turkey shows opened in the city auditorium, the royalty of Turkeydom on display drew the eyes of the world's Turkey men to this city." Another phrase read, "400 of the highest priced turkeys in the United States and Canada," "104 STATIONS ON BROADCAST OF EXHIBITION."[7]

Phrases from the *Grand Forks Herald's* front pages relating to international events were not at all promising, but instead disturbing. "Hitler Orders Duce to Keep Peace for Year." "Premier Mussolini was reported authoritatively Friday to have refused to 'retreat one inch' from Fascist colonial demands on France in his talks with British Prime Minister Chamberlain in Rome." Reports from Rome were that "the British premier had left Mussolini no doubt that Britain . . . supported French resistance to any territorial demands." Adolf Hitler had "stepped into Prime Minister Chamberlain's appeasement visit to Rome by urging Premier Mussolini to keep the peace for one year before demanding fulfillment of Italy's 'natural aspirations.'" "Hitler's repeated request was interpreted as indicating that

[7] Other headlines read as follows: "Every section of the United States and Canada was represented as breeders and show officials finished banding, labeling and cooping approximately 400 strutting gobblers." The *Herald* further noted that "[a]lso exhibited here, and on display for the first time in the history of the world, were three specimens of the new Royal Palm breed."

the Fuehrer was not ready to make payment to Mussolini for his support at the Munich conference" that "partitioned Czechoslovakia and marked a milestone in European history."

More news from the *Herald* indicated that live turkeys began arriving in Grand Forks Saturday night, January 14, "by train and truck from all corners of the country."

Other headlines brought readers' thoughts back to the home front, announcing that live turkeys began arriving in Grand Forks Saturday night, January 14, "by train and truck from all corners of the country." Included were pens of White Holland and Narragansett, entered by Mrs. August Swenson, of Gilby, North Dakota. Oakdale Turkey Farm, Kensington, Minnesota, entered twenty Black turkeys and eight Narragansetts. John Martinson, also of Kensington, entered six Blue Slate. Hugo Meyer, Cold Camp, Missouri, attending the show for the first time, entered both Bronze and White Holland birds. Mr. and Mrs. John O. Allen, Radium, Minnesota, entered three Bronze. Exhibitors from Canada included Ed Milrae, Oak River, Manitoba, who entered six Bronze. Portage La Prairie, Manitoba, was represented by two breeders, Mr. and Mrs. H. M. Laugheed and Mrs. Lily Wallace with three Bourbon Reds.

Mr. and Mrs. Al C. Johnson, Wayzata, Minnesota, "wrote a new chapter in All-American Turkey Show history . . . when their yearling Bronze tom was adjudged grand champion of the show." The second consecutive Grand Championship for the Johnsons was "the first time in the 16-year history of the show that a grower" had "captured two grand championships." The Johnsons' Bronze yearling tom was the show's Grand Champion in 1938. The Johnsons' yearling tom was also the Bronze champion in 1939. Mr. and Mrs. Albert Payne, Towner, North Dakota, received the 1939 Master Breeder Award. One of their birds, a Narragansett yearling hen, was named the show's Reserve Champion.

"Also exhibited here, and on display for the first time in the history of the world," announced the *Grand Forks Herald*, were "three specimens of the new Royal Palm breed developed by Enoch E. Carson of Lake Worth, Florida." Mr. and Mrs. Carson intended to fly to Grand Forks for the All-American Turkey Show.

Royal Palm turkeys, according to *The Livestock Conservancy*, are small, but strikingly attractive, with a coloration that is difficult to perfect.

The first ones in America appeared in a mixed flock of Black, Bronze, Narragansett, and wild turkeys on Enoch E. Carson's farm in Lake Worth, Florida, in the 1920s. Writing in *Feathered Magazine* in 1931, a breeder noted that turkeys "of this type of coloration do crop up by chance where different color varieties are crossed . . . but it takes years to perfect their markings."[8]

Judges who examined the three Royal Palm turkeys exhibited at the 1939 All-American Turkey Show cited their "remarkable beauty and body formation" and commented that a slightly bigger size, which would be "easy to obtain, was the only factor lacking." The three on display were, they said, about the size of wild turkeys. Other birds on display at the 1939 All-American Turkey Show were wild turkeys, exhibited by Marvin M. Magnus, Sterling, North Dakota. The Glendalough Game Farm, Battle Lake, Minnesota, had an exhibit of pheasants.

More than 350 "choice dressed birds," described by show officials as "the best array of dressed turkeys" entered in recent years, were displayed in the All-American Turkey Show's cold storage room in the City Auditorium's balcony. Ranging in weight from nine to thirty-six pounds, the dressed birds were entered by breeders from Washington, Oregon, Minnesota, Missouri, Nebraska, Iowa, North Dakota, South Dakota, Michigan, New Hampshire, Montana, and Manitoba, Canada.

Mr. and Mrs. John O. Allen, Radium, Minnesota, also made history at the 1939 All-American Turkey Show. For the fourth time in the previous eleven years, one of their dressed birds, a heavy young Bronze tom, was named the Grand Champion in the dressed division. The Allens also won the Sweepstakes Championship for their display of dressed turkeys, and one of their Bronze was also Breed Champion.

Forty-one Special Pack or Single Pack turkeys, suitable for gifts or for the "fancy trade," were also displayed in the auditorium's cold storage room. Special Packs were featured for the first time at the 1938 All-American Turkey Show. Weighing between nine and thirty-four pounds, the birds were entered by exhibitors from nine states. So comparable were the

---

[8] Similar to a European variety of turkey called the Pied, Crollwitz, or Black-laced White, Royal Palm turkeys are white, with a sharply contrasting, metallic black edging on the feathers. The black saddle provides a sharp contrast to the white base color of the body plumage. Tails are pure white, each feather having a band of black and a white edge.

Top: The dressed turkey division of the 1939 All-American Turkey Show, shown with the judges in the City Auditorium's refrigerated area. Bottom: Showroom scene at the 1939 All-American Turkey Show. The dressed division was located in the refrigerated section at the far left, behind the glass. *American Turkey Journal*, October 1939.

forty-one birds in quality, appearance, and packaging that judges declared ties and split ten awards among eighteen entries.[9]

When the judges had finished their work, the exhibit was opened to the public for viewing. Those wanting to purchase Special Packs as souvenirs or to give as gifts could submit bids on their choice. The bids would be opened when dressed birds were auctioned on Friday afternoon, the last day of the show.

An exhibit of canned turkey, "another show specialty" and "a 1939 innovation," was open for the first time on Wednesday morning, January 18. Entries were pints or quarts of turkey with or without bones, turkey sandwich spread, giblets, and turkey broth. Mrs. Al C. Johnson, Wayzata, Minnesota, received first in the turkey without bone class, and Mrs. Roy Utne of Ortley, South Dakota, received first in the turkey with bone class.

A smoked turkey was another first for an All-American Turkey Show, exhibited by Mrs. Al C. Johnson. For display, Mrs. Johnson boxed the smoked turkey in Single Pack style. The smoked turkey "received high praise" from "visitors and turkey experts and the praise was backed by rapid disappearance of the meat."

More than 350 "choice dressed birds," Special Pack turkeys, canned turkey, and smoked turkey showed the turkey industry was changing. Consumers' tastes were changing. Consumers were eating more turkey, but they wanted it in different forms, and they did not care whether the turkey's color had once been bronze, white, black, slate, or bourbon red. Breeders were adapting to the change.[10]

Radio personality Everett Mitchell's broadcast of the popular *National Farm and Home Hour* from the floor of the Grand Forks City Auditorium on Thursday, January 19, and his opening the broadcast with his

---

[9] Mrs. Roy Utne, Ortley, South Dakota, received first and third. Mrs. George Fowler, Hoberg, Missouri, second. Fourth went to Mrs. Jim Martinson, Kensington, Minnesota, and fifth to the Broadway Turkey Farm, Yakima, Washington. Seventh place was shared by Broadway Turkey Farm and Mrs. Don English, DeSmet, South Dakota. Eighth place, another tie, was shared by Mrs. W. J. Janda, St. Hilaire, Minnesota, and Shisler's Sheep and Turkey Farm, Aitkin, Minnesota. Those in tenth place included Mrs. Henry W. Domes, Richreall, Oregon; Mrs. O. O. Kreuger, Morris, Minnesota; Mr. and Mrs. J. M. Olson, Webster, North Dakota; and Mrs. Elvin Risbrudt, Dalton, Minnesota.

[10] Prominent breeders who had consistently taken top prizes with their live turkeys were also receiving top prizes for their dressed birds and Special Packs—among them Mr. and Mrs. Al C. Johnson, Mr. and Mrs. John O. Allen, Mr. and Mrs. Roy Utne, Mr. and Mrs. Henry W. Domes, and Mrs. C. H. Folz.

signature phrase—"It's a bea-u-u-tiful day"—was the most memorable feature of the 1939 All-American Turkey Show. It was also among the most memorable events occurring in Grand Forks, North Dakota, in the period before World War II.

Starting in 1930, with Everett Mitchell as announcer, the National Broadcasting Company's *National Farm and Home Hour* was the first regularly broadcast, nationwide show. Designed to appeal to rural Americans, the show featured farm news and events, agricultural and home advice, educational talks, and music.

In 1932, in the depths of the Depression, Mitchell met a man on the train one morning who said, "We're doomed. God has forsaken us." Mitchell pondered the man's statement, thinking that it was actually a nice day, even though it was dark and raining. He could walk and he could see sunrises and sunsets. When he went on the air, he exclaimed, "It's a beautiful day in Chicago."

Some critics advised Mitchell to drop the line, but the station received thirteen thousand calls and letters requesting that he continue to use the line, and it became his signature phrase. During World War II, when United States radio stations were not allowed to broadcast the weather, Mitchell received special permission to continue telling about "the beautiful day in Chicago."

The phrasing of accounts in the January 19, 1939, issue of the *Grand Forks Herald* is a measure of what Everett Mitchell's broadcasting the *National Farm and Home Hour* from the floor of the Grand Forks City Auditorium meant for the All-American Turkey Show and for the Greater Grand Forks area.

The *Herald* headline was "U.S. 'Hears' Turkey Show," followed by "104 STATIONS ON BROADCAST OF EXHIBITION." "Page, Burton and Hackett Among Speakers on Program." "MUSIC UNITS ON LIST." "Everett Mitchell in Charge of Radio Chain Feature Here." With the broadcast, "Grand Forks went on the national map of agricultural promotion 'with a bang,' in a half-hour nation-wide radio broadcast over 140 stations of the National Broadcasting Co., during its daily Farm and Home hour."[11]

[11] *Grand Forks Herald*, January 19, 1939.

This was "the peak of publicity Grand Forks and this section has obtained through the medium of this premier turkey exposition of the United States, conveying with telling emphasis to millions of listeners just how important is this annual event." The broadcast "portrayed to the country as a whole the significance of the show in the development of the turkey industry," and "it also gave convincing demonstration to citizens of Grand Forks of the true value of the show in reflecting favorably on the community and the state."

The article noted that "[c]onstantly in the background during the broadcast was the gobble-gobble of the prize turkeys, their proud challenges echoing through the crowded city auditorium and out over the national network as the introduction to the broadcast before Mitchell uttered a word." The "usual introductory number of the Farm and Home hour," and "The Stars and Stripes" was played by the University of North Dakota band, under the direction of Professor John E. Howard.

Then came Mitchell's "emphatic assurance" that "it's a bea-u-u-tiful day in Grand Forks." From that "effective start at 11:30 A.M. until the closing strains of *Yankee Rhythm* by the Central high school orchestra under Leo O. Haesle at 12:04 P.M., the program moved rapidly and without a hitch."[12]

The article described how "[e]very inch of standing room on the main floor of the auditorium and all gallery seats were taken long before the time the broadcast was scheduled to start. Thousands of others in Grand Forks heard the broadcast over their home, office or car radios." Everett Mitchell, "in a tour of the rows of exhibit coops," interviewed a number of turkey breeders on the air.[13]

Mitchell interviewed Albert Payne, recipient with his wife of the Master Breeder Award in 1939. Mitchell took note on the air of Mr. and

---

[12] The University of North Dakota band played twice during the broadcast. Other music was provided by Central high school's seventy-five Centralian Singers, directed by Arthur W. Seith, and by thirty-three University of North Dakota Madrigal Singers, directed by Professor Hywel C. Rowland.

[13] On air, Judge M. C. Herner, Winnipeg, Manitoba, "described the values of the new Royal Palm toms from Florida," exhibited at the All-American Turkey Show "for the first time anywhere in the world." Mrs. William Eddie, Northwood, North Dakota, described her Narragansetts, and J. M. Olson, Webster, North Dakota, his White Hollands. Judge Frank E. Moore, from the North Dakota Agricultural College in Fargo, commented on the Bronze variety, and Mrs. Ole Nelson, Kensington, Minnesota, described Black turkeys. Mrs. Otto Thieke, Beardsley, Minnesota, introduced Mitchell to her Red Bourbons.

All-American Turkey Show went on the air in January 1939, when Everett Mitchell broadcast the NBC *National Farm and Home Hour* over 116 radio stations from the floor of the Grand Forks City Auditorium. Image from January 19, 1939; used with permission from Grand Forks Herald.

Mrs. John O. Allen, Radium, Minnesota, Sweepstakes winners in the dressed division, and of Al. C. Johnson, Wayzata, Minnesota, who with his wife had exhibited the show's Grand Champion in 1938 and again in 1939. Mrs. August Swenson, Gilby, North Dakota, perhaps made the best impression on the air when she "demonstrated turkey calls and received a roar of gobbles from the 400 birds on display."

C. Dyke Page, All-American Turkey Show president, and turkey judge Walter Burton of Arlington, Texas, offered lengthy on-air comments. Page began by welcoming the *Farm and Home Hour* to "this turkey marketing center of Grand Forks and this great agricultural state of North Dakota on this first broadcast of an exclusive turkey show over the National Broadcasting Company network."

"The great Northwest was," Page said, "a leader in the production of fine turkeys, with North Dakota second, and our sister state of Minnesota third, in turkey production in the United States. Turkeys are peculiarly adaptable to this region." He continued, "The dry climate, longer hours of sunshine, and early, cool falls cause rapid growth and early fattening for market, and the quality is such that Northwestern turkeys invariably command premium prices on the eastern markets." The turkey industry, Page pointed out, made a significant contribution to the nation's agricultural economy with its annual production of seventeen million birds worth fifty million dollars.

It was "perfectly natural," Page said, "that Grand Forks should be the home of the first exclusive turkey show ever held anywhere, the All-American, a show that has grown in prestige and importance to the industry continually since its founding." He continued, "Now in its 16th year, the All-American is all that its name implies. It attracts high-quality pure-bred turkeys from as far distant as New Hampshire and Florida, from Oregon and California, Colorado, Texas, and Missouri, and from all the turkey growing states in between and from the provinces of Canada." Some of the "fine breeding birds" exhibited at the 1939 All-American Turkey Show, Page pointed out, were valued at one thousand dollars.

"The All-American Turkey Show," Page said, was "a centralizing agency, serving all phases of turkey growing." Its Education Sessions advised breeders on how to improve quality, provide better care, and conduct more profitable operations. "Backed financially by Grand Forks businessmen," he said, "and by industries allied to turkey growing," the All-American Turkey Show was "one more demonstration of the effective alliance of farm and city in a mutual effort to improve the conditions of both."

"The next time you sit down to a turkey dinner," Page advised listeners, "remember that the bird in front of you is a better turkey than you could get 15 years ago, and that turkey shows like the All-American are largely responsible for this fact."

Judge Burton said that he had judged turkey shows all over the United States for twenty-five years, and that what had "taken place here at the All-American Turkey Show at Grand Forks" was "a marvel of turkey breeders everywhere." He had judged in Grand Forks several times, and he had left each time "believing the limit in improvement of the show and the quality of the bird exhibited had been reached," but, on each return, he had found "new features presented in the show itself and always higher quality in the birds to be judged." As for the quality of the turkeys exhibited at the All-American Turkey Show, he said, there were no better ones to be found anywhere.

Nor, Burton said, had he seen better cooperation at a turkey show than in Grand Forks. Everyone was "turkey minded during show week, with the business interests, civic clubs and even society being interested in the show and its activities so capably directed by manager Hackett." Wherever he had judged, Burton said, it was the "highest ambition" of exhibitors "to someday exhibit at the All-American." He did not blame them. No other turkey show offered so many classifications of awards and so many special prizes, many of them originating at the All-American Turkey Show, including the Master Breeder Award, which was "considered the highest attainable in turkey show awards."

The *American Turkey Journal* noted that the sixteenth Annual All-American Turkey Show was believed to be the first ever to be covered by a nationwide radio broadcasting hookup and broadcast to stations all across the United States and Canada. "The gobbling of the turkey toms served as an unusual and pleasing background" for the program, "which did so much to enhance the popularity of the turkey and the industry, as well as to advertise that Greatest of all Turkey Shows, the All-American," which had the distinction of being "the first of its kind to be held anywhere in the world."[14]

January 19, 1939, was—as noted in the *Grand Forks Herald*—a "RED LETTER DAY" for the turkey industry in many ways, but especially because the industry had been recognized as "worthy of a place" on the *Farm and Home Hour* in this the "FIRST NATIONAL TURKEY BROADCAST."

"Show officials Win Praise for Radio Program" was another article heading in the *Herald* that day. Officials had received telegrams from listeners all over the United States. Mr. and Mrs. E. J. Severson of Grand

[14] *American Turkey Journal,* March 1939.

Forks, en route to Texas, sent word from Omaha, Nebraska: "Listened to the turkey show Farm and Home program. Sounded fine. Wonderful advertising for Grand Forks. Felt as proud as the grand champion of the show."[15]

Mrs. M. D. Butler, also of Grand Forks, sent a telegram from Seattle, Washington, where she was visiting. "Congratulations on your wonderful show. Heard gobblers here. Program came in wonderfully. Glad North Dakota won good recognition."

John Hesheth, president of the Grand Forks Chamber of Commerce, paid "special tribute" to C. Dyke Page and to W. W. Blaine, secretary and treasurer of the All-American Turkey Show, for the work they had done "in completing arrangements for the broadcast." The value of the broadcast to Grand Forks "from a publicity viewpoint cannot be accurately estimated," he said, but we know that it did "the turkey show and the city great good."

Any event of the All-American Turkey Show occurring after the noon hour on January 19 was pale by comparison, but the show did not close until Saturday afternoon, January 21. On Friday, "the shrill cries of an auctioneer punctuated the sale of all canned turkey and special packs" and show officials contracted for the sale on consignment of some three hundred dressed birds.[16]

Forty of the 350 dressed birds were purchased by visitors to the show, and 250 pounds of dressed birds were ordered by the Grand Forks Elks Lodge. The remaining dressed birds were purchased by the Red River Produce Co. The adult Bronze tom, the first in its class in the dressed division and exhibited by Broadway Turkey Farms of Yakima, Washington, went to the World's Poultry Congress in Cleveland, Ohio, to be put on display.

The sixteenth Annual All-American Turkey Show, described as "decidedly one of the most successful shows in the history of turkey competition," closed Saturday afternoon, January 21, 1939, and turkeys were released to owners at 3:00 p.m. The "work of crating and shipping the turkeys started immediately." Many of the "high bred aristocrats" were crated

---

[15] *Grand Forks Herald*, January 19, 1939.

[16] Cellophane-wrapped Special Pack birds sold for 35 cents to 65 cents per pound. The top price of 65 cents per pound was paid for a bird exhibited by Mrs. Roy Utne, of Ortley, South Dakota. Forty-eight quart jars of canned turkey sold at 12½ cents to $1.25 per quart.

for shipment to new owners, as many of them had been purchased as breeding stock. Outstanding toms had sold for $30 to $50 apiece. Because heavy snow and strong winds were making highway traffic difficult, many exhibitors who had trucked their turkeys to the show decided to wait for better weather and improved road conditions before leaving Grand Forks.

The seventeenth Annual All-American Turkey Show was held during the week of January 15–20, 1940. When the show opened on Monday morning, the temperature was -17°F, and the wind was blowing from the northwest at 12 MPH. Temperatures did not moderate. The temperature on Wednesday morning was -20°F with light snow. The "stinging cold" was "calmly described as normal winter conditions" by weathermen. William Huggins had begun arranging the turkey coops the previous Friday, and workers were busy decorating the City Auditorium.

Dates and features of the seventeenth All-American Turkey Show had been announced two months earlier in the *American Turkey Journal.* "For the 17th time," read the announcement, "the far famed All-American Turkey Show" had "bright prospects for making good its motto of 'CONTINUED PROGRESS!'"[17]

The show "boasted" of offering a wider range of competitions and more awards than were offered at any other show, and "every obligation" was "fully discharged" on the show's final day. "The educational and social programs for the week" could not "be duplicated," with the "'Homecoming' atmosphere pervading the whole gathering from first to last, and climaxed in a grand Banquet, Program and Ball on Thursday night of the week."

Held "in the heart of a great turkey producing section" and "conducted by turkey breeders," the All-American Turkey Show represented "every phase of the turkey industry" more completely than was true of any other turkey show in the country. The *American Turkey Journal* described the All-American Turkey Show as "a school, an 'Institute of Higher Learning' in Turkey education for large and small growers."[18]

The "SHOW WINDOW" of the turkey industry, the All-American Turkey Show offered opportunity to the "smallest and least pretentious"

---

[17] *American Turkey Journal,* November 1939.

[18] *American Turkey Journal,* March 1940.

growers by allowing them to attend, make comparisons, exchange ideas, attend educational programs, and receive advice on management, disease control, and feeds. Breeders could also learn how to improve their turkeys by observing judges "table handling" the birds and by taking note of what judges considered defects.

The dressed bird exhibit, which had become a regular feature of the All-American Turkey Show, attracted "wide attention," according to the *American Turkey Journal*, by adding to the competition and interest. The exhibits had become "the greatest source of influence in the improvement of turkey type," and breeders who did not enter their birds in the dressed competition were missing "a good bet." The "SPECIAL oven-ready single pack" and the "canned turkey competitions" had also become popular features of the All-American Turkey Show.

Belying the bright prospects, the boasts, and the high expectations, events occurring in 1940 in the turkey industry and in international affairs did not augur well for the All-American Turkey Show's future. "A war just as grim and far-reaching in its implications as those raging in Europe and Asia," noted the *Grand Forks Herald* that year, was "being fought out in every factory in North America." It was "the War of the Substitutes—nylon against silk and rayon, Plastacele against wood and sheet metal." So far, the chemists had "won one decisive victory after another."[19]

Their "new brain children, conceived in retorts and crucibles," were being used in thousands of ways "never dreamed of" before. "Plastics," as the new materials were called, could be used to produce telephones, steering wheels, radio cabinets, safety glass, buttons, and "thousands of other products." Plastics could be cast into rods, sheets, or tubes and then sawed, machined, turned, and cut into whatever final shape was desired.

Plastics were popular because they combined advantages of their own with those of natural materials. It was like rubbing Aladdin's lamp. If a material was not resistant to electricity, for example, "add a pinch of this and a dash of that," and the resulting plastic was "positively shock-proof." If one wanted the new material to repel moisture, one took "a slightly different 'rule' out of the recipe book," and the new substance was waterproof. "No wonder" this was "the Chemical Age!"

---

[19] *Grand Forks Herald*, January 12, 1940.

Plastics, the article continued, had "rooted themselves in the shade and darkness of the worst depression years," and they had grown by "leaps and bounds" since. They had "barely scratched the surface" of their possibilities, however. When plastics had "finally come to grips with the Maginot Line of the natural materials" in automobile bodies, the war would "enter its final stage." The War of the Substitutes meant "vast social changes" as well, and industries would be "caught in the backwash of this epoch-making war."

The War of the Substitutes was not the only one covered in the *Grand Forks Herald.* Finnish marksmen had "virtually annihilated" large groups of Soviet Russian troops, shooting them in the air as they attempted to land by parachutes behind Finnish lines on the Karelian Isthmus. Russia was massing troops for an attack on the Mannerheim Line on the Karelian Isthmus, but the Finns wanted to prevent a decisive encounter before spring.[20] In their resistance, the Finns were relying on the "twin scourges" of the worst winter in twenty-five years and temperatures as low as -54°F. The "strategy of the Finnish woodsmen" had also proven successful.[21]

War news also included the observation that the "Rome-Berlin 'axis' was rusting in the wake of the Nazi-Soviet alliance." Italy, therefore, was strengthening its position in Europe by diplomacy, trade agreements, and force of arms. By concluding a defensive alliance with Hungary, annexing Albania, improving trade relations with London, and courting Russian friendship by seeking an adjustment of Balkan territorial disputes, Italy had become a new hub in Europe's balance of power.

Although the European war was only in its fifth month, having begun in early September 1939 when Germany invaded Poland, countries were already experiencing shortages and governments were instituting rationing. Germany, according to the *Herald*, had placed "restrictions on practically all basic necessities of life." Even horses and cows had to "have cards to be able to stick their noses in the feed trough."[22]

---

[20] *Grand Forks Herald,* January 12, 1940.

[21] Finnish military installations were completely hidden in the forests, and one could drive hours without seeing them until halted by a sentry who slipped out from the cover of the woods. The Russian troops were at a disadvantage because they were poor skiers and lacked mobility and information about the terrain.

[22] *Grand Forks Herald,* January 13, 1940.

German people were subjected to two meatless days per week and limited to a little over one pound of meat per week, when they could purchase it at all. Germany's seafood supply had been virtually cut off by the British blockading ships in the North Sea. Those Germans with money could obtain meat, sugar, and other products from Hungarian smugglers, but "Nazis," noted the *Herald* article, "don't drink coffee." There was none to be had. A substitute for coffee, made from barley and with a dash of coffee scent added, tasted "somewhat like the real article," but lacked "its kick." Beer was still available, but it was "not the beer it used to be."

The British government had placed restrictions on butter, bacon, ham, and eggs. Even the king was limited to four ounces of bacon, four ounces of butter, and twelve ounces of sugar per week. As in Germany and Italy, gasoline was rationed in Britain, and its price placed it out of reach of most motorists.

Neither its isolationist policy nor its neutrality regulations granted the United States immunity from the war or from its ramifications. Early in January 1940, French military commissioners at St. Louis, Missouri, arranged the purchase of 1,100 horses from America's western plains. The horses were intended as mounts for French soldiers serving on France's border with Germany.

A *Grand Forks Herald* article reported on the flight of twin-motored American-made bombers to Pembina, North Dakota, near the border of the United States with Canada. The two bombers, the first of a contingent produced by Lockheed for sale to Canada, were flown from California to Omaha, Nebraska, and they passed over Grand Forks on their way to the border. Because of neutrality regulations, the bombers had to be paid for and titles had to be transferred to Canadian authorities before they could leave the United States. Neutrality regulations also required that the bombers had to be taken across the border on the ground.[23]

---

[23] *Grand Forks Herald*, January 16, 1940. Two large, smooth "flying fields" had been constructed, one on either side of the international border, separated only by a narrow strip of virgin prairie. "The huge black camouflaged bombers" landed at the Pembina airport, and the motors were left turning over in the -20°F temperatures while two of the pilots cleared customs. The pilots then flew the bombers to the landing field on the United States' side and taxied to the international border. Motors still turning over, the planes were pulled across the border by a team of horses. After the bombers were on the Canadian airstrip, the pilots once again took to the air and flew the planes to Winnipeg. The bombers were to be used to train Canadian bomber pilots.

Sharing pages in the *Herald* with the Lockheed bombers, the All-American Turkey Show opened with more than seven hundred live and dressed birds exhibited by breeders from more than a dozen turkey-raising states. Breeders and turkeys competed for more than $1,000 in cash prizes and for a long list of awards provided by businesses and organizations in Grand Forks. As with previous shows, the 1940 All-American Turkey Show had a number of new features, not all of them, unfortunately, boding well for the show's future. There were as many birds in the dressed division, for example, as there were in the live bird division.[24]

The All-American Turkey Show was on the national airwaves again in 1940. Station WDAY in Fargo, North Dakota, the National Broadcasting System's northwestern station, carried a special broadcast of the show on Thursday, January 18, from 1:00 to 1:30 p.m. Station KFJM, the University of North Dakota station, carried broadcasts of the show each day from 11:45 a.m. to noon. The station also broadcast highlights of the 1940 show at other times.

The 1940 Homecoming Banquet was held at the Hotel Dacotah at 6:30 p.m. on Thursday, January 17. Always the highlight of a show's social activities, the banquet in 1940 featured an innovation. All-American Turkey Show "pioneers" were guests of honor and seated at a special table. The pioneers were those who had exhibited at one or more of the first five All-American Turkey Shows, "at a time when it was truly the first exclusive turkey show ever held anywhere." Of the more than two dozen old-timers, only Mrs. C. H. Folz of Drayton, North Dakota, and Alfred Malmberg, of Crookston, Minnesota, had exhibited at every one of the All-American Turkey Shows since their inception in 1924.

A first for this All-American Turkey Show was as unwelcome as it was unexpected. Pleading the press of other business, George W. Hackett announced that he was retiring as manager. The heart and soul of the

---

[24] Mr. and Mrs. Frank Ralston of Crystal, North Dakota, set a single-entry record at the 1940 All-American Turkey Show with their thirty-one Bronze turkeys, the largest single-breed entry in the show's seventeen-year history. One of their turkeys, a yearling Bronze, was named the show's Grand Champion. Mr. and Mrs. Roy G. Utne of Ortley, South Dakota, displayed seventeen birds of a single class, and one of their Bronze adult hens was named the show's Reserve Grand Champion. The Utnes also received the Master Breeder's Award and a gold medal.

All-American Turkey Show, Hackett had been with the show for all of its seventeen years, five years as judge and twelve as manager.

Page, All-American Turkey Show president, spoke true when he described Hackett as "unquestionably the best known turkey man in the United States and Canada." It was through Hackett's efforts, Page said, that the All-American Turkey Show had "grown in its prestige to its present commanding position as the top-ranking turkey show" in the country.

As a token of their appreciation for Hackett's long association with the All-American Turkey Show, exhibitors presented him with a gift. He received a travel bag on behalf of the All-American Turkey Show Association. Although he would no longer serve as manager, Hackett said, he would continue to work with the show in any way he could. Frank Moore, of the North Dakota Agricultural College in Fargo, was named acting manager to replace Hackett. He was later elected manager.

The seventeenth annual All-American Turkey Show featured a new class in the dressed division, the utility bird class. In this class, for the first time in the show's history, crossbred birds were allowed to compete. In other competitions, turkeys had to be those varieties admitted to the *Standard of Perfection*. In the utility bird competition, entries were judged by their market value.

The All-American Turkey Show in 1940 also stressed the Special Pack and the canned turkey competitions. The popularity of these divisions was evident in the number of entries and awards and in the prices received at auction. In the Special Pack division, awards were given down to twentieth place and down to the twelfth place in the canned turkey division.

Attendance on Friday, January 19, was higher than on any other day of the show, probably because of the auction that afternoon of the Special Pack and canned turkey exhibits. The champion Special Pack was purchased by P. M. Onstad, manager of Grand Forks Mercantile, for $1.00 per pound. Records show that "other fancy packed animals were sold at prices ranging down to 38¢ per pound." Onstad also purchased the blue-ribbon winner in the canned turkey exhibit for $2.10. Prices for the other canned exhibits ranged down to sixty cents.

Exhibitors attempting to bring their turkeys as close as was humanly possible to the specifications in the *Standard of Perfection* may have been dismayed at some of the seventeenth All-American Turkey Show's features: dressed birds and live birds about equal in number; a new utility class, non-Standard birds; and thirty-two Special Packs and as many or more canned turkey entries.[25]

Breeders may have been given pause when they read the description of the 1940 All-American Turkey Show's dressed division's Grand Champion in the February 1940 issue of *Turkey World*. The Grand Champion was a heavy, young Broad Breasted Bronze hen weighing 18¾ pounds. The hen had been shipped to the show by the "well-known Pacific Coast breeders, Mr. and Mrs. Art Hamilton, Chehalis, Washington." This bird, the description went on to say, "was truly a specimen of Broad Breasted Bronze that did credit in every way to this popular breed. It was also champion Broad Breasted Bronze."

This non-Standard hen sold at auction on the last day of the show for $3.25 per pound, reportedly the highest price ever paid for a dressed turkey. The Dr. Salsbury Laboratories of Charles City, Iowa, was the buyer. The Reserve Champion in the dressed division, a Standard-bred Bronze young hen, was exhibited by the Allen Turkey Farm, Radium, Minnesota. At auction, the Hubbard Milling Co. was the high bidder for the Reserve Champion at sixty cents per pound.

The War of the Substitutes, according to the *Grand Forks Herald*, was "being fought out in every factory in North America." Armed with retorts and crucibles, chemists had won one victory after another. The skirmishes between Standard-variety turkeys and those of the Broad Breasted varieties were occurring in the dressed divisions of turkey shows, in the meat aisles of grocery stores, and in well-placed ads in the *American Turkey Journal*.

[25] Exhibitors may also have noted that many breeders whose live turkeys had often been cooped in the Court of Honor were also entering exhibits in the Special Pack and canned turkey divisions, including Mrs. William Eddie, Mr. and Mrs. Roy G. Utne, Mr. and Mrs. Jim Martinson, Grace Baxter, Mrs. Frank Ralston, and Mr. and Mrs. Albert Payne.

The 1940 All-American Turkey Show's Grand Champion in the dressed division was a heavy, young Broad Breasted Bronze hen weighing 18¾ pounds. The bird was described as a "beauty" and as a specimen that "did credit" to what was becoming a popular turkey variety. Also doing credit to the variety, the Grand Champion was sold at the highest price ever paid for a dressed turkey.

At the market, in the display case, or on the rack, according to the *American Turkey Journal*, consumers cared little for *Standard of Perfection* specifications, feather color and shading, and conformity, and they were the "court of last resort." Consumers wanted turkeys that appealed to them for size, shape, and finish. Butchers, wanting to satisfy their customers, did not explain the differences among turkey varieties; they stocked the birds that they knew would appeal to customers and keep them coming back to the store.[26]

That Broad Breasted breeders by 1940 were pressing their advantages was evident in their ads in the *American Turkey Journal*.[27] Of the breeders advertising breeding turkeys in the October *American Turkey Journal*, one-half were offering utility, meat-type varieties.

In the same issue, George W. Hackett reprinted an article from *California Turkey News*. The article noted that the average turkey marketed in 1929 weighed 13.2 pounds. By 1939, the average weight was fifteen pounds, and increase of nearly two pounds in eleven years. By 1940, it seemed, Broad Breasted breeders had succeeded in taking much of the turkey industry's high ground. Those raising and exhibiting Standard variety turkeys, armed with little more than feather color, shading, and pattern, were fighting a rear-guard action.[28]

Always quick to sing the praises of Standard-bred Bronze, Mrs. W. J. Janda was secretary of the All-American Bronze Club. "It is true," she ad-

---

[26] *American Turkey Journal*, October 1940.

[27] The following were representative: "Hamilton's 'Perfect Meat Type' Broad Breasted Bronze. Won All-American Grand Champion. 7000 beautiful 'Perfect Meat' type breeder hens for 1940. Blood tested, 100% free of pullorum." "Lusby Turkey Farm, Owenton, Ky. Ralph Lusby, Mgr. Lusby small type Whites. New small type, quick maturing Whites with broad breasts, in great demand for quality broilers as well as family type market birds. Ideal for year round sales." "Fowler's Valley View Bronze. The Broad-Breasted Type-best for meat and exhibition. Well-developed, blocky-type breeding toms at reasonable prices. Hoberg, Missouri." "Amos' Famous Shelton Bronze Broad Breasted meat type, Russellville, Missouri."

[28] *California Turkey News* was the official publication of the California Turkey Growers Association.

mitted, "that you cannot eat feathers," but feathers were important. Feathers told the "true story" of the bird. "A highly colored bird," she believed, showed good breeding, and large, wide feathers showed "a thrifty bird." Narrow, loose, dull-colored feathers indicated that, at some time during the bird's life, something had gone amiss—poor breeding, sickness, improper feeding—all making the bird unfit for breeding.

"We each strive," Janda wrote, "to raise our chosen breeds true to color, type and marketing value." Were there not "a standard to go by, turkey raising would be a dull occupation." Because breeders could continue to raise turkeys "true to breed" and still make money with them, there was no need to produce birds of the Broad Breasted varieties, having no regard for Standard color and type.

Mrs. Ole C. Nelson, Kensington, Minnesota, was the Narragansett Club secretary. She cited a letter from Edna and Maude Sheckler, Nevada, Ohio, who had been raising Narragansett turkeys for years. The Shecklers had raised other varieties, but, for many reasons, Narrangansetts were their favorite, preferred "for their beauty," because there was nothing more beautiful or striking in the show room than a display of pure-bred Narrangansetts. A person could "see their beauty and fine qualities from the farthest corner of the show room without the aid of a bright light or the sunshine." At the shows, more visitors each year came to admire Narragansetts "for both type and feathers." And, there was nothing "more gratifying than to look over the green fields" and feast ones eyes "on a flock of pure bred, high class Narragansetts."[29]

By reason of being editor of the *American Turkey Journal*, Hackett earned the last word. Trends in the turkey industry were changing, he realized. The unwarranted "utility" craze had swept the country (turkey parts, canned turkey, smaller turkeys, smoked turkeys), robbing "the breeder of standard turkeys of much of the business rightly due him."

Breeders of Standard varieties had to assume some of the blame because, too satisfied with things as they were, they had not heeded the

---

[29] Narragansetts had the essential qualities of a good market type bird, having a good size and exceptionally deep breasts and an abundance of white meat. High in the production of eggs of good fertility and hatchability, Narragansett's matured early and were easy to dress.

advice that would have allowed them to maintain their position in the industry. “The dressed turkey shows,” Hackett reminded breeders, had “provided the means” for them to prove the quality of their birds, but “few took advantage of it and most of those who did, only halfheartedly.”[30]

“Standard breeders,” Hackett advised, “may as well adjust themselves to the fact that the new turkey” was “here to stay.” Breeders were raising fine flocks of Broad Breasted turkeys and the way seemed “to be open” for the admission of the new turkey to the *Standard of Perfection*. No one knew whether the utility birds would remain popular, but “the mass productionist,” the “common raiser of turkeys,” and “the consuming public” had to be satisfied, or “lasting popularity” was “out of the question.”

Hackett saw no reason, however, “for standard breeders to be discouraged.” He advised them to continue to “make every effort to improve type,” because “the better color of standard birds” would “always be in demand by the average turkey raiser.”

---

[30] *American Turkey Journal*, October 1940.

CHAPTER 15

Can One Desire Too Much of a Good Turkey Show?

(with apologies to William Shakespeare, *As You Like It*, 4:1)

GIVEN INTERNATIONAL AND NATIONAL events and many of the features of the eighteenth All-American Turkey Show, held January 20–25, 1941, breeders of Standard turkey varieties could be forgiven had they become disconcerted. Beginning shortly after its inception in 1924, the All-American Turkey Show enjoyed being the Turkey World's center of attention, and, for an entire week every year, accounts of the show dominated the pages of the *Grand Forks Herald*. In January 1941, accounts of international and national events, many of them unsettling, took precedence, and the newspaper's coverage of the All-American Turkey Show was markedly reduced from what it had been in previous years.

Perhaps disconcerting as well, flock owners, season after season, had bred their birds to meet, as nearly as was humanly possible, the specifications demanded by the *Standard of Perfection*, including feather color, pattern, and shading. Implemented for the first time at the eighteenth All-American Turkey Show were changes in the way that judges evaluated a bird's characteristics, awarding more points to meat type and marketability and fewer points to plumage.

Taken together, news of international and national events and some features of the eighteenth All-American Turkey Show did not bode well for the future of Standard turkey varieties nor for that of the All-American Turkey Show. Three days before the opening of the eighteenth All-American Turkey Show, the *Grand Forks Herald*'s January 17, 1941, headline was, "U.S. Navy Will Be Outnumbered If Britain Loses, Knox Asserts." Frank Knox, a Republican serving as President Franklin Roosevelt's Secretary of the Navy, told members of the House Foreign Affairs Committee that the country's navy would be "heavily outnumbered by the fleets of the Axis powers if British sea power should be destroyed," and he urged passage of Roosevelt's lend-lease bill as being in the country's "best national interests." The United States, Knox advised committee members, "needs time to perfect its defenses," because the two-ocean navy construction program then under way "would take six years to complete." By 1943, Knox warned, the US would be outnumbered by the Axis powers in every fleet category, from battleships to submarines. "Only Britain and its fleet," Knox noted, could "give us that time, and they need our help to survive."

Another front-page article on the same day announced that guardsmen were being mustered into service. "The boys of Company M," the Grand Forks machine gun unit of the 164th Infantry, would be mustered into military service on February 10, 1941, and sent to Camp Clairborne, Louisiana. On February 25, 6,300 guardsmen from Minnesota; 3,500 from Iowa; 1,250 from South Dakota; and 1,300 from North Dakota would also leave for Camp Clairborne. North Dakota's 188th Field Artillery Regiment, of which Battery F of Grand Forks was a part, would leave for Fort Warren, Wyoming, on April 1, 1941.

More alarming were the headlines on the fourth day of the All-American Turkey Show: "U.S. Britain Together Can't Win War, Should Negotiate Peace, Lindy Says" and "British Aid 'Mistake': Air Invasion Impossible Famous Flier Asserts." Colonel Charles A. Lindberg warned that the combined strengths of the United States and Britain could not win the European war "on the present basis," and he urged the two nations to pursue a "negotiated peace." Lindberg also believed that "the stand of 'the American people' in favor of aiding the British 'was a mistake.'"

Readers of the *Grand Forks Herald* were reminded of the war in other ways as well. The headline on Saturday, January 18, 1941, was, "Crowds

Throng Capital for Precedent-Breaking 3rd Inauguration of F.R." The lead article opened with the words, the "bustle of carnival crowds overshadowed somber defense preparations Saturday in the nation's capital, splashed with banners and bunting for the precedent-shattering third inauguration" of Franklin D. Roosevelt. "Virtually the entire inaugural parade" was "a sober reminder of the preeminent place of defense in the government's activities." The parade included only three civilian units—two hundred "green-clad CCC boys, as many marchers from the National Youth Administration, wearing red, white and blue mackinaws, and a WPA delegation in working clothes."

The other parade units were "military in nature-khaki-garbed infantrymen; row after row of blue jackets; brightly-uniformed marines"—signs of the country's growing military strength—and battalions of cadets and midshipmen from West Point and Annapolis. General George C. Marshall, Army Chief of Staff, was grand marshal. Batteries of big guns, light and heavy tanks, scout cars, and motorcycles rolled along the street and 280 military planes flew overhead.

War may have been distant for turkey raisers in 1941, but they were, nevertheless, directly affected by it. Turkey and breeders were experiencing a labor shortage. Young, able-bodied men were being drafted or their guard units were being activated. Workers could receive higher wages in defense plants than turkey raisers could afford to pay, given the prices they received for their turkeys. Working on WPA projects, on which hours were shorter than those on farms, also paid better. A writer for the *American Turkey Journal* observed that "things are badly out of balance and the answer is not in sight."[1]

As insignificant as it may have seemed to those unfamiliar with the turkey industry, turkey raisers needed aluminum leg and wing bands to mark birds in their breeding flocks. In 1940, the turkey industry had used 150,000 pounds of the bands, but, aluminum was a vital defense material,

---

[1] *American Turkey Journal,* September 1941.

and it was in short supply. In 1941, the Office of Production Management (OPM) allowed poultry raisers only 11,000 pounds of aluminum to be used exclusively for wing bands. The OPM released aluminum for this special purpose because eggs were deemed important defense foods, and poultry raisers were encouraged to produce more.

According to the *American Turkey Journal*, "real turkey breeders" sought to improve the quality of their turkeys by using only the best birds for breeding. They were advised, however, to have the breeding birds selected and banded by inspectors authorized and licensed under the American Poultry Association Flock Inspection Program. These inspectors handled thousands of birds per year, they knew what the *Standard of Perfection* specified for each turkey variety, and, with their years of experience, they knew how to apply the standards.[2]

The cost was minimal, five to six cents per bird, including the cost of the aluminum bands. Besides improving the quality of the flock, inspecting and banding was a guarantee to prospective buyers that the turkeys were birds of a variety listed in the *Standard of Perfection*, and banded birds commanded a higher price as breeders. Having breeding turkeys banded also brought flock owners national recognition, especially when they advertised their breeding stock in turkey journals. Finally, having the best turkeys in their flocks inspected and banded allowed turkey raisers to take a greater, rewarding interest in their flocks.

More encouraging to turkey raisers than many of the defense measures undertaken by the federal government was the *American Turkey Journal* notice that "vast military establishments" were being built all over the country. According to reports from quartermaster procurement officers, "hundreds of thousands" of United States marines, soldiers, and sailors would "eat a lot of turkey."[3]

Turkey meat was inexpensive in comparison to other meat and "this new and hungry market" for the "great American turkey" was expected to have a positive effect on the prices paid for turkeys. An estimated 33.5

---

[2] *American Turkey Journal*, October 1941.

[3] *American Turkey Journal*, February 1941.

million turkeys would be produced in 1941, a record. Nevertheless, the United States Department of Agriculture was recommending a 10 percent increase in turkey numbers in its effort to increase food production for defense purposes.

On Friday, January 17, 1941, a day when the *Grand Forks Herald* carried mostly unsettling news accounts of war on its front page, an abbreviated announcement informed readers that more than three hundred coops, "destined to house the finest turkeys in the United States for a week, sprang into being at the city auditorium . . . as William Huggins and his crew prepared the building for the All-American Turkey Show" that would open Monday at 10:00 a.m.

"From Virginia and Oregon," noted a *Journal* writer, the turkeys came, "from Manitoba and Texas, Washington and Ohio, Kansas and South Dakota, and many other states in between, more than 300 of the best quality turkeys in the land to compete at 'The Turkey Classic of America,' the 18th annual All-American Turkey Show at Grand Forks, North Dakota."[4]

Actually, there were 335 live entries and 273 dressed entries from eleven states and one Canadian province—Washington, Virginia, Ohio, Oregon, Texas, Montana, Nebraska, Kansas, South Dakota, North Dakota, Minnesota, and Manitoba. The number of entries was down from the 1940 All-American Turkey Show because of market conditions, snow-blocked roads, and the Armistice Day storm, November 11, 1940.

The Armistice Day blizzard that struck the Upper Midwest in 1940 was the type of storm of which legends are made. Autumn weather had been deceptively mild, and on the morning of November 11 the temperature was 55°F at Chicago and 54°F at Davenport, Iowa. By early afternoon, temperatures approached 65°F over much of the region.

During the day, however, conditions quickly deteriorated across much of the Midwest, with sleet, heavy rain and snow, a tornado, gale-force winds, and sharply dropping temperatures. An intense low-pressure system, tracking from the southern plains northeastward into western Wisconsin, pulled moisture up from the Gulf of Mexico and a cold arctic air mass down from the north.

---

[4] *American Turkey Journal,* February 1941.

The result was a raging blizzard lasting into the next day. Snowfalls of up to twenty-seven inches and winds of 50 to 80 MPH produced twenty-foot snowdrifts. Sharp temperature drops across Nebraska, South Dakota, Iowa, Minnesota, Wisconsin, and Michigan produced cold so severe, according to survivors, that breathing was difficult. The cold seared lungs like a red-hot knife, and the air was so heavy with moisture it was as thick as syrup.[5]

Because Armistice Day was a holiday, many businesses and schools were closed, and people were outside and away from home. Along the Mississippi River, several hundred duck hunters took advantage of the ideal hunting weather. When the storm struck, many sought shelter on small islands in the river, only to be stranded when gale-force winds swept water over the islands. The water froze, and heavy snow reduced visibility to zero.

Unprepared for cold weather, many hunters froze to death and others drowned. Some, fortunate enough to survive, lost hands or feet because of severe frostbite. In Watkins, Minnesota, two people died when a passenger train and a freight train collided in the blizzard. In Lake Michigan, sixty-six people drowned when three freighters and two smaller boats sank. A total of 145 deaths were blamed on the storm, including 49 in Minnesota, 13 in Illinois, 13 in Wisconsin, and 4 in Michigan.

Across the Upper Midwest, hundreds of abandoned vehicles clogged roads and highways in twenty-foot snowdrifts; passenger trains were stranded; roads and highways were closed; newspaper deliveries were halted; telephone, telegraph, and power lines were down; and business, home, and farm buildings were damaged or destroyed.

The Armistice Day storm was a seminal event with lasting economic and cultural impacts. Before the storm, Iowa was a leading fruit-growing state, second only to Michigan in apple production. The blizzard and ice storm killed hundreds of fruit trees, devastating Iowa's apple industry

---

[5] A few days earlier, on November 7, a strong weather system moving into the Pacific Northwest had taken down the Tacoma Narrows Bridge in Washington. Referred to as Galloping Gertie because of the way it undulated in the wind, the bridge was an engineering wonder and, before its collapse, the third longest suspension bridge in the world. By November 10, the storm system had moved across the Rocky Mountains and into Colorado, before curving northeast into the country's midsection. By the time the storm was centered over Lake Superior, the barometer reading had dropped to 28.57 inches of mercury.

and changing the state's landscape. Planting new orchards is expensive, and trees do not produce fruit for several years. In 1940, the threat of war was growing, and the country was facing hard times. Apple growers transformed their orchards into fields and began raising corn and soybeans. Turkey breeders were also affected. In Minnesota alone, 1.5 million turkeys died.

The turkeys exhibited at the 1941 All-American Turkey Show were the six varieties listed in the *Standard of Perfection*—Bronze, Narragansett, White Holland, Bourbon Red, Slate, and Black. For the first time in the history of the show, twenty-one turkeys of the Broad Breasted variety were allowed to compete in the live division. In the dressed division, for the first time, Commercial Box Packs—boxed dressed turkeys packed by turkey produce houses and ready for the market—were introduced.

"With the advent of the 1941 show," the All-American Show Association announced in the *American Turkey Journal*, it was "hoped to make of the All-American, more and more, a 'turkey industry exposition.'" Continuing to cover all phases of both live and dressed turkey competition, the All-American Turkey Show "hoped to attract the entire turkey industry: dressing, shipping, marketing, advertising and activities aimed at increasing national turkey consumption." The "ultimate goal" was the "thorough and complete coverage of all angles" of the turkey industry. "Continued Progress" was still the All-American Turkey Show's slogan, and, as in the past, "this great show" would "be found in the fore" in everything that made "for better turkeys."[6]

This "First and Foremost of the World's Exclusive Turkey Shows" in 1941, it was announced, would offer a large number of awards, including $1,500 in cash and Special Prizes. "To perfect 'team work,'" which was "the purpose of the management," a number of "working committees" were named, including the Show Advisory Committee, the Educational Program Committee, and the Annual Banquet Committee.

The Turkey Hen Club would have a rest room, as in the past, and an information booth "where anyone desiring information regarding the show program" could obtain it. The Turkey Hen Club would also pro-

---

[6] *American Turkey Journal*, December 1941.

Turkey Hen Club Nest, in which women exhibitors could seclude themselves and rest during the week of the All-American Turkey Show. *American Turkey Journal*, December 1941.

vide a reception committee. To insure that all visitors and exhibitors were made to feel welcome, the reception committee members would see to it that everyone was "well cared for."

The Court of Honor, "always an attractive feature" of the All-American Turkey Show and "an educational demonstration as well," would be on the auditorium stage. "In this unique arrangement of highly decorated coops, fifteen of the 'honor' representatives of the various breeds" were displayed "under proper labels and decorated with the respective trophies" they had won.

Anticipating the conditions and circumstances that might prevail in January 1941, the All-American Show Association decided to open the 1941 All-American Turkey Show to the public for only five days, officially closing it on Friday evening, January 24, 1941, instead of on Saturday afternoon as in the past.

"All roads of Turkeydom" led to Grand Forks, North Dakota, in January 1941, and the *Grand Forks Herald* introduced a new feature, a "Minute Editorial," on its front page. "Community chests (the patted kind)," read the editorial,

> should swell with pride in the ranking of the All-American Turkey Show as the greatest event of its kind in the world in quality of entries.
>
> This is no mere Chamber of Commerce boast, but the considered judgement of an internationally known poultry expert—Harry M. Lamm of Adams Center, New York, president of the American Poultry Association and a judge at the show in Grand Forks this week.
>
> THE GREATEST SHOW IN THE WORLD and right here in our own backyard. Certainly every resident of this community should see it. It is entertaining, educational, deserving of the fullest recognition—and the admission is FREE.
> Go today![7]

An article on Armistice Day 1940 in *Newsweek* gave added reason for community chests (the patted kind) in Grand Forks, North Dakota, to swell with pride. "Recognition by such magazines as this," noted George Hackett, *American Turkey Journal* editor, emphasized "the importance of the turkey industry" and of the All-American Turkey Show. The article included a picture of the 1940 All-American Turkey Show's Grand Champion.[8]

"Election year or no election year," according to *Newsweek*, "every January the best walking Thanksgiving dinners in the United States and Canada compete at Grand Forks, N.D. for the North American blue ribbon. This year's winner . . . owned by Mr. and Mrs. Frank Ralston of Crystal, N.D. was judged the most perfect of 341 turkeys but will grace no Thanksgiving table. He is valued at $2,500 and doesn't care whether

---

[7] *Grand Forks Herald,* January 22, 1941.

[8] *Newsweek,* November 11, 1940.

Thanksgiving comes on the third Thursday in November or the first Tuesday in May."[9]

Giving the 1940 All-American Turkey Show's Grand Champion its due, Harry Yoder described the bird in the February 1940 issue of *Turkey World*. The Grand Champion, he wrote, was a yearling Bronze tom and "most worthy of this high award. He was marked beautifully, his type second to none, his carriage and general behavior was a sight to gladden the heart of any true turkey lover. He was 'King of the Show' and conducted himself in like manner."[10]

The Premium List for the 1941 All-American Turkey Show, one of the longest in the show's history, was published at the end of 1940 in the *American Turkey Journal*. Awards consisted of $1,500 in cash and fifty trophies and Special Prizes, plus merchandise awards of worming pellets, feed, other products, and merchandise certificates of $1.00. [11] There were silver loving cups, suitably inscribed, for the champions of each turkey variety, but only one Grand Champion award, one Reserve Champion award, and one Master Breeder award.[12]

Special Prizes included five pairs of beautiful, hand-turned candlesticks of walnut and curly maple, donated by the *American Turkey Journal*. The candlesticks would be awarded to the five "lady exhibitors coming from the greatest distance" and representing the Bronze, Narragansett, White Holland, Bourbon Red, and Black turkey varieties. The All-American Bronze Club would provide pairs of hand-turned bookends for the Best Breeder's Display of live Standard Bronze and for the Best Standard Bronze dressed bird.

---

[9] The reference to Thanksgiving's being observed on the third Thursday in November was because in 1939 President Franklin D. Roosevelt had, by proclamation, moved the date of Thanksgiving to the third Thursday from the fourth Thursday in November. In 1940, twenty-six states, including North Dakota, announced that they would observe Thanksgiving on November 21.

[10] Yoder was advertising manager of *Turkey World*, a monthly magazine serving the turkey industry. The journal, with an international subscription list, was published at Mount Morris, Illinois. Subscription prices were $1.00 per year in the United States, $2.00 for three years, and $1.50 per year in Canada and other countries, $3.50 for three years.

[11] *American Turkey Journal*, December 1940.

[12] The Master Breeder gold medal, provided by the First National Bank in Grand Forks, North Dakota, would be inscribed with the winner's name. The Master Breeder Award was the most coveted of all prizes. The recipient was determined by judges' comparisons of the winning Breeders' Displays of each of the varieties in the Standard classes.

A portion of the prizes offered at the 1940 All-American Turkey Show. The battery-powered Farm Radio—the show's most coveted prize—is prominently displayed in the center of the top row. *American Turkey Journal*, November 1940.

The Poppler Radio Award, provided by the Poppler Piano and Furniture Company, was perhaps the best Special Prize of the 1941 show. This battery-operated Model 105 B Philco Farm Radio, with its beautiful walnut cabinet, was "one of the finest farm radios" on the market. Equipped with four new-type, low-drain tubes, the radio's 1½-volt battery would last for 750–1,000 hours. This "fine trophy" would be awarded to the Best Display of live Standard turkeys entered by an exhibitor from Traill, Steele, Grand Forks, Nelson, Walsh, Pembina, or Cavalier County in North Dakota, or by an exhibitor from Norman, Polk, Red Lake, Pennington, Marshall, Kittson, or Roseau County in Minnesota.

To appreciate what this fine trophy meant to exhibitors is to realize that in 1941 a radio for many families would have been a luxury. For turkey raisers, isolated on their farms and lacking electricity, a battery-operated radio may have been the sole means by which they could obtain news of happenings outside their communities.

The 1941 All-American Turkey Show, fortunately, enjoyed a generous amount of time on the airwaves. KFJM, the University of North Dakota radio station, aired a broadcast from the showroom floor every day at 11:45 a.m. and again at 5:30 p.m., under the auspices of the Hubbard Milling Company of Mankato, Minnesota, and the Archer-Daniels-Midland Company of Minneapolis, Minnesota. Station WDAZ of Fargo, North Dakota, the NBC affiliate, staged a one-half hour broadcast Thursday, January 23.[13]

More than a dozen commercial firms had booths at the eighteenth All-American Turkey Show, attesting to the show's reputation among businesses serving the poultry industry.[14] Dr. Salsbury's Laboratories of Charles City, Iowa, displayed a transparent hen, depicting the bird's internal organs. Consolidated Products Company of Danville, Illinois, manufacturers of Semi-solid Buttermilk feeds, featured what was described as a "giant brilliant yellow revolving barrel as the eye-catcher" in its exhibit. The North Dakota State Mill and Elevator, a long-time supporter of the All-American Turkey Show, always attracted attention with its Dakota Maid products.

Offering services rather than products, the Greater North Dakota Association—the statewide Chamber of Commerce—and the North Dakota Poultry Improvement Board, both of Bismarck, North Dakota, had attractive displays. The Railway Express Agency, St. Paul, Minnesota, returned as usual with exhibits and personnel who cooperated with show organizers to assure that valuable turkeys moved into and out of the City Auditorium efficiently. Turkeys were promptly delivered to the auditorium from all incoming trains and, when the show concluded, the Railway Express Agency established a branch office in the show room for billing

[13] C. Dyke Page, All-American Turkey Show president, acting as Master of Ceremonies, introduced the show's judges—Harry M. Lamm, of Adams Center, New York; P. M. Pierce, of Denver, Colorado; and M. C. Herner of Winnipeg, Manitoba. Page also introduced a number of exhibitors and other notables. Music was furnished by the Centralians, singers from Grand Forks Central High School.

[14] Northwood Grain and Seed Company and Northwood Hatchery, both of Northwood, North Dakota, exhibited their products. The Archer-Daniels-Midland Company, Minneapolis, Minnesota, displayed the company's wheat germ oil. Hubbard Milling Company, Mankato, Minnesota, was a regular at the All-American Turkey Show with its display of Sunshine feeds. Purina Mills of St. Louis, Missouri, reqularly supported the show, and Vic Cranley, its representative, was noted for the cold watermelons he furnished for the Hen Club Picnic in Riverside Park every summer.

and weighing the turkeys that were then taken directly and promptly to trains for transport to buyers or exhibitors' farms.

Among the 1941 All-American Turkey Show's highlights was a sumptuous turkey dinner held in the office building of the North Dakota State Mill and Elevator at 1:00 p.m., Thursday, January 24, at the invitation of C. A. Harvey, the mill's feed department superintendent. About eighty exhibitors attended this event. Men were treated to cigars and cigarettes, and women were presented with white canvas aprons.

The State Mill and Elevator, "makers of fine Dakota Maid feed," also raised turkeys in its experimental flocks at the mill. Some of the finest birds were dressed and served at the dinner, together with all the trimmings, and everyone expressed their pleasure when telling about the fine roast turkey they had been served. The food was so delicious, the guests found it difficult to decide which dish was best. The Jell-O salad, however, won favor, with its unique dressing made with sugar, flour, butter, and fresh lemon. Guests requested the recipe, which the cook shared.

The *Turkey World Journal* was represented by Harry Yoder, combination advertising manager, subscription agent, editorial expert, and photographer. Yoder commented on the eighteenth All-American Show's promising features. Deb: the paragraph continues with "This report and ends with ever attended "This report," he noted, was "being written in the quiet of my office late at night. I'm alone and away from the hustle and bustle of the great All-American Turkey Show that has just drawn to a close at Grand Forks, N. Dak. I can look back tonight, and after due and calm thought, truthfully say that it was the greatest and the finest All-American I have ever attended."[15]

Although not the largest of the All-American Turkey Shows, he admitted, anyone who had spent a week at the 1941 show would agree that "there are finer things than mere size alone." Yoder referred especially "to competent management, to friendliness, to keen competition, plus fair play, and above all a spirit of welcome to one and all that met you at the door on opening day and went with you all week on every occasion and

[15] *Turkey World*, February 1941.

penetrated every meeting and social function. That's what makes this show so different from all others!"

Yoder commented that when he had arrived for the 1941 show on opening day, he had been "warmly greeted and extended every courtesy" by show officials. It had been "a treat," he wrote, "to see how the cooping in the live department had been rearranged to provide more commercial exhibit space on the main floor, and how the large stage had been utilized to accommodate more birds and actually improve the general appearance of the show and how the new arrangement worked like clock-work with the judges and their many assistants."

Judge M. C. Herner, of Winnipeg, Manitoba, Yoder noted, had worked closely with Judge Harry M. Lamm, president of the American Poultry Association. The judges at the All-American Turkey Show, as show manager Frank E. Moore, so aptly put it, were "supreme court judges," and "their combined work and judgement" was "needed and necessary in a show of birds that in many, many cases" had previously been exhibited at other leading shows.

Yoder congratulated Mr. and Mrs. Frank Ralston of Crystal, North Dakota, for the many awards they had received at the 1941 All-American Turkey Show, including the Master Breeder Award, which they "so justly deserved" for the years they had been "loyal, hard-working turkey folks" and All-American Turkey Show exhibitors. The Ralstons' adult Bronze tom was the show's Reserve Champion, as well as its Champion Bronze. Their young Bronze tom and hen were selected as Champion tom and hen. The Ralstons also received the Turkey World silver trophy.

The 1941 All-American Turkey Show's Grand Champion, as Yoder described it, "was a striking Narragansett yearling tom shown by the well-known turkey people, Mr. and Mrs. Albert Payne, Towner, N. Dak." The tom "was a champion in every sense of the word and a credit to the popular Narraganset breed. He was also champion Narragansett." Mr. and Mrs. Otto Thieke, Beardsley, Minnesota, took Champion Bourbon Red with their adult tom, "as well as many other honors in the live and dressed divisions."

Herbie Olson, Bergen, North Dakota, exhibited "one of the most beautiful adult White Holland hens" Yoder had ever seen. The hen was named Champion White Holland. Mrs. W. T. Hall, of Denton, Texas, exhibited the Champion Black, and Jim Martinson and Son, Kensington, Minnesota, showed the Champion Slate. The champions of each variety and the Grand Champion and the Reserve Champion, Yoder wrote, "occupied coops in the famous court of honor."

The turkeys in the dressed division, Yoder believed, made the division "one of the very finest *quality shows* ever staged at the All-American." Judge P. M. Pierce, assisted by J. A. Amble and Allen Nixon, "had the entire dressed exhibit in its usual good form. It was well laid out and proved to be most popular with show visitors." The division's Grand Champion, a Broad Breasted young hen weighing 17½ pounds, was exhibited by Mr. and Mrs. Art L. Hamilton of Chehalis, Oregon. They had also shown the Grand Champion in the 1940 dressed division. "There's no doubt about it," Yoder wrote, "[T]he Hamiltons just know how to produce champions, as is proven by their consistent winnings at many major shows."

The dressed division's Reserve Champion, a "most refined specimen," was shown by Mr. and Mrs. John Allen of Radium, Minnesota. Yoder congratulated the Allens and commented that he was looking forward to future shows in which, he was certain, the Allens would "win even higher honors."

On Friday afternoon, in Yoder's words, "some mighty fine birds" from the dressed division were auctioned for "uniformly good prices" by Martin Potratz of Dr. Salsbury's Laboratories. The Hamilton Grand Champion was purchased by M. M. Oppegard, *Grand Forks Herald* publisher, for $1.75 per pound. Oppegard's paper, Yoder believed, with its "columns of publicity and news," was "responsible in a large manner for the continuous success of the show."[16]

"No report of the All-American," Yoder acknowledged with good reason, "would be complete without a brief mention of the many and varied social activities that often start with breakfast and on occasion ended about the same time!" He continued, "On Tuesday evening, a get-together

[16] The Hubbard Milling Company, Mankato, Minnesota, paid $1.00 per pound for the Reserve Champion, and the same company purchased the Champion Bourbon Red. W. G. Micklin, manager of the Hotel Ryan, was the high bidder for the Champion Narragansett, and the Red River National Bank bought the Champion White Holland. The First National Bank purchased the Champion Black.

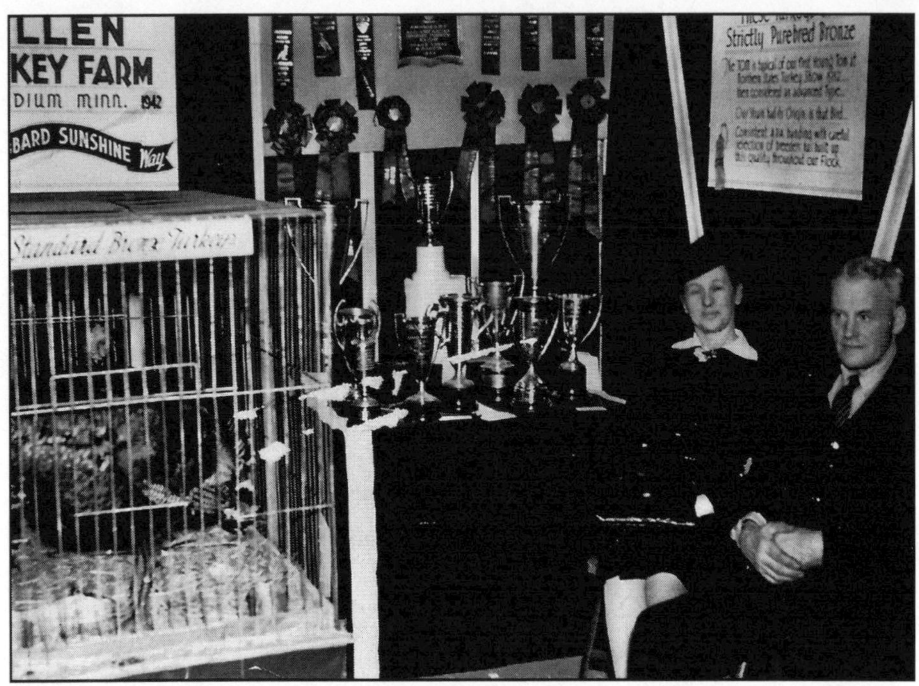

John and Amanda Allen of Radium, Minnesota, at the 1942 All-American Turkey Show with a display of their trophies and a pen of their Standard Bronze turkeys. Image courtesy of Jane and Richard Peterson, Cannon Falls, Minnesota.

luncheon was held at the Hotel Dacotah. Here various breeders, judges and commercial folks were introduced and a social evening followed."

During the week, Turkey Club members held their meetings, and on Wednesday evening, January 22, the Hen Club members held their annual banquet at the Hotel Ryan. Yoder observed that, as usual, "the ladies had a fine time and an opportunity to renew acquaintances and make new friends."

On Thursday evening, January 23, Yoder attended "the grand homecoming banquet" in the Hotel Dacotah's Garden Room. About two hundred people attended the dinner, at which C. Dyke Page, *American Turkey Journal* publisher and All-American Turkey Show president, presided.

Page, wrote Yoder, "ran the program with his usual technique" that always included "a few laughs and lots of fun for all." Yoder was "greatly enjoying" Page's calling on various guests to sing verses of songs, until, "out of a clear sky," Page invited him to sing a verse. Yoder admitted that he "couldn't carry a tune in a bushel basket," and that is where "the other 199 people present enjoyed themselves!"

An orchestra from the University of North Dakota, directed by John E. Holland, provided dinner music, and Bill Pond favored the group with a cornet solo. All of the judges gave short talks, and other visitors and show officials made brief remarks. The talks were interspersed with entertaining numbers by Central High School students.

Yoder reported that Judge George W. Hackett received "a warm welcome from his many friends," and Frank Moore, show manager, made the major awards "in a most fitting and appropriate manner." Moore pledged to conduct future shows "to the best of his ability," and those who knew him did not doubt that he would keep his word, and, with Moore as manager, they looked forward "to the finest shows in the history of the industry."

Following the banquet and the program, "a splendid 5-piece orchestra" provided dance music that pleased everyone's tastes. "Old time waltzes were favorites, but there was apparently no dance too fast or too new for these fine fun-loving turkey folks to enjoy." William Page, Grand Forks County Agent, attended the banquet with his wife. They danced until after 11:00 p.m., and, Page observed, "most of the crowd spent much time in [the] saloon."

"As the dance ended," Yoder wrote, "so must this brief summary of the eighteenth All-American. If my comments have given . . . a picture of a splendid show, highly educational, combined with friendly social activities, I shall feel my job is well done until another year rolls around. Yes, I'm planning on seeing all of you again in 1942."

Other features of the 1941 All-American Turkey Show, most of them new, were not so much promising as they were predictive. The features all reflected the changes occurring in the turkey industry. They all reflected that the All-American Turkey Show had succeeded in encouraging flock owners to improve the quality of their turkeys and to produce more of them. The features all reflected that the All-American Turkey Show had succeeded in encouraging people to eat more turkey and that consumers' tastes were changing.

In sum, the features all reflected that the All-American Turkey Show may have succeeded too well in accomplishing the purposes for which it was intended, and that it stood to be done in by its success.

For long, a criticism of judging by *Standard* was that they placed too much emphasis on plumage color, pattern, and shading. Implemented for the first time at the 1941 All-American Turkey Show, the new *Standard* reduced the number of points for feathers from forty to thirty-two and added eight points to a turkey's market value, that is, to meat and weight. (Two turkeys exhibited at the 1941 All-American Turkey Show weighed over forty pounds.)

Perhaps in response to the new judging Standard, manager Frank Moore introduced another innovation at the 1941 All-American Turkey Show. In previous shows, most birds in each class had been appraised and placed by a single judge. Moore's innovation was that the birds in each class were appraised and placed by two, sometimes three, judges.

The Canned Turkey exhibits had become one of the most popular features of the All-American Turkey Show, and, in 1941, exhibitors in this division were required to follow a set of rules recently established by the Home Economics Department of the North Dakota Agricultural College. An exhibit had to consist of four pint or four quart jars of turkey, and the Home Economics Department advised exhibitors to can the turkey using the pressure cooker method.[17]

The Special Single Pack division at the 1941 All-American Turkey Show drew seventeen entries. A Special Single Pack was a dressed turkey attractively packed in a handsome box and suitable to give as a gift. Commercial Box Packs, new to the dressed division in the 1941 All-American Turkey Show, created "unusual interest," an interest, Yoder believed, that would "undoubtedly . . . grow rapidly in years to come." A Commercial Box Pack contained six to eight dressed turkeys attractively packaged for

---

[17] The turkey had to be canned by the exhibitor; the canning method had to be stated on the label; four ways of preparing the jar's contents for the table had to be described on the label; no jar could be entered in both single and collection exhibits; no jar could be opened for judging; and no jar would be accepted for judging unless it had the possibility of winning an award.

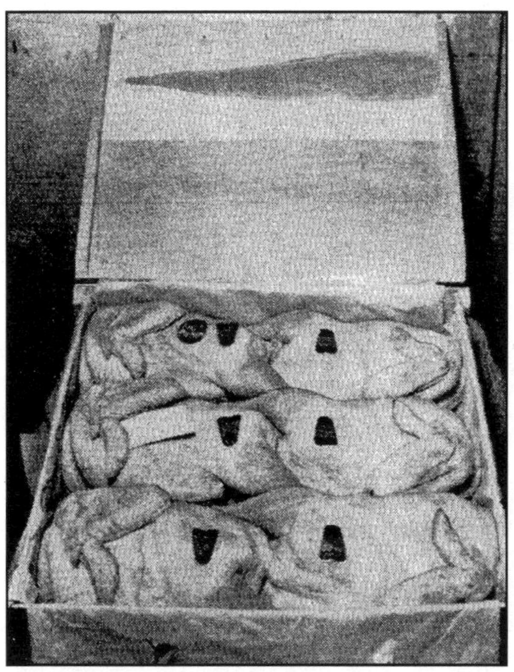

The Commercial Box Pack competition was an All-American Turkey Show innovation. Shown here is the Grand Champion Commercial Box Pack of the 1942 show, exhibited by Armour Creameries, Fargo, North Dakota. *American Turkey Journal*, February 1942.

the purpose of demonstrating "how to make the most out of turkey shipments in the line of merchandising appeal."[18]

The Commercial Box Pack division drew fifteen entries with a total of ninety turkeys. Exhibitors included Cudahy Packing Company; Armour Packing Company; Fairmont Creameries; Land O'Lakes Creameries; Sunnyside Farms of North Platte, Nebraska; Larn Gates of Aurora, Kansas; and A. W. Hoffman of Aitkin, Minnesota.

The Cudahy Packing Company's pack of six young hens was the Commercial Box Pack Grand Champion and also the best young hen Commercial Box Pack. Fairmont Creameries' Commercial Box Pack was the young tom champion. All Commercial Box Packs scored above ninety points in judging. In Harry Yoder's opinion, the Commercial Box Pack division "was a splendid educational feature," and he advised breeders to take note of it.

[18] There were three categories in Class I: Young Hen Pack: 8-10 pounds, 10-12 pounds, and over 12 pounds. Class II, Young Tom Pack, had two categories: 16-18 pounds and 18 pounds and over.

When assessing the prospects for the All-American Turkey Show and Standard turkey varieties, the numbers of birds entered in the 1941 All-American Turkey Show were telling. There were 335 entries in the live division and 273 entries in the dressed division. Of the 273 entries in the dressed division, however, 15 were Commercial Box Packs containing 90 birds, bringing the number of turkeys in the dressed division to 363. At the eighteenth All-American Turkey Show, dressed birds outnumbered live birds by a significant margin, and dressed turkeys do not have feathers.

The eighteenth All-American Turkey Show's most controversial feature was in the live division. "For the first time in the history of this show," noted Harry Yoder in the February 1941 issue of *Turkey World*, "a class was created for live Broad Breasted Bronze." Those with exhibits in this class represented six states and, in Yoder's estimation, they were "some of the finest and best-known Broad Breast breeders in America today! It was a quality class and one of which any show could be most proud."[19]

Harry Lamm of Adams, New York, president of the American Poultry Association, served as judge at the 1941 All-American Turkey Show, but he also attended the show in order to observe the reaction of exhibitors to the new variety. Breeders "of this strain of turkeys," Lamm remarked in his featured address, "had something that other breeders so far have not accomplished." They were producing what consumers wanted in a turkey: more white meat.

Lamm advised, however, that the term "bronze" should not be used when designating the new variety because "at the present time," the variety lacked "uniformity as to feather markings." The American Poultry Association set the standards by which turkeys were judged and, he believed, the Association would "naturally go slow" about setting standards for the new variety until it had become "more thoroughly established."

Acknowledging the controversy between breeders of Standard turkeys and those breeding the Broad Breasted variety, Lamm observed that even breeders of Standard turkeys had developed types with broader breasts and more compact bodies than were specified in the *Standard of*

[19] The Broad Breasted variety, a new type of meat turkey in which plumage and coloring had not yet been standardized, had been developed in the western part of the United States by pioneer raisers such as Dave Cooper of Mt. Hood, Oregon, and A. L. Hamilton of Chehalis, Washington.

Perfection. He also pointed out that the new scoring system for turkeys called for more emphasis on marketability and conformation than before and less emphasis on plumage. Perhaps, he said, the controversy between Standard and Broad Breast breeders could be dealt with by breeders of Standard varieties "taking over the new type" by breeding larger and meatier turkeys of the Standard varieties.

Partial to Bronze turkeys and an authority on their coloration, Hackett offered critical comments on the recently developed Broad Breasted Bronze turkey. Writing in the *American Turkey Journal,* he commented that breeders of the Broad Breasted Bronze variety had developed the variety without regard to any definite standard or even any agreement on specifications. Breeders had bred only for size and marketability, not for conformity or color. As a result, there was no uniformity to the turkeys produced by different breeders.[20]

Broad Breasted Bronze turkeys should be judged, Hackett insisted, as all turkeys of established varieties were judged—by type and conformation—and there had to be agreement on color and description so that the birds could meet a standard and compete with other Standard varieties. Breeders developed turkey varieties by cooperating with nature, through careful selection of breeding stock for a definite purpose.

When any turkey variety was perfected to the point where it would reproduce a large percentage of offspring reasonably true to the type and color desired and agreed on by breeders, those setting standards had to fit a standard to the variety. It was contrary to the laws of nature and the desires of the breeders, Hackett pointed out, to fit the turkey into an existing standard.

Breeders of the Broad Breasted variety of Bronze, Hackett wrote in an earlier issue of the *American Turkey Journal,* had appropriated the name Broad Breasted Bronze for their birds. They could not in fairness, however, do so, because American Poultry Association regulations did not allow any poultry variety to be divided under one name. The word Bronze was specific to a single turkey variety, and it could not be used in designating another variety. Hackett admitted that it was of "untold advantage to this new turkey" to use the Bronze name with the "broad breast" prefix, there-

[20] *American Turkey Journal,* June 1941.

by "trading on the prestige of the most advertised and most popular of all varieties," but "the ethics and fairness" were other issues.[21]

The only recognized breeds of poultry in the United States and Canada, Hackett pointed out, were those that qualified and had been approved by the American Poultry Association through its nearly seventy years of service to agriculture. All the Standards that had been developed were the work of "far-sighted men of brains and integrity."[22]

Those breeding broad-breasted turkeys wanted their variety included in the *Standard of Perfection*, and they wanted to compete in turkey shows and compete for awards. It was their right to do so, Hackett admitted, but they had to adopt a name for their variety different from Broad Breasted Bronze. He offered a number of suggestions such as Cornish, Throssel, Oregon, and Washorgon.[23]

At the 1941 All-American Turkey Show, fifty turkey breeders—members of the All-American Bronze Turkey Club and members of the International Narragansett Turkey Club—unanimously approved a Resolution protesting the use of the "usurped" name of Bronze, with the prefix "broad breast," to designate the new turkey variety. The usurping of this name, according to the petition to the American Poultry Association, was "obviously detrimental to the well-established standard Bronze Turkey and to all other standard turkeys."[24]

---

[21] *American Turkey Journal*, March 1941.

[22] It was the right and the duty of those promoting the new turkey variety to propose and secure standards satisfactory to those interested in the new variety. But, this had to be done consistent with existing Standards and in conformity with the recognized process of Standard making. Surely, Hackett wrote, there was some color scheme that was consistent and desirable for the broad-breasted variety, but this was up to those breeding them to agree on.

[23] *Cornish*, because the birds somewhat resembled the Cornish fowl and the foundation stock was reported to have come from England. *Throssel*, to honor Jesse Throssel, the man who had first produced the broad-breasted turkey variety. *Oregon*, because broad-breasted birds were developed in the state after being introduced from British Columbia, Canada. *Washorgon*, to accord to the states of Washington and Oregon the credit for developing the new variety.

[24] "Therefore, Be It Resolved," stated the petition, that the name Broad Breasted Bronze be "disapproved, discouraged and not accepted" for these reasons: a) The new turkey variety did not qualify as a Bronze turkey under the specifications given in the *Standard of Perfection*, and the term "broad breast" was a term used by the *Standard* to describe shape for all standard turkeys. The term, therefore, could not properly be used or accepted as part of the name of a turkey variety. b) The new turkey was a distinct variety and should be given its own appropriate color description and weights, consistent with other turkey varieties, and breeders should follow the regular procedure prescribed by the American Poultry Association for admission to the *Standard*. In the best interests of the turkey industry, the American Poultry Association should not depart from its current plan of standards whereby a single scale of points and a single shape description applied to all varieties, thereby making judging consistent.

The Resolution, petitioners acknowledged, was not offered in criticism of or antagonism to any variety, but in a spirit of fairness to all varieties and breeders, "to preserve the beauty and utility with which the turkey is particularly endowed and to preserve the achievements of master breeders past and present."

George W. Hackett, nationally recognized authority on turkeys and with years of experience in all aspects of the turkey industry, was uniquely qualified to assess the Broad Breasted Bronze issue as it had developed by 1941. Editor of the *American Turkey Journal*, he used the journal as his forum. In the journal, he reported on his experience at the Ohio State Fair the previous August. For years, the Ohio State Fair had led all others in the country in numbers of turkeys and exhibitors.[25]

In 1941, only 150 turkeys were exhibited, slightly more than one-half the usual number, so few that adult and yearling birds were combined in a single class. Bronze entries "fell flat." Only six Bronze turkeys were exhibited. Numbers were reduced in part because of the labor shortage, but also because many breeders who had once exhibited large numbers of birds had disposed of their flocks for "business reasons."

"Talking with some of the Bronze breeders," Hackett wrote, "it was frankly stated that the current propaganda for the new western turkey had severely cut sales of Bronze breeding stock to the point where advertising through showing was not profitable." Propaganda, he admitted, was a powerful factor in the turkey industry as it was in war, and "just what the eventual result" would be was "unpredictable."

The trend of the times, Hackett acknowledged, was "unquestionably" toward the best type of carcass of the right size that could be produced economically, but Standard color, he knew, need not be sacrificed in order to achieve it. By their own admission, breeders of "the new western" turkey cared "nothing for color" nor for the beauty and uniformity of a turkey flock. With a singleness of purpose, they had developed a bird with more meat.

The same was true of those "western breeders" who produced hatching eggs in order to provide large numbers of poults to "big production"

[25] *American Turkey Journal*, September 1941.

turkey raisers. Although the turkeys were lacking in uniformity of color and evenness of type, the "mass productionists," Hackett's term, were satisfied with the new variety as it was.

Hackett acknowledged that the broad-breasted turkey was gaining in popularity. Because there was no way of knowing, however, whether the popularity would last, he advised that "all the better breeding flocks of standard turkeys should be preserved and no recession permitted on *Standard* color." It would be a "tragedy," he warned, to "throw away" the results of careful breeding that had established the present Standard turkey varieties in preference to anything that had "yet or is likely to appear."[26]

Sobering national and international events and for-the-first-time features of the 1941 All-American Turkey Show aside, judges Harry M. Lamm, P. M. Pierce, and M. C. Herner were alike in commenting on "the remarkable high quality" of the turkeys exhibited. Show officials, manager Frank E. Moore, president C. Dyke Page, and secretary W. W. Blaine, agreed that the eighteenth All-American Turkey Show had been "one of the most successful ever held here." Hackett had "nothing but praise" for Frank E. Moore, who had succeeded him as manager, and, in his opinion, the eighteenth had been "one of the best in the long history of the All-American."

[26] *American Turkey Journal*, April 1941.

CHAPTER 16

The Last Turkey Show

(with apologies to Larry McMurtry, *The Last Picture Show*)

ACCORDING TO THE *American Turkey Journal*, thirteen states and three Canadian provinces "defied Emperor Hirohito and all-out war to help stage a great 1942 All-American." Scheduled for January 12–16, 1942, the nineteenth was the first All-American Turkey Show held during a war, and the show's management was prepared for almost anything to happen, "all of it bad."[1]

At the close of the show on Friday, January 16, 1942, manager C. Dyke Page commented that the nineteenth had been "one of the finest shows we have ever had," and he was "very gratified" that this had been possible even though people were "more interested in the war than in anything else." With Japan's attack on Pearl Harbor on December 7, 1941, the United States was at war, as was evident in the features of the *Grand Forks Herald* sports pages, comics, fashion news, beauty columns, and ads.[2] An Associated Press release warned brides that there would

---

[1] *American Turkey Journal*, February 1942.

[2] On Monday, January 12, 1942, opening day of the All-American Turkey Show, a column was headed, "Joe Passes Final Exam for Army." Boxer Joe Louis, the heavyweight champion, had undergone one and one-half hours of physical tests for induction into the army at Fort Jay, Governor's Island, "along with some 400 other selectees." On Wednesday, January 14, he took the final step at Camp Upton, Long Island, the step that changed him to Private Joe Louis Barrow. The Gasoline Alley comic strip had already enlisted, with its characters serving as recruiting agents. "Gosh, Tops," Skeezix gushed when he greeted his friend, home on leave and in an army uniform, "you look swell! Army life agrees with you." Tops replied that he was "having a swell go of it, an' learnin' things too." Skeezix said he was glad and that he was nearly due for service himself. "Choose well your Easter bonnet," women were advised, "because it may have to last a long, long time." Hat manufacturers reported that because the major part of their supplies of straw braid had been shut off because of the war in the Pacific, "the straw hat won't survive another season." Patricia Lindsay's "Beauty and You" column included a picture of a young woman's hands holding cribbage cards above the caption: "Games played at home have become popular for the duration. Keep your hands as lovely as possible to add to general morale."

soon be "no more dinners out of tin cans," because the country had to conserve its "dwindling tin supplies." "If Mrs. Newlywed" was smart, according to the AP release, she would "rustle up a cook book and start studying" because the Office of Price Administration had "an order in the works, prohibiting the use of tin for packaging virtually everything that can be marketed satisfactorily in some other type of container or in no container at all."[3]

At the new year's beginning, the *Grand Forks Herald* informed readers of several measures the federal government was taking in the interests of national defense. To create a large land force, the army induction rate was expected to triple, and many once-deferred men would be reclassified and considered fit for service. "The immediate goal was a hard-hitting land force of four million men" that would be "sent to Britain and anywhere else the high command deemed advisable."[4]

Sugar rationing, the article advised, was being considered in order to forestall "an unwarranted hoarding epidemic." The United States House of Representatives had approved a daylight-saving bill providing a "uniform one-hour advance of the nation's clocks." The Senate version of the bill approved a two-hour advance. Once enacted, the measure would remain in effect until six months after the end of the war. Motorists were advised to save on tires. Most of the country's rubber came from the Dutch East Indies and Malaya and, with the war, rubber supplies had been cut off. Motorists had one of two choices—take care of the tires on their vehicles or put their cars away for the duration.

---

[3]Included in the OPM's order would be such "old reliables" as pork and beans, spaghetti, and canned meat. Beer and dogfood would also be affected. Dogfood output alone averaged 650,000,000 one-pound cans per year.

In a quarter-page ad, Butter-Nut coffee drinkers were urged to "save tin" and "Help National Defense" by buying Butter-Nut coffee in usable one-quart glass jars. Regular Mason lids fit the jars and the jars could be used to can summer produce or they could be donated to charitable organizations.

The R. J. Reynolds Tobacco Company's ad for Camel cigarettes included an attractive woman holding a carton of Camel cigarettes with a "To" shipping label prominently displayed. "SEND HIM A CARTON OF CAMELS," readers were urged, "Your dealer has a special wrapping and mailing service to save you time and trouble." An inset showed three servicemen—Army, Navy, and Marine—all smiling and holding cigarettes in their lips. According to the ad, "Actual sales records in Post Exchanges, Sales Commissaries, Ship's Stores, Ship's Service Stores, and Canteens show the favorite cigarette with men in the service is Camel."

[4] *Grand Forks Herald*, January 9, 1942.

Accounts of the war dominated the pages of the *Herald*, and those from the Pacific War Zone were grim. One January headline was "Japs Attack East Indies." Front-page columns had headings such as "Japs Renew Attack in Philippines," "U.S. Dutch Fliers Bomb Ships in Indies," and "Gain First Footholds as Dutch Battle Furiously."[5]

"Japan's all-out gamble for a quick sweeping victory before the United States can take the offensive," according to an Associated Press release, "carried her troops dangerously closer to Singapore Monday, touched off a preparatory artillery battle along the entire Philippine front and stirred up a hornet's nest of resistance in the Netherlands East Indies. The outcome of the land, sea and air battles for vital outposts in the Dutch Archipelago was still in the balance; the fighting was mounting in bloody, destructive fury. For assistance, the Dutch had United States warships and American and Australian planes."[6]

Among the states and Canadian provinces defying Emperor Hirohito to make the nineteenth All-American Turkey Show another success were California, New Hampshire, Washington, Missouri, Minnesota, North Dakota, South Dakota, Nebraska, Iowa, Wisconsin, Montana, Oregon, Colorado, Virginia, Alberta, and Saskatchewan. "When the smoke had cleared away," noted the *American Turkey Journal* in the next month,

---

[5] *Grand Forks Herald,* January 12, 1942.

[6] An Associated Press release on January 14, 1942, reported that General Sir Archibald P. Wavell, Commander-in-Chief of the new allied Far East command, had arrived in the Dutch East Indies and had launched a series of "dynamic counter-blows against Japanese invasion forces." Dutch troops were moving into action on the border of the "'White rajah' Kingdom of Sarawak, where Japanese troops had seized the capital, Kuching, and most of the northern territory." On the Philippine Front, General Douglas MacArthur's American and Filipino defense forces had beaten off two new Japanese attacks, inflicting "heavy losses." MacArthur's troops had "suffered only comparatively small casualties."

For those following the war in the Pacific Theater and dependent on the region for their supplies and raw materials, the headline in the January 19, 1942, issue of the *Grand Forks Herald* was particularly alarming, "JAPS GAIN IN MALAYA." The paper's lead column was headed, "British Facing Bitter Fight to Hold Singapore."

The Australian Imperial forces, according to Clyde A. Farnsworth, Associated Press War Editor, had "turned back waves of Japanese attackers in the fierce defense of Singapore, Fulcrum of the United Nations 'defense system'" in the region. "With the Japanese still advancing," reported Farnsworth, "Malaya's steaming jungle land, which the British had looked upon as a sort of natural Maginot line for Singapore, had taken its place as another outworn defensive concept and the battle for Singapore Monday became a test of fighting men and their weapons. Some of the fiercest hand-to-hand combat in the history of war was in store, if not already in progress, along the narrowing, fluid line of defense in Johorestat, well within 100 miles of the pivotal naval base which the British call their Gibralter of the Orient." Within days, having breached the "jungle land Maginot line," Japanese forces were within fifty miles of Singapore.

entries were received from a territory stretching from coast to coast and border to border. The entries were "a symbol of these stirring war times that this great group of breeders were prepared to carry on and do their part in America's war effort, no matter what dark days might be ahead." Given that entries were received from all parts of the country, competition was expected "to be keen in most of the classes."[7]

When the civic auditorium's doors were thrown open to the public on the morning of January 12, 1942, 315 turkeys "of proud lineage and individual perfection strutted and preened their feathers in the coops of the nineteenth annual All-American Turkey Show."[8]

Challenged by the controversial Broad Breasts, the Standard varieties had their staunch advocates. "Again," reported Mrs. W. J. Janda, secretary of the Bronze Club, in the *American Turkey Journal*, "our beautiful Standard Bronze were tops." The "handsome Bronze yearling tom" exhibited by John O. Allen was "one of the most outstanding Grand Champions" she had ever seen. Not only was he "marked beautifully, but his body type and conformation were superb." The Risbrudts of Dalton, Minnesota, won the "much coveted Master Breeders Award" with their "unusually splendid entry of Standard Bronze" that were also "of the best market type."[9]

Showing the quality that had been developed over the nearly twenty years of the All-American Turkey Show, the White Holland yearling hen of Herbie Olson of Bergen, North Dakota, was "an unusually high quality specimen." The Champion young tom, shown by Mr. and Mrs. Albert Payne, Towner, North Dakota, was another of their "quality Narragansetts." The Champion young hen, exhibited by Mrs. Anna Brackett, Richland, Missouri, was a "beautiful Bourbon Red."

---

[7] *American Turkey Journal*, February 1942.

[8] Numbered among the 315 turkeys were 96 Bronze, 83 Narragansett, 36 White Holland, 26 Black, 23 Bourbon Red, 6 Slate, and 45 Broad Breast. With 14 Blacks and 12 Narragansetts, Oakdale Turkey Farms, Kensington, Minnesota, had the largest entry. J. J. Olson, Beltrami, Minnesota, entered 21 Narranganisetts, and Mr. and Mrs. Alfred Payne, Towner, North Dakota, 16. Herbie Olson, Bergen, North Dakota, exhibited 15 White Hollands, Mrs. William Eddie, Northwood, North Dakota, 12 Narranganisetts, and Alfred Malmberg, Crookston, Minnesota, 12 Bronze. William Kemp, Mohall, North Dakota, entered 24 Broad Breasts and Alvin Eastvold, Mayville, North Dakota, 2.

[9] *American Turkey Journal*, February 1942.

Harry Yoder noted in *Turkey World* that the show's Champion Broad Breast was "a striking yearling tom" exhibited by M. M. Lyons of Triple-B Turkey Farm, Portland, Oregon. Lyons, "one of the older and better-known Broad Breast breeders on the West Coast," said that having his tom chosen as the show's Champion Broad Breast "was well worth the trip of several thousand miles."[10]

A pair of young turkeys, a tom and a hen, exhibited by Judge M. C. Herner of Winnipeg, Manitoba, generated "considerable comment among the experts" at the 1942 All-American Turkey Show. The birds, called McHerners, were crossbreeds. Herner bred the McHerners from five Standard turkey varieties to produce a suitable small turkey for the market.

George W. Hackett, Wayzata, Minnesota, headed the judging staff in the live division at the 1942 All-American Turkey Show. Hackett had judged at the show for many years before becoming manager, "and his work at the tables was a pleasure to watch because he knew what he was looking for." Hackett was assisted at the judging tables by M. C. Herner and Cleve Angen, Portland, Oregon.

Professor O. A. Stevens, of the North Dakota Agricultural College, judged the 407 entries in the dressed division, 267 individual dressed birds and 140 birds in Commercial Packs. It was telling that at the 1942 All-American Turkey Show the entries in the dressed division handily outnumbered the entries in the live division. The sixteen Commercial Packs, exhibited by seven well-known packers, for Harry Yoder were a show in themselves.

"Professor O. A. [Stevens] (and his cigar)," commented Yoder in *Turkey World*, "were on hand early and late!"[11] Stevens "very interestingly" described "conformation," "body weight," "type," "bloom," "finish," and "dozens of other terms that all go into the make-up of a prize-winning dressed bird." There was "no question," Yoder affirmed, that Professor Stevens knew dressed turkeys, and he also knew "how to place the awards."

---

[10] *Turkey World,* February 1942.

11 *Turkey World,* February 1942.

Live Entries Close
JANUARY 5
Dressed Entries Close
JANUARY 12
1 P. M.

ENTRY BLANK

ALL-AMERICAN TURKEY SHOW

Grand Forks, North Dakota

January 12-16, 1942

ENTRY NO.

For Secretary's Use Only

NAME OF EXHIBITOR..........

STREET OR ROUTE.......... POST OFFICE..........

EXPRESS OFFICE.......... STATE..........

I make the following entries, subject to the rules published in your Premium List:
Enclosed find remittance for live bird entries (entry fee $2.00 per bird, $4.00 for pen)..........$..........

Enclosed find remittance for dressed bird entries (Fee 50c per bird). If you wish to omit entry fee on dressed birds herewith, we will deduct same from receipts from sale of your birds....$..........

TOTAL $..........

Drafts, Money Orders and Checks should be made payable to W. W. BLAIN, Treasurer.

ENTRY FEE $2.00 PER BIRD
ENTRY FEE $4.00 PER PEN

Address Entries and Communications to
W. W. BLAIN, Secretary, Grand Forks, North Dakota

This Column for Secretary's Use Only	Check In	Check Out	All-American Seal Band	NAME OF TURKEYS Such as Bronze, etc., In the space below	Owner's Band No.	SEX Designate by X mark in proper column						Entry Fee
						Adult Tom	Y'rling Tom	Young Tom	Adult Hen	Y'rling Hen	Young Hen	

Breeders Display

Consists of 6 birds—one in each of single classes—4 of which must have been bred by exhibitor

Does entry qualify? (YES) (NO) No extra fee required Extra entry blanks on request

Pens

(Consists of 1 Tom and 2 Hens)

COOP	CHECK		BREED	Old or Young	EXHIBITOR'S BAND NOS.			Entry Fee
	In	Out			Tom	Hen	Hen	

Sale Prices and Band Numbers May Be Listed Here.

Dressed Turkey Department

ENTRY FEE 50c PER BIRD

This Column for Secretary's Use Only	Check In	Number of Birds	BREED OF TURKEYS	Class	Exhibitor's Weight When Shipped	Weight in at Show	Entry Fee
			Y. Toms				
			Y. Hens				
			Old Hens				
			Old Toms				

The Entry Blank for the 1942 All-American Turkey Show. The features are worth noting. Supplement to the *American Turkey Journal*, December 1941.

The Grand Champion dressed turkey was exhibited by R. H. Jandebeur of Sunny Slope Farm, North Platte, Nebraska. "This perfect specimen," as Yoder described, "a Broad Breast old hen, weighing 22 pounds was indeed worthy of this, the highest honor bestowed in this division." To win against almost three hundred birds from more than a dozen states and two Canadian provinces "should make any breeder most proud." The Reserve Champion was also a Broad Breast. "This fine specimen was a young hen weighing 18 pounds," entered by Mitchell Turkey Ranch, Marysville, Washington.[12]

Turkeys, live and dressed, were the primary features of the All-American Turkey Show, the reason for its being. By show procedures and long practice, turkey entries were registered, caged, exhibited, judged, and placed—all by variety and class. Other show features were informative, educational, supportive, and entertaining, but they were subsidiary to the turkeys.

Just as the *Grand Forks Herald* by January 1942 was publishing fewer pages in each issue—sometimes only ten, likely to support the country's defense efforts by conserving material—the nineteenth All-American Turkey Show had fewer features, perhaps because exhibitors and visitors were sobered by the realization that the country was engaged in an all-out global war that was likely to be of long duration. Accounts of the show in the *Herald* were few, brief, and to the point; show features, for the most part, were limited to those that were practical and educational.

No accounts of judges from southern states commenting on North Dakota's winter weather and their donning long woolen underwear to ward off frigid temperatures; no droll stories of prominent judges and exhibitors recounted by judges in their characteristic Texas drawls; no accounts of All-American Expresses transporting turkeys by truck to Grand Forks from Philip, South Dakota, in the dead of winter; no scandal sheets

12 In the Commercial Pack division, the Grand Champion was a box of heavy hens, 10–12 pounds, entered by Armour and Co. The company's young hen entry, 8–10 pounds, also placed first. Peterson Biddick Co. of Crookston, Minnesota, took second place in the young hen class, and Red River Produce Co. of Grand Forks, North Dakota, took third. Armour and Co. took first place in the young hen, heavy class, 10–12 pounds; Fairmont Creamery second place; and Cudahy Packing Co. third. Cudahy Packing Co. also took first place in the heavy tom class.

published by self-appointed investigative reporters who discovered nudist colonies under the judges' noses in the auditorium balcony; no brightly colored revolving barrels drawing attention to commercial exhibits; no exhibits showing in graphic detail the workings of laying hens' reproductive systems; and no gossipy accounts of properly behaving but dour tom turkeys.

An article in *Turkey World* described the 1942 All-American Turkey Show's commercial exhibits as numerous, interesting, educational, and "a prominent factor" in the show's success "as usual." The Railway Express Agency's attractive booth was manned for the entire week by a large number of company representatives.[13] Purina Mills of St. Louis, Missouri, had its usual booth and displays, with the "popular Victor J. Cranley in charge." Cranley supplied the cold watermelons for the Hen Club July picnics in Riverside Park. Boote's Hatcheries, Worthington, Minnesota, advertised its "fine 1942 crop of poults" and "took many orders." *Turkey World*, Mount Morris, Illinois, had a booth, with the "always welcome" Harry Yoder on hand to greet friends and "do anything to aid the show's success." Mrs. William Huggins was in charge of the *American Turkey Journal's* table.[14]

Two exhibits, neither of them commercial, were among the All-American Turkey Show's most popular attractions in 1942. Ida Bisek Prokop, of Prokop Studios, Lidgerwood, North Dakota, titled her exhibit *Dakota Prairie Pictures*, works of art that would grace the table of any turkey grower. Prokop incorporated turkey and North Dakota game bird feathers into pictures and in the designs on decorated trays and other

---

[13] *Turkey World,* February 1942.

[14] The North Dakota Mill and Elevator's booth, with manager C. A. Harvey in charge, featured a large display of Dakota Maid Feeds. Minnkota Power Cooperative of Grand Forks, North Dakota, had a display of electric farm equipment, including an ultraviolet water sterilizer. The Consolidated Products Company, Danville, Illinois, displayed its Semi-Solid Buttermilk Feeds, and Salsbury's Laboratories, Charles City, Iowa, showed its line of products. The Hubbard Milling Co., Mankato, Minnesota, had its usual booth, with the famous T. C. "Doc" Haney and L. A. Hanson, who "extolled the merits of Hubbard's Sunshine feeds" and staged a radio broadcast over Station KILO every noon during the week, "adding much to the show's publicity." Farm Owners Mutual Insurance Co. of St. Paul, Minnesota, had a representative on hand to explain the company's turkey insurance program that had been particularly successful in Minnesota. Grand Forks Glass and Paint Co. of Grand Forks, North Dakota, offered samples of the company's new poultry litter and shared the names of the area's distributors.

objects. Another impressive exhibit was a display of the awards won by turkeys from the Allen Turkey Farm in the country's leading shows, along with a pair of Bronze turkeys from the farm.

On Thursday afternoon, January 15, a thirty-minute program was broadcast from the showroom over thirty-two stations of the North Central Broadcasting System of the Mutual network. The program was arranged by Station KILO, and manager C. Dyke Page served as Master of Ceremonies. Page introduced the winner of the Master Breeder Award and the owners of the live and dressed bird Grand Champions. Hackett spoke for the show's judges.

Using a portable microphone, Page interviewed several exhibitors as he walked down the auditorium's aisles. As further described in *Turkey World*, Mrs. August Swenson, Gilby, North Dakota, "rendered her inimitable turkey calls" that met "with a hearty response from the attending turkeys." Music was provided by the Centralian Singers and a Quartette of Boy Sopranos from Grand Forks Central High School and by the University of North Dakota band, directed by John Howard.

Counted as strengths, the All-American Turkey Show enjoyed strong local support, and individuals from Grand Forks and the region served as officials and on the board of directors, some of them for many years. At the 1942 All-American Turkey Show Association business meeting, held during show week, C. Dyke Page, who had previously served as show president for ten years, accepted reappointment as manager for another year when Frank E. Moore resigned from his university position.[15]

Moore's new position would be coordinator for the National Poultry Improvement Plan (NPIP) that had become operative in 1935. The objective of the NPIP—a voluntary industry, state, and federal cooperative testing and certification program for poultry breeding flocks, baby chicks, poults, hatching eggs, hatcheries, and dealers—was to improve poultry and poultry products throughout the country.

---

[15] Page was named manager for the 1942 show when Frank Moore resigned from the North Dakota Agricultural College to accept an attractive offer from the Bureau of Animal Husbandry in the United States Department of Agriculture in Beltsville, Maryland, and Washington, D.C.

A director for several years and always "a good booster for the show," H. H. Herberger was elected president of the All-American Turkey Show for 1943. Engaged in the ready-to-wear business in Grand Forks, Herberger was "prominently connected" to all civic activities and his name was "a real addition to the All-American roster."

Although not a real turkey man, unless the turkey was on a platter, Herberger was reportedly a "live wire," and he would "put his shoulder to the wheel and add plenty of pep to the organization," something that was "always needed." Herberger and Page, it was believed, would "make a good team."[16]

Turkeys, exhibits, radio broadcasts, and business meetings were only part of what the All-American Turkey Show had become by 1942. It was, in Hackett's words, the "Annual Homecoming" of a "large family of turkey folks" who had common interests. As Yoder wrote in *Turkey World*, the All-American Turkey Show probably had more social affairs directly connected with it than any other show in the country.

This neighborly attitude was in evidence at the 1942 All-American Turkey Show, Hackett believed, and "rightly recognized" in the many informal luncheons and teas, in the Turkey Hen Nest (into which some men ventured), in Breed Club meetings, in the annual Hen Club banquet at the Hotel Ryan, and in the annual Homecoming banquet held in the Terrace Room of the Hotel Dacotah, followed by the annual dance. These, Hackett observed, were "but a few of the enjoyable social features of the All-American that have a wide appeal to turkey growers and those in allied industries."[17]

Hackett also described the 1942 All-American Turkey Show Homecoming Banquet as an "outstanding affair" with the largest attendance in

---

[16] Also a director for many years, John O. Allen, Radium, Minnesota, was elected vice-president of the All-American Turkey Show. Mr. and Mrs. John O. Allen were, "without question, outstanding breeders" and they had shown and won awards with their Bronze turkeys at the All-American Turkey Show for more than seventeen years. W. W. Blaine was elected Secretary-Treasurer, and George W. Hackett was appointed Senior Judge for the 1943 show. Of the fifteen elected to the Board of Directors, seven were from Grand Forks and five were from Minnesota. South Dakota, North Dakota, and Manitoba were each represented by one director.

[17] *Turkey World*, February 1942.

many years. C. Dyke Page, toastmaster, introduced a number of dignitaries and judges.[18] Yoder, who reported on the evening's events, wrote that Page had "made one remark which should be long remembered": "Regardless of national and international conditions, we are still going to promise you another turkey show in 1943." Yoder, "like all those present," believed that "this $80,000,000 turkey industry of ours shall continue to prosper and grow during the coming year and that there will be a need for another get-together at the '43 All-American, where we may all again talk turkey, enter friendly competition, and leave Grand Forks, already making plans for next year's bigger and better show."[19]

The evening's entertainment was furnished by a group of students from the University of North Dakota under the direction of John Howard, director of bands and orchestras. Singers included songstress Margaret Grundy and three young ladies forming the Gamma Phi Trio. Gordon Erickson played solos on a vibraphone, and Miss Beryl Willis played cowboy songs on her mandolin. Immediately following the "sumptuous banquet," Yoder wrote, tables were cleared away "and the annual Turkey Trot Band swung into action." The dance orchestra was made up of university students who played "the real 'danceable' kind of music."

Yoder also reported on another measure of what the All-American Turkey Show had become by 1942, a show in which dressed exhibits outnumbered live exhibits 407 to 315. The dressed turkey division had to be in a refrigerated enclosure, and, he wrote, "The All-American has one of the most roomy and handy dressed departments of any of the shows in the country."

---

[18] Dignitaries included J. Earl Cook, Extension Poultryman, from the North Dakota Agricultural College, Fargo; Elmo Ellingson, Deputy Executive Secretary of the North Dakota Poultry Improvement Board, Fargo; Walter J. Hunt, Assistant to the Director, Department of Agricultural Development, Northern Pacific Railway, St. Paul, Minnesota; and F. J. Higginson, representing the Alberta Turkey Breeders Association and Fieldman for the Alberta Department of Agriculture. Judges included O. A. Stevens, North Dakota Agricultural College, Fargo; M. C. Herner, Winnipeg, Manitoba; George W. Hackett, Wayzata, Minnesota; and Cleve Angen, Portland, Oregon. Local dignitaries introduced included H. H. Herberger, newly elected president of the All-American Turkey Show; John O. Allen, vice president; W. W. Blaine, secretary-treasurer; and attorney Harold Shaft, president of the Grand Forks Civic and Commerce Association.

[19] *Turkey World*, February 1942.

The room had to be kept "at a very low temperature," which was easy to do in Grand Forks when the shows were held in January and the temperatures were often -30°F to -40°F for the entire week. As Yoder described it, the top of the auditorium balcony was covered with a heavy canvas tent and the outside windows were left open to "provide a natural and splendid method of refrigeration. Few shows could depend so much on nature's help unless located in a northern climate!"[20]

It was becoming more common at turkey shows, Yoder observed, that the winning birds in the dressed division were auctioned off at the end of the show. The first time this was done at the All-American Turkey Show, the Grand Champion had sold for eighty cents per pound, "and everyone present 'oh'd' and 'ah'd' at the high price." Since then the price per pound had gradually increased "through competitive bidding," and the auction had become one of the most popular features of the show.

In 1942, the auction, held Friday afternoon, January 16, "attracted a splendid crowd of real buyers who in many cases came with blood in their eyes and folding money in their pockets." The Grand Champion, a Broad Breast old hen weighing twenty-two pounds and owned by R. H. Jandebeur of Sunny Slope Farms, North Platte, Nebraska, was the first bird auctioned. Bidding started at $1.50 per pound, and the bid quickly went to $5.00. When the bid reached $6.00, according to reports, the auctioneer and bidders had sweat "rolling off" them. The bird was "knocked down finally at $7.40 per pound," totaling $162.80, a record price for Champion dressed meat. The International Livestock Show's record price for Champion Beef that year did not come even close to that.

M. M. Lyons, Triple-B Turkey Farm, Portland, Oregon, had the winning bid. Lyons purchased the dressed Grand Champion for Governor Charles Sprague of Oregon, who graciously presented it to Governor John Moses of North Dakota. Lyons, Yoder believed, "was to be commended for his fine sportsmanship in making such an event a thing of national interest."

The dressed division's Reserve Grand Champion, a Broad Breast young hen weighing eighteen pounds and exhibited by Mitchell Turkey Farm, Marysville, Washington, brought $1.25 per pound. In "a most

[20] *Turkey World*, February 1942.

The Grand Champion dressed turkey at the 1942 All-American Turkey Show, a twenty-two-pound Broad Breast, sold for $7.40 per pound, the highest price ever bid for prizewinning dressed meat at a livestock show. Auctioneer M. B. Potratz, on left, hands the bird to M. M. Lyons, Portland, Oregon, who had the winning bid. *American Turkey Journal*, February 1942.

friendly and worthy gesture," the *Turkey World* article continued, the bird was presented to W. C. Micklin, manager of the Hotel Ryan, show headquarters for the week.[21]

The winning Special Pack single exhibits, entered by A. W. Hoffman and Son, Aitkin, Minnesota, brought seventy cents per pound and were purchased by M. M. Oppegard, publisher of the *Grand Forks Herald*. Other Single Packs, eleven in all, brought sixty-five to eighty-five cents per pound. The auction of dressed birds, including those in the Commercial Pack division, netted "well over $500, and was one of the most profitable and popular events of the show."

The auctioning of prizewinning dressed birds, noted the *Grand Forks Herald* in its Friday edition, "marked the closing hours" of the annual All-American Turkey Show, and "at 6:00 p.m. today the live exhibits will be released to owners, and the show will have passed into history."[22] The last phrase was prophetic.

At the annual Homecoming banquet, when Page promised "another turkey show in 1943," he could not have known that within six months he would be dead of complications following surgery for a perforated ulcer. The Page Printing Shop was closed, and the business was sold. In February 1942, ten years to the month from when its first issue was published, the *American Turkey Journal* was sold to *Turkey World*. George W. Hackett, citing health reasons and the press of work, had resigned as editor. In 1943, the All-American Turkey Shows were cancelled for the duration, with the expectation that when the war was over they would be held again. They were not.

With the nineteenth show in January 1942, the All-American Turkey Show passed into history.

---

[21] Breed champions also brought top prices, 50 cents to $2.20 per pound. The Bourbon Red Champion, owned by Mr. and Mrs. Otto Thieke, Beardsley, Minnesota, was purchased for $17.00 by the Hubbard Sunshine Feed Co. of Mankato, Minnesota. The company paid $2.20 per pound, for a total of $34.10, for the Bronze Champion, displayed by the Allen Turkey Farm, Radium, Minnesota. J. C. Sherlock, of East Grand Forks, Minnesota, paid 50¢ per pound for the Champion Black, entered by Mrs. W. T. Hall of Denton, Texas. The Hubbard Sunshine Co. also purchased the Champion Narragansett, exhibited by Mr. and Mrs. Albert Payne, Towner, North Dakota, for $1.00 per pound.

[22] *Grand Forks Herald*, January 16, 1942.

AMERICAN TURKEY JOURNAL SOLD

The announcement is made of the sale of THE AMERICAN TURKEY JOURNAL to Turkey World, of Mount Morris, Illinois, effective at once. This February 1942 issue is the last to be published under its old ownership.

All unfilled subscriptions will be filled by Turkey World on a pro rata basis and the subscription of those taking both magazines will be extended on the same basis. Involved in the transfer are the name, good will, subscription lists, advertising contracts, back number files, cuts and engravings of the American Turkey Journal.

A Statement by the Editor

GEORGE W. HACKETT

THE CURTAIN

Ten years ago this month The American Turkey Journal made its bow to the turkey public. Without plans for any "conquest" of the field, or of great financial gain, it was launched to fill what we then thought was a crying need in the turkey business for a magazine of the kind we hoped to create.

And now, as this last issue goes to press we are reminded of what we said in our opening editorial ten years ago..."that our policies would always be progressive and not aggressive, that we would be conservative in estimates and statements, and in all things apply the Golden Rule." As we review our work over the years we believe we have kept faith with that creed. It has been a labor of joy to work with and for that fine aggregation of wonderful Americans known as the turkey growers.

Our many friends will find their interests well taken care of in Turkey World and I bespeak your cooperation with them in every way that may help the industry.

For myself, I am NOT retiring from turkey work, as later announcements will disclose. And so I say "good bye" from these pages for the last time with all good wishes and a promise of continued support to the industry.

Sincerely,
GEORGE W. HACKETT

Wayzata, Minn.

A Statement by the Publisher

Before relinquishing ownership of THE AMERICAN TURKEY JOURNAL I would like to express my warm thanks to the many friends and supporters THE JOURNAL has garnered in its 10 years of existence. Your patronage and your friendly helpfulness would fill many pages in any book of memories.

I believe the best interests of every phase of the turkey business are in fine and capable hands in Turkey World and I know they will fully merit your every confidence and support.

C. DYKE PAGE

Grand Forks, N. Dak.

Ten years to the month after its first issue, the *American Turkey Journal* was sold to *Turkey World.* *American Turkey Journal*, February 1942.

CHAPTER 17

Wonder Women of Turkeydom

(with apologies to Princess Diana of Themyscira, aka Wonder Woman)

A HISTORY OF THE All-American Turkey Show would be incomplete did it not include a chapter on the women, who, with their turkeys, made the All-American Turkey Show into what it became between the years 1924 and 1942. On family farms, raising turkeys was women's work, and 80 percent of the turkeys exhibited at the All-American Turkey Shows were raised by women. Women, constituting the majority of those attending the Education Sessions at the All-American Turkey Shows, were the ones asking questions, sharing experiences, and offering advice.

Wanting to improve the breeding birds in their turkey flocks and thereby increase their chances of having their turkeys being cooped in the Court of Honor, it was women who observed the judges at the judging tables as they "scanned each bird for a discolored feather, a misplaced quill or an improper conformation." And it was women, who, between 1924 and 1942, brought the Standard turkey strains as close to perfection as was humanly possible.

These women were appropriately described as businesswomen, conducting "one of the important industries of the Northwest." Their status as businesswomen was acknowledged when, during the week of the All-American Turkey Show, they were luncheon guests of the Grand Forks Business and Professional Women's Club and of the city's service clubs. They were also invited, together with their husbands, to tour the North Dakota State Mill and Elevator, to observe the mill's experimental turkey flocks, and to learn about the mill's line of turkey feeds. As businesswomen, knowledgeable about market trends and consumers' preferences,

Idaho woman camouflaged by her turkeys in 1940. During years of drought and Depression, turkeys such as these were often the only source of income on family farms. Photo used with permission of Ansgar E. Johnson, National Geographic Creative.

they began exhibiting dressed turkeys, canned turkeys, and Special Pack turkeys at the All-American Turkey Show. Accepting the challenge presented by the development of the Broad Breast turkey variety, they bred their Standard variety birds to have larger frames that carried more weight and meatier breasts.

The women were also farm wives, identified as such by invariably being referred to by their husbands' names. During periods of drought and years of crop failure, their turkeys were often the family's only source of income. Having fed all season on little more than grasshoppers, bugs, and weeds, the turkeys, when sold in the fall, provided the money with which to pay bills and make mortgage payments.

In previous chapters, these women's contributions to the turkey industry and to the All-American Turkey Show were treated only in passing. In this and following chapters, three of these women's most significant contributions will be given their due: the variety or breed clubs, the Turkey Hen Club, and the Hen Club picnic.

In *The Turkey: An American Story*, Andrew Smith wrote that the American Poultry Association's recognition of poultry breeds encouraged the development of turkey shows in which flock owners could compete to determine how closely their birds came to meeting the rigid standards specified in the Association's *Standard of Perfection*. Turkey clubs were also developed to foster interest in the six varieties or strains recognized in the *Standard of Perfection*: Bronze, White Holland, Narragansett, Bourbon Red, Black, and Slate. These clubs, George W. Hackett believed, were of prime importance to the improvement of turkey varieties and to the country's turkey industry. Each club had a president, vice president, and secretary-treasurer—almost always a woman—and women made up most of the clubs' membership. The clubs met once per year, during the week of the All-American Turkey Show, to elect officers and to plan activities for the upcoming year.

In Hackett's opinion, the progress, standing, and value of any turkey club depended more on the secretary than on all the other clubs' members combined. The secretary had to have a "thorough understanding" of the turkey strain the club sought to promote, a "general knowledge" of the turkey industry, and a keen appreciation for "turkey folks."

Andrew F. Smith, *The Turkey: An American Story* (Champaign: University of Illinois Press, 2009). Image located at Saveur.com (The Food Series).

In addition to these attributes, the secretary had to have "a deep interest in the work." Few club members, he believed, realized or appreciated the amount of time and effort the secretary devoted to furthering the club's interests, all while caring for her family and her own turkeys, recruiting members, collecting dues, conducting correspondence, soliciting news items, and providing Club Notes for each issue of the *American Turkey Journal.*

Club Notes, a feature in the issues of the *American Turkey Journal* until 1942, when the *Journal* ceased publication, were the principal means by which members of each "turkey clan-family" kept in touch with one another. These Club Notes were for each of the Standard varieties, for breeders of Bronze turkeys, another for Narragansett breeders, and yet another for those breeding Bourbon Reds. The Notes were awaited as eagerly and read as avidly by club members as were the columns in their local newspapers.

The Club Notes reveal more about the women who raised turkeys than do newspaper accounts, Hackett's editorials, or Yoder's columns. The Club Notes include comments on weather and crops, grain harvests and potato digging, turkey flocks, aged mothers who had fallen and broken hips, daughters who had had tonsil operations, relatives who had died, houses that had burned, those mending after auto accidents, those who had married or had babies, those who had gone on vacations, those canning quarts of tomatoes and pickling pecks of cucumbers, and those who had tried new recipes.

Particularly poignant are the letters received from club members in 1941 and 1942 who admitted dreading going to their mailboxes each day for fear they would find letters informing them that their draft-age sons were being ordered to report for military service. Unfortunately, only a sampling of excerpts from the Club Notes can be reproduced here, only enough of them to tantalize and serve as examples. We begin with excerpts and summaries from Mrs. W. J. Janda, St. Hilaire, Minnesota, secretary of the All-American Bronze Turkey Club and among the most conscientious and hardest working of the club secretaries. Her Club Notes in the *American Turkey Journal* reveal how much she was involved in the country's turkey industry, how insistent she was that breeders continue raising Standard varieties, and how deeply she cared for "turkey folks."

***

July 1934: "We enjoyed a visit by Mr. and Mrs. Edward Youngren, Northgate, Minnesota." They had a nice flock of Bronze poults and had lost only two in 1934. Only two weeks to the Grand Forks picnic. Hope everyone will attend and return home "with so much enthusiasm" that they would work to get more club members. Two AA toms would be offered as prizes-one to the member getting the largest number of new members and one to the person getting the most old and new members. Remember to submit dues of $1 per year.

***

August-September 1934: Janda had received a letter from Rolla Henry, Mercer, Missouri, one of "our turkey bachelors." Besides "a Bronze turkey fancier," he was also a "flower enthusiast." Henry ended his letter with "the nicest compliment," writing that he would always be a Bronze Club member. Mrs. J. W. Walker, Williamstown, Missouri, had "just recently recovered from an auto accident and had the sad misfortune of losing her baby daughter." Club members extended their sympathies to her and her husband.

Mrs. Janda ended her Club Notes with a plea for news items and new members "to make our club outstanding."

***

April 1935: "Greet Miss Marion Joan, new assistant secretary to the All-American Bronze Club," born on February 22, 1935. Although not big (she weighed 6½ pounds), she was as "plump as a partridge." Marion Joan was sure to learn all about turkeys, but, even though she had been born on February 22, she would be taught more modern methods "than to use the hatchet."

Mother and daughter were doing fine, and, "through it all," Mrs. Janda had not missed a month submitting Club Notes to the *American Turkey Journal.*

Club members were submitting their season's mating lists. Many of them had won awards at the All-American Turkey Show, attesting to fine breeding.

Mrs. S. Birk, Maxbass, North Dakota, had raised two hundred turkeys from ten hens—"more than most of us can do."

Mrs. Janda wished that more members would use the Club emblem on their stationery and mailing lists. She ended her notes with a "Plea for more members—they would be welcome."

***

May 1935: Mrs. Janda had not seen any of "our turkey folks" for a while, so welcomed a visit by Mr. and Mrs. John O. Allen, Radium, Minnesota, and Mr. Ward Claney, Stephen, Minnesota. Both the Allens and the Claneys were "head over heels" in their turkey work and their efforts to raise only the best of Bronze birds.

Mrs. Winn, Ashland, Virginia, had been ill most of the fall and winter, but she was well on the way to recovery. Janda had missed her "newsy letters." Janda wished good luck to Mrs. J. W. Walker of Williamstown, Missouri, on her large sales of eggs and poults. Mrs. Walker was a "hard worker" and a "good booster" of the Bronze turkey.

Mrs. Janda received "so many letters of congratulations" on the birth of her new daughter. "Marion Joan is some girl, and growing like a weed." She keeps her mother busy.

Mrs. Janda urged club members to submit news items and helpful ideas on raising turkeys and to submit club dues.

***

Many people had asked Mrs. Janda what benefits they would receive by joining the Bronze Club. In June 1935, she pointed out that it was a specialty club to promote the production of Standard Bred Bronze turkeys. The only way to do this was to exhibit at shows. People could not appreciate the quality of Standard Bred Bronze, she wrote, until they had seen them for themselves. A portion of the money received in membership dues was spent for premiums at the largest shows in order to reach more Bronze producers. A Club rule was to offer prizes at shows in those states in which the Club had ten or more members.

In other Club news, she promised to send a list of members to those asking from whom they could purchase breeding stock so they could begin raising Bronze turkeys, and the Club enrolled its first member from Kentucky—the Lusby Turkey Farm, Owenton, Kentucky, Mr. Ralph Lusby, Manager.

Mrs. Janda was pleased to receive an invitation from the American Poultry Association to present a paper on Bronze turkeys at the sixth annual convention of the American Poultry Association at Danville, Illinois, August 6–8.

Mrs. Janda made note of reports indicating a smaller turkey crop in 1935, and she advised readers to feed their birds well so they would command better prices. "With other meats so high the turkey should be on a par with them in price, not mentioning the delectable eating in comparison to beef and pork." Breeders should be "talking turkey for the table," she said, rather than talking of breeding birds. She had spoken with many in her town who had not eaten turkey meat in years and most said they did not care for it because it is so dry. But, they had not eaten turkeys raised in confinement and fed on wholesome grain rather than on bugs and grasshoppers.

"Why not try to sell turkey meat as well as 'show birds,' breeders, etc." The Bronze was a good bird for every purpose. "Young hens are just the size for a small family, while our plump young toms are the delight of hotel chefs and for a large dinner party. Anyway, let's sell our Bronze."

***

March 1936: Mrs. J. W. Walker, Williamstown, Missouri, reported "a long cold drive home from the All-American (Jan. 1936) with a couple of accidents that could have been quite serious." Once, their car tipped on its side into a snowbank when they were passing a snowplow. The other accident occurred when a snowplow hit their car and spun it to one side. No one was injured in either accident.

Mr. and Mrs. Roy Utne, Ortley, South Dakota, wrote that they had arrived home safely from the All-American Turkey Show (January 1936), but they had been forced to drive one hundred miles out of their way. They had gotten to within ten miles from their home when they were stopped by a huge snowdrift. They had to detour fifty miles to get around the drift. They had, however, enjoyed the show and they planned to attend in 1937.

Mrs. Janda reminded club members that dues are due each January—twenty-eight members renewed their memberships at the All-American

Turkey Show in January 1936. "I will," she wrote, "appreciate hearing from you all."

***

April 1936: Mrs. Janda noted that she had received letters from members all across the country: Mr. Alonzo H. Card, Lisbon, Maine; Ralph Lusby, Owenton, Kentucky; Mrs. Ernest Thompson, Bardsdown, Kentucky; Ed Mybrea, Oak River, Manitoba, Canada; Mrs. William Dumbrill, Charleswood, Manitoba, Canada; Robert Perry, Straughn, Indiana; Mrs. G. C. Brenzel, Ollie, Montana; Howard Tanner, Gettysburg, South Dakota; and Mrs. Roy Utne, Ortley, South Dakota.

***

June 1936: Mrs. Janda had received the announcement of the marriage of Vernon Revier of Olivia, Minnesota, to Miss Margaret Brenner on April 29, 1936. Revier was one of the "famous Revier Bros. of the Minnesota Turkey Farm." She extended best wishes to the new Mr. and Mrs.

***

April 1937: Mr. Roy Youngren, Northcote, Minnesota, remitted his dues and included a letter. He enjoyed the Club Notes "so very much" and looked forward to them each month. "It seems so nice," he wrote, "to read about the different turkey folks, even though I don't know them, but when you see their winnings you think of them as your personal friends." Youngren closed his letter with "Greetings to you and the club."

Mrs. Janda had received a letter from Mrs. Mack Burnett. Mack was "improving nicely." His bones were "commencing to knit" and they were "hardening," so he would soon "be able to be moved around." Other club members wished Mack well and hoped that he could soon leave the hospital.

It was time to pay club dues, she reminded readers. "Please remit so we can carry out our club work." Mrs. Janda reported that she was preparing a pamphlet on the Bronze variety. Copies of it would be free for the asking.

***

August 1937: Mrs. Janda had received another letter from Mrs. Mack Burnett. Mack was back at St. Luke's hospital in Fargo, North Dakota, for an operation on his ankle. He had been home from the hospital for a few days and doing nicely. "But he was too enthusiastic about his ability to walk with crutches and broke a small bone in his ankle, thus making it necessary to have the operation." Mrs. Janda expressed "regret hearing of this accident again but trust he will recover soon."

Mrs. Janda had received a letter from Glenn Clay, Meyers, Kentucky, a 4-H Club member who had raised four hundred young turkeys. All were doing well. He had attended the 4-H Club Junior Week at Lexington, Kentucky, where he was honored by being elected president of the Kentucky Association of 4-H Clubs and presented with a gold medal for "the most outstanding 4-H boy in the state. (My, but we are proud of you Glenn, and hope that someday you may obtain the supreme honor of all, president of our country!)." Glenn attributed "a large part" of his success to his turkey project.

***

April 1938: Mrs. Janda wrote, "St. Patrick's Day here in northern Minnesota, having ideal spring weather, snow all gone, had our first rain. Turkey hens are laying, and we are looking for the first robin to verify that spring is here to stay. We had a beautiful mild winter, and now an early spring, hoping for an ideal summer with lots of sunshine and moisture and lots of beautiful Bronze to make us completely happy."

Many members had not yet submitted dues. Again, she appealed to the readers, "Join us in our work and help us boost the Bronze."

***

June 1938: "We are again having more than a reasonable share of moisture. Red Lake River is the highest it has been for years and today it commenced to rain again so we may have a small flood." Janda wondered about "you turkey folks at Aitkin . . . Were any of you marooned by the flood?"

***

September 1938: Janda reported that "Mr. and Mrs. Roy Utne, Ortley, South Dakota, had a fairly good grain crop, but rain was badly needed to save the corn. Mrs. Godfrey Morris, Neche, North Dakota, wrote that her turkeys were growing "like weeds," but crops had been badly damaged by hail. Mrs. Mae E. Driscoll, Henning, Minnesota, a club member and All-American Turkey Show director, had spent an "enjoyable afternoon" visiting Judge Hackett at his home in Wayzata, Minnesota.

***

October 1938: Mrs. Janda had received a number of letters, some of them from members new to the club.[1]

***

November 1939: Mrs. Janda wrote, "I was deeply shocked to hear of the death of Mrs. E. D. Grant of Glyndon, Minnesota. Especially so as I had not heard of her illness." Janda noted that Mrs. Grant had been a charter member of the club, always "so interested" in the club's affairs and she took "great pride" in the Bronze breed. "We will miss her so much at our turkey gatherings" and her customers will miss her helpfulness in providing good breeders. "Personally I will miss her deeply as we have been friends so many years. She was a quiet, sweet and understanding personage. Our deepest sympathy goes to Mr. Grant and family in their loss."

Janda chastised show managers for not serving turkey more often at the "turkey gatherings." Turkey was, perhaps, served at the banquet, but other meat was served at other times. She had attended the World Poultry Conference in Cleveland, Ohio. Everywhere, there were eggs and poultry, but, those who wanted something to eat had to eat hamburgers and hot dogs. The way to a man's heart was through his stomach, she believed, and the best way to educate the public to eating more eggs and poultry was to serve eggs and poultry.

Turkey was served at the World Poultry Conference to the 250 people at the conference banquet, all they could eat, but the 100,000 people

---

[1] S. Ching of 939 7th Ave., Honolulu, Hawaii, the first member from a "foreign country"
Mrs. William Wright, 57 14th St., Wheeling, West Virginia, the first member from that state
Mr. Wilford Charron, Jr., Norfolk, Connecticut, the first member from Connecticut
H. H. Hill of Hill Turkey Farm, Youngstown, Missouri
E. Edwards, Mt. Bethel, Pennsylvania, a new member
Joe Kline, Mgr. of Willow Springs Farm, Mt. Morrison, Colorado
Mrs. O. J. Vinji, Church's Ferry, North Dakota
Mrs. John W. Walker, Williamstown, Missouri

each day who viewed the exhibits had to eat hamburgers and hot dogs. Why not, asked Janda, have poultry products for people to eat, and use up some of the surplus turkeys, chickens, and eggs?

***

January 1940: Mrs. Janda wrote that she was sorry to hear that Mack Burnett, Cummings, North Dakota, was again at St. Luke's hospital in Fargo for an operation on his knees and may not be home for Christmas. "We are hoping that this will be Mack's last painful operation and that he will recover soon."

She continued her report of member news with regret over "the passing of our friend and club member Al C. Johnson," a faithful booster of Bronze. "Our sympathy is extended to Mrs. Johnson and Lois."

Janda had "the loveliest Christmas surprise," she wrote; Mr. S. Ching of Honolulu, Hawaii, had sent her "the nicest box of jams, jellies and preserves made of native Hawaiian fruits." She thanked Mr. Ching, and remarked, "We will be thinking of you enjoying the sunny climate of Hawaii while we are battling snow storms."

***

April 1940: "We regret the passing of Mr. F. E. Murphy, owner of Glendalough Turkey and Game Farm, Battle Lake, Minnesota," Mrs. Janda wrote. She reminded readers that Murphy raised only Standard Bronze turkeys, and Bronze breeders were "deeply indebted" to him for his work with the Standard Bronze variety.

***

May 1940: Mrs. Janda reported that Mrs. Martha B. Walker, Williamstown, Missouri, was feeling "real well," gaining strength, and busy with her turkeys. Her new home was completed—new furniture, electricity, "and all." Mrs. Walker had been ill most of the winter, "but we all hope she will soon be her usual self."

Janda's Club Notes for May affirmed that the Standard breeds are still "the backbone of the turkey industry and are the basis from which all turkeys are judged. Let us keep boosting our Standard Bred Bronze as we always have." Over the past twenty to twenty-five years, other types had come along—Bronze-Bourbon Cross, baby beef type, and other kinds, but Standard breeds were still holding their own and the others are just a memory. "Let's raise Standard Bronze."

***

September 1940: Mrs. Janda was embarrassed because, for the first time in the eight years since the Bronze Club had been organized, she had nothing to submit for the Bronze Club Notes in the August issue of the *American Turkey Journal.* Club members had not sent her anything that would interest other club members, such as information about themselves, turkeys, turkey equipment, or feeding programs. If club members did not care to write letters, postcards, she suggested, would do.

***

April 1941: The area had been hit by a "disastrous" snow and windstorm that struck in "full fury" at 9:15 p.m. Saturday, March 15, 1941. It was "an inferno" of wind, snow, and fire. Half the town burned. Snow drifts, six to twelve feet deep, prevented help from getting through. Over sixty people were stranded in town. "Turkey losses, Mrs. Janda reported, were greater than they were in the Armistice Day storm in 1940. The storm struck at night, after a warm day, and turkeys were roosting outdoors. The storm came up so fast, the turkeys could not be gotten into shelters. Any bird or human caught in the storm smothered."

***

June 1941: In her Club Notes, Mrs. Janda included a line from John Ruskin's *Modern Painters*: "Color is the most sacred element of all visible things." Would not the world be drab and uninteresting, she asked, without color, were all animals, birds, flowers, and clothes the same color? "Flowers could be just as sweet smelling if all were the same color, but wouldn't it be monotonous to be looking at one color all the time? Isn't it a great pleasure when the poults start getting their second feathers, to see the Bronze appear on the back, then in fluff and breast, and what a gorgeous sight when the poult is fully feathered to see it in all its full glory of gorgeous Bronze? Yes, 'color is the most sacred element of all visible things.'"

***

September 1941: Mrs. Janda was pleased to have a visit by Mr. and Mrs. Mack Burnett and daughter the first part of August. The Burnetts had toured northern Minnesota and were on their way home. "Mack is getting along fine" and could get by without crutches when in the house. It had been five years since his accident, and he had spent most of the time in the hospital. "But we are glad he is recovering. (By the way, we want to extend our best wishes to Mr. and Mrs. Willard Burnett, who were married in June)."

Mr. and Mrs. Roy Utne, Ortley, South Dakota, had written that crops were fair. A nine-day heat wave had rushed ripening and quality was not the best.

Rolla Henry, Mercer, Missouri, wrote that he could not find regular help; dependable hired hands were scarce because of the draft. Because of drought conditions and temperatures since July 23 of 93°F to 105°F every day, crops and pastures were, he reported, "burning badly." Vegetable and berry crops had been good. The oat crop had been fairly good and quality was excellent. Harvesting had finished. Henry was all for the Standard Bronze. The birds had every quality suited to be a "perfect market bird, besides their beautiful color. (I am all for good colored Bronze)."

***

October 1941: The Standard Bronze turkey "of today," wrote Mrs. Janda, "is the ideal turkey in all respects." She had raised them for twenty-four years and had worked through several "remodelings" of the breed. The Bronze was an ideal bird for size and conformation. They were quick maturing, with fine-grained flesh, and a large amount of white meat. "Besides their exceptional market qualities," she said, "you also have the satisfaction of having beautiful birds to work with."

***

January 1942: Mrs. Janda urged club members to attend "the big grand 'round-up' at the All-American, Grand Forks, Jan. 12–16." "Let us," she wrote, "make this the biggest and best Bronze show held in recent years." Important matters would be discussed at the annual club meeting, held during the week of the show. "Let's have a big attendance as we are always assured of a grand time at the All-American." Club trophies would be awarded in both live and dressed classes. "They are worth competing for."

She needed letters telling of members' experiences, to make the Club Notes more interesting. It was embarrassing, she admonished, if there were no Club Notes in each issue of the *American Turkey Journal*, and it was also embarrassing if the club secretary had nothing to write about but herself. Neither did credit to the Bronze Club.

***

Those partial to the Narragansetts were as taken with their variety as those who favored the Bronze were taken by theirs. Until elected vice president, Mrs. William Eddie, Northwood, North Dakota, was the Narragansett Club secretary. Mrs. Ole C. Nelson, Kensington, Minnesota, was elected secretary in 1936 to succeed Mrs. Eddie. Club Notes from both women follow.

August–September 1934: Mrs. Eddie wondered why only four Narragansett Club members had attended the Hen Club picnic the previous July. "That looks bad, folks, and all I can say is that you don't know what you missed." Not even the club president, George Gilbertson, had attended. "I hope he has a good alibi. He can't have roads and weather for an excuse this time. If more members did not attend gatherings, it would appear that the club was not "very active."

Mrs. Eddie mourned Mrs. Parr's misfortune of losing all her poults in a brooder house fire. "She has our deepest sympathy," Eddie wrote.

Mr. Olin wrote that he had four hundred to five hundred poults "all doing nicely," Eddie reported. Some dark night, she joked, "I am coming around."

Mrs. Eddie reported she'd had a "three-legged poult hatch. It was doing nicely" until something killed twelve of her turkeys, including her poult.

Because she was still waiting for "a line or two" from members, she wondered if Narragansett breeders had raised any birds in 1934. If they had, "they are very quiet about it." She was busy writing the Club Notes, caring for someone who was ill, entertaining company, and feeding a large harvest crew.

***

April 1935: Mrs. Eddie was preparing a "descriptive, illustrated circular" to send to those inquiring about Narragansett turkeys, to make good the club motto: "The coming breed."

***

June 1935: Mrs. Eddie's first poults were hatched April 23. From the first twenty eggs she gathered, eighteen hatched, and she had sixteen left as of her writing. More had hatched since. "All peppy and growing fast."

She urged club members to attend the Hen Club picnic on July 7. She was certain that club members had "some real interesting things to tell" about their poults, so they should send them to her for the Club Notes. "Wishing you the best of hatches. Will see you all at the picnic."

***

March 1936: Mrs. Ole C. Nelson, Kensington, Minnesota—taking up the Club Notes responsibilities— included a comment from Mrs. William Eddie, former club secretary, that it had saddened her to turn the books over to Mrs. Nelson because she had so much "enjoyed the touch through correspondence" with club members. Mrs. Eddie would miss this, but she hoped to keep in touch. She urged members to help Mrs. Nelson by sending in material for the Club Notes "to swell the place we have in the *Journal*." Mrs. Nelson reminded club members of the challenge issued by Mrs. Eddie: "Let's try and get one letter off to our secretary every month." That would be good—but do more than that. Bring in one new member every month, she encouraged. "Come folks, let us do this and watch our club grow!"

Jim Martinson, Kensington, Minnesota, a new club member, had spent several weeks visiting his aged parents in Norway. He had enjoyed his boyhood occupation of fishing. He planned to be home in March and back to raising "quality Narragansetts."

A well-known breeder in Tangent, Oregon, A. D. Hudson, was a new club member. He invited Narragansett Club members to visit him and he was sure that they would have a good time together.

Mrs. Nelson had received a letter from Mrs. A. H. Muedeking, Tracy, Minnesota. She was not a club member, but Vice President Eddie would write and "ask her to join our happy throng." Mrs. Muedeking had purchased two toms, one Narragansett and one Black. She named them Ole and Nels, after club secretary Mrs. Ole C. Nelson.

At the time of writing, Mrs. Nelson had heard on the radio that another storm and cold wave were on the way. They had been snowed in for a month already, and they were waiting for a break in the weather. It had been "a very hard winter" for the turkeys and she hoped for a change. She closed with "I shall be waiting for those letters and those new members."

***

June 1936: Mrs. Nelson had visited the George Gilbertson home at Garfield, Minnesota. Mrs. Gilbertson was "somewhat crippled up as she had stepped on a rusty nail while mowing the lawn the day before. We sincerely hope she will soon be better again." Nelson had been busy constructing a rock garden and lily pool in her backyard. It was hard, but interesting, work. She urged members to write and to submit dues, "remember we have to keep our club growing larger and better every month."

***

April 1937: "I just came in from picking turkey eggs so I should be in the right mood for writing 'Club Notes.'" The snow was melting, and the hilltops were getting bare, Nelson noted, but there were still plenty of drifts.

She had received a letter from Mrs. G. C. Brenzel of Ollis, Montana, who had about decided not to raise turkeys in 1937 because it looked "very much" as if the area would have another dry season. Mrs. Nelson hoped that they would receive moisture before time to seed crops.

She had also received "a long enthusiastic letter and club dues" from Mrs. Clara Fero, Whitewater, Wisconsin. Fero had not had a good year in 1936 because she had broken a bone in her foot in April. "Imagine a turkey woman with a broken foot! Some class! But I am alive and raised 900 turkeys last year . . . I am trying to get spring housecleaning done before turkeys come."

Mrs. Nelson had not received applications from any new members in March, so club members would have to work much harder in the coming month "so our club would continue to grow."

***

April 1938: "Today, March 20, is an ideal spring day." The snow was almost gone, Mrs. Nelson observed, and farmers were getting ready to start in the fields "now anytime."

Everything pointed to an early spring. Warm weather had helped turkey raisers "in these northern" areas get some early eggs. The club had prepared a folder on Narragansetts, and she had already received several requests for it.

***

September 1938: Mrs. Nelson assumed that club members had been too busy to write letters. She had been busy too, but harvest and most of the threshing were done, so the big rush was over for a while. Crops were good, and "the corn is looking fine, but we are badly in need of rain or it may not fill out properly."

***

March 1939: Mrs. Nelson had received a nice letter from Mr. and Mrs. Albert Payne, Towner, North Dakota, thanking the Nelsons for "the lovely mantel clock" they had won at the January 1939 All-American Turkey Show. It would be a reminder "of the splendid spirit of cooperation that is exhibited" at the shows to further the progress of the All-American Turkey Show and the Narragansett Club. The Paynes had also won the "Pyrex console set" at the January show for the Best Breeder Display. "It had "already been put in service. It is as useful as it is beautiful." The Paynes thanked the Narragansett Club for "this lovely trophy—also for the array of beautiful ribbons."

***

November 1939: Mrs. Nelson urged members to submit annual dues right away because there would be extra expenses in 1939, the treasury was "rather low," and the club wanted to do its share at the leading shows.

***

March 1940: Mrs. Nelson concluded her Club Notes with, "Last of all I wish to mention that we got quite a wonderful valentine this year. It was a baby daughter, Marlys Jean. She is the only girl in our family, so needless to say we are all immensely happy."

In the same issue of the *American Turkey Journal*, Editor Hackett congratulated Mr. and Mrs. Ole C. Nelson on "their superb valentine," Marlys Jean, born on February 14, 1940. "This is just as ordered," he wrote, because the couple already had "two fine boys about 6 and 8 years old," and "are they ever proud."

***

July 1940: "This month the mailman has failed to bring me many letters from our club members," so Club Notes are brief, Mrs. Nelson chided. She hoped that members would make up their minds "to send a few lines to yours truly," or she would follow the suggestion of Mrs. Albert Payne of Towner, North Dakota, whose husband was club president, that we write about our baby daughters. "Marlys Jean was 4 months old the 14th of June, and weighed 16 pounds at that time. I had her out to see the turkeys for the first time a few days ago, and she surely looked rather surprised."

***

September 1940: Mrs. Nelson was busy the day she wrote because the threshers were there and it was entry day for the Douglas County Fair. Everyone was rushing because they wanted to attend the fair. This was her son's first year in 4-H, and he was showing a baby beef. The crops looked good, and they would have more feed than they had had in years. The corn should "turn out O.K. also." She had to close because it was "time to see that the threshers get lunch." She urged members to send her letters.

***

March 1941: Mrs. Nelson had received a letter from club vice president Mrs. William Eddie, Northwood, North Dakota. Because of the weather, Mrs. Eddie had been forced to remain in Grand Forks for several days after the All-American Turkey Show in January, "and still the roads were blocked when she went home so she had to go 7½ miles by team."

***

April 1941: North Dakota and Minnesota had experienced another "record-breaking storm on March 15." Mrs. Nelson was out in it, so knew "how terrible it really was." The family had been visiting a neighbor that evening and had started for home. They had gotten only as far as their driveway—eighty rods from the house—when the car stalled. They attempted to walk, but, with the wind so strong and the cold so intense, they realized that they could not make it. She and the children remained in the car, and the men, by keeping to the fence to find their way, got to the barn.

They hitched up the horses and returned to the car for Mrs. Nelson and the children. "It is very sad to read about the storm and the many people who froze to death because of it," she wrote. Turkey losses were also high.

***

October 1941: Mrs. Nelson had received an interesting letter from Mrs. Albert Payne, Towner, North Dakota. I wish, she wrote, that "you could see our Narragansetts now." They make "a pretty sight for those of us that are admirers of this breed." When the Paynes walked through their flock, the turkeys came to meet them, flapping their wings and perking up their heads. "We enjoy walking through the flock, stopping here and there to admire feather markings, nicely built blocky birds and prospective show birds." It was interesting "to watch them blossom out in full beauty and maturity."

Mrs. Nelson had also received a welcomed letter from Mrs. Eric Norrie, Crookston, Minnesota, who was convinced that the Narragansett was the best variety. Bronze were good, she admitted, but "the Narragansetts are more beautiful to me."

***

December 1941: Mrs. Nelson had received a letter from club vice president Mrs. William Eddie, always a booster for the variety. "My Narragansetts," she wrote, "mean to me what a good captain's ship means to its captain. Easily handled, reliable and keeps its course." Given their "wonderful disposition," Narragansetts were easy to handle and, because they were not flighty, they gained weight faster. "They are reliable," in that if given proper care and feed during the growing season, they were ready for market early in the fall. "They are not disposed to roam."

She had raised them for ten years and never had any of them visited a neighbor. "Narragansett are very beautiful. Their colorings of steel gray, black and white, together with their red heads, make a very attractive bird and in a flock they are as beautiful at a distance as at close range." No wonder breeders maintain the claim that Narragansetts "yield the best returns in both pleasure and profit."

***

January 1942: Mrs. Nelson noted that there was always a keen competition among the clubs—which had the most members, which had the most new members, which had the most members at the shows, and which variety had won the most awards.

She urged club members to attend the All-American Turkey Show, January 12–16, 1942, and to exhibit their birds. "Let us all do our part to make the Narragansett exhibit the largest and best in the history of the show."

Late off the mark at the All-American Turkey Show when compared with the Bronze and the Narragansetts, the Bourbon Reds soon demonstrated that they were favored by many breeders. An entry in the *American Turkey Journal* noted that the National Bourbon Red Club had enjoyed a revival of interest in 1934 under the able "secretaryship" of Mrs. Homer Stone, Oregon, Wisconsin, and "great progress" was predicted for 1935.[2]

Mrs. Stone was succeeded as secretary by Mrs. Emma Snyder, Perrysburg, Ohio. In her Club Notes for April 1937, Mrs. Stone noted that it was the spring rush period of getting ready for poults. She urged members not to forget the Bourbon Red section of the *American Turkey Journal.* Letters, she noted, were "always welcomed and anticipated from Bourbon breeders."

April 1938: Mrs. Stone had received club dues from two new members, both from Canada: Mrs. Jessie Marten of Wapella, Saskatchewan, and Mr. Ralph Libby, Ayers Cliff, Quebec. Both would serve as vice presidents of their clubs.

***

June 1938: Mrs. Snyder admitted that May was the busiest month of the year. "But," she noted, "what a thrill we get when we open those brooder-house doors every morning and find those little fellows all up and busy! Then, to watch them develop, and to compare the results of this pen and that pen. Already, we can see outstanding poults among them."

***

[2] *American Turkey Journal,* April 1935.

September 1938: Mrs. J. Oliver McMorris of East Greenwich, New York, had written that because of their spell of wet weather they could "scarcely get oats dry enough to thresh." Oats were a good crop for the area, but army worms had destroyed much of the crop, entire fields in some places. The corn looked good and the turkeys were doing well. She had weighed a young Bourbon tom on August 10. At 3½ months old, it weighed 9½ pounds. Her Bourbon Reds were showing their colors. Mrs. Anna Button, club vice president, of Glasgow, Kentucky, had sent some news items. Mr. and Mrs. V. O. Roney, Gallatin, Tennessee, had both had the flu. Their turkeys were doing well, up to the time the couple had become ill.

Mrs. Clara Fero's husband had been very ill and in the hospital. She was kept busy caring for eight hundred "lovely turkeys" and visiting her husband in the hospital each day.

Mrs. John Griffeth, Madison, Missouri, a long-time club member, wrote that "chiggers have been bad here this year. If they worry the turkeys as much as they do me, I don't know how they can stand them." Wheat, corn, and oats were too low in price to sell. Wheat cost $1 per bushel to produce and it was selling for 55 cents. They would feed it, in order to get back the cost of production, but there would be no profit.

***

June 1939: Mr. King Kohl, Cleveland, Ohio, had sent a $5 check as a donation to the club. If nineteen more members gave $5 each, he would donate another $5, and if $100 was raised altogether, he would donate another $5. Mrs. Snyder sent a copy of Kohl's letter to every club member. She also urged members to send her news, because she did not want "a vacant space" in the Bourbon Red Club portion of the Club Notes in the *American Turkey Journal.*

***

September 1940: Mrs. Snyder had one son working away from home. Another son and a daughter were in summer school, working on Master's degrees. Only one son was at home to help with the work. The July notes for the *American Turkey Journal* were due just when her favorite aunt, eighty-two years old, was very ill. Mrs. Snyder was asked to help during the last days of her aunt's life. She had not received any news for the July notes anyway. "I did not intentionally shirk my task," she wrote. Beginning September 3, all four of her children would be working away from home. “May I rely on letters being sent to our department?”

“Grains are turning out very well in our community. Wheat is running as high as 40 bushels per acre, and of very good quality; oats yielding up to 75 bushels per acre. Wheat is 72¢ per bushel, oats 28¢, and corn now is more than at this time last season.” On the night of August 20, the temperature was 56°F.

She closed her message with, “Please don’t make us beg for notes—surprise us.”

***

August 1941: Mrs. Snyder’s message began on a personal note. “I wonder how many other folks fear to go to the mail box because of calls for the army. When three sons are awaiting calls, one is quite anxious.”

She added, “Trouble has hit us lately.” Her mother, eighty-six years of age, had injured a hip in a fall, and Mrs. Snyder had to care for her, while at the same time caring for her turkeys. “Our house caught on fire. Our daughter had a tonsil operation, and three of our relatives passed away. So we have had no time to spare. Others, I imagine, have experienced their share of events too.”

Thankfully, she also had “a happy bit of news to report.” Mr. and Mrs. Arlo Polloch, Delphi, Ohio, announced the arrival of “a charming new little model named Dennis Tyrone, born on May 14. Congratulations, proud parents!”

Sadie B. Caldwell, known for her Bourbon Reds, advertised her turkeys in the *American Turkey Journal* as “Sadie’s Bourbon Beauties.” Located in Broughton, Kansas, she was the secretary of the Royal Turkey Club, the club that sponsored the American Royal Turkey Show held in Kansas City.

Caldwell’s Club Notes were as newsy as those of other club secretaries, but hers were often the more interesting, because she occasionally laced them with humor.

April 1937: “Another month has rolled by, and what a busy one! Everyone seems too busy to write much,” and she had not had time to answer letters.

Hugo Meyer from Lake of the Ozarks County reported a nine-inch snowfall in a recent storm, this in mid-March and in that part of the country. The storm slowed egg production. Turkeys, Caldwell observed, were "the most temperamental creatures" one could imagine.

***

April 1938: A Mrs. Wolfe wanted advice on how to "outwit the egg-hunting crow. Any ideas, anyone? Crows watch for eggs every spring, and it is difficult to get them before the crows did."

Caldwell was looking for letters from club members.

***

June 1938: "With deepest regrets comes news of the passing of Erle Smiley, May 5th. Nicely recovering from a recent operation, death came as a result of a blood clot which stopped his breathing. We can scarcely believe that we shall no more be privileged to hold turkeys across the table for his judging. Quiet and unassuming as was his manner, his services were repeatedly demanded in countless poultry shows throughout the United States and Hawaii alike. Kindly and helpful always. Never flustered. Win or lose we parted, knowing something of the depths of a loyal and generous friendship. We looked forward to many future meetings of this kind. Greater tributes than these we know none."

"We express our heartfelt sympathy to Mrs. Smiley and the kiddies of whom he so frequently spoke," and also to those who worked with him in his hatchery and "his wonderful water gardens." In parting, after the All-American Turkey Show in January 1938, he had said, "Somehow, sometime, somewhere, we'll meet again. And so, Dear Friend, though no words can express the depth of loss we feel through your absence, yet we like to think of you as being there among some other water gardens such as you so loved to work among while here. And we know no good-byes."

***

September 1938: Many had been on vacation recently, Sadie Caldwell wrote. "Mine was to the cool Colorado mountains, so it seemed really hot coming back to Kansas. But, it was nice to see how rapidly the turkeys had grown during my short absence." There had been rain, so it was not so bad, "even in Kansas." Caldwell was pleased that the Lucien Maags of

Savannah, Missouri, had paid her a visit the previous week. The Maags were enroute home from their vacation, and it was "so pleasant to visit with them." Caldwell looked forward to learning to know them better at the turkey shows. "They were getting a bit homesick for their Bourbons."

She urged club members to submit their dues of $1 per year, to provide premiums at the turkey shows.

***

March 1939: Caldwell had received a letter from Mr. and Mrs. Odell Dyers. One night, stray dogs had gotten into their pens and had killed twelve hens. Their dogs had put up a fuss, but Odell was sick and Velma was afraid to go out in the dark by herself.

"There's a man for you. Laid up when a woman needs him most. Maybe it's best I'm an old maid," Caldwell wrote. "I'm used to going out after the turkeys in the dark whenever necessary. It's really the same place as in the daylight, once you're used to it."

***

June 1939: "Once more it is the good old summer time, and just had a couple of good rains to break our Kansas drought," wrote Caldwell.

It was warm enough to go without socks. Caldwell saved enough by going sockless to pay her 1938 club dues of $1 per year, and she urged other club members to submit their dues, with the reminder that dues were used to provide premiums for the shows.

For several days, she had had "some very lovely and rare Iris, the first blooms from plants received last season from Judge Smiley's gardens, and from the Sass Bros., the latter coming as a token of deep friendship to our mutual friend." The flowers were "a most pleasant reminder of a most pleasant friendship with Judge Smiley."

***

March 1940: Secretary Caldwell had gotten a letter from Mrs. Wolfe. Because they had had a bad winter, the turkeys kept trying to crowd the horses and cows out of the barn. "Bill wasted a lot of language on them, but they gave as good as he sent! Didn't we have a honey of a winter? I froze both ears. They didn't come off, but they swelled up and looked like a couple of wild goose plums!"

***

April 1940: As Sadie Caldwell was writing, it was trying to rain. She wanted to get some pasture ground seeded to oats for her poults and she wanted to get her garden plowed. She was two weeks late. Wheat looked good.

She was sorry to hear that Mrs. Wallar had had "quite a time" with the flu the previous winter, but her turkeys were doing well.

Roy Utne, a South Dakota breeder, had called on Rolla Henry, and they had visited several members' flocks in Missouri. "There's something special," Caldwell wrote, "about the congenial feeling turkey breeders and growers feel among themselves. They will get together for a good time in spite of everything."

***

May 1940: Secretary Caldwell had discovered that black-strap molasses was good for turkeys, and they liked it. She put it in their drinking water once per week. A "splendid conditioner," it got the poults started to eat sooner. Her hens chose nests that were difficult to get to, a couple high overhead in one of the farm buildings. She had to climb a ladder to get to the eggs. Thankfully, the hens had not started to fly up into the trees to lay their eggs in crows' nests. "And how those crows do watch any hen that commences to look around the hedge rows" for a nesting spot.

***

July 1940: Secretary Caldwell had received letters from the Freemans and the Turners, the only ones she had received in the past month. "The Freemans were busy with their turkeys, music lessons, 4-H and Farm Bureau Club work and what-not. And, I still recall the vast numbers of cattle I saw (mostly well-bred White Faces) in the hills and pastures as I went to visit in their home last fall. Didn't know Kansas had so many hills."

Mrs. Turner wrote, "I am taking a correspondence course from K. U."—she teaches, too, Caldwell added—"in addition to taking care of the poults and other work." Caldwell knew that everyone was busy, but she urged members to "write a letter for the August notes. I'll really expect all of you to, as it's been so long since most of you have." She also reminded them to submit their 1940 club dues.

The area had experienced only a few warm days. "You'd have thought by the necessary blankets again last night that you surely were up in the mountains vacationing."

***

September 1940: "I can't recall having 'fired' the gang, but somehow, I don't get much news from you for this column."

The month before, Caldwell had written about "how hot Kansas was" and that she felt like moving out. "Now I'll tell you how really lovely it is—frequent rains, everything green (I have to mow the lawn again) and pleasantly cool. Slept under covers last night. Rains every day." The turkeys were growing and getting accustomed to "water on their heads and take it like troupers."

Clair Bidleman had written that someone had stolen one hundred of his turkeys. "Isn't there someone to invent some sort of poison for that type of thief? Stray dogs seem to like to watch mine—but, so do I."

***

March 1941: Mrs. Lloyd Fuller of First View, Colorado, had sent pictures of their new home, "which, combined with a knowledge of the hospitality of her family makes us all wish we might drop in on them." They had enjoyed several trips to the mountains in their new car, and their turkeys were doing well.

***

June 1941: Caldwell invited people to stop for a visit when they went through Kansas on summer vacation. She was surprised to learn that an Ohio club member had been drafted and was in training in Fort Riley, Kansas, just thirty miles from her. "For once, I'm glad I'm not a man," she said, "for how I would hate to have to give up my flock for even one year's time—and we hope it won't be longer than that." Friends of John Ward, of Delaware, Ohio, would be pleased to know that he had acquired "a beautiful Kansas tan while in training here. He hasn't forgotten how to feed turkeys either."

"And now I know you'll all sit down—or do it standing if you prefer —but do it, please—and give me all the news from your turkey farm for July. If you won't write, you'll just have to listen to my own meanderings, I suppose."

***

August 1941: "We all have our problems," Caldwell began, "but when together with other turkey folks—gain a valuable lesson. Others have their own problems and sometimes we are relieved that we don't have all the troubles. Others have some too."

Editor Hackett included a notice in the August 1941 issue of the *American Turkey Journal* that on July 22, 1941, Miss Sadie Caldwell was married to Dave L. Lloyd at Topeka, Kansas. The couple would live at Broughton, Kansas. Caldwell—a long-time prominent figure in the turkey industry, familiar to her readers through her Club Notes in the *American Turkey Journal*—received the journal staff's "hearty congratulations. We hope that turkeys will not lose out under this new partnership."

Turkeys may not have lost out under the new partnership of Sadie and Dave Lloyd, but the Club Notes section of the *American Turkey Journal* must have. After her name was changed, it seems that Sadie B. Lloyd no longer submitted American Royal Club notes.

Once second in preference to Bronze, White Hollands gave place to the Narragansetts among breeders, and the White Holland, Slate, and Black Clubs were neither as large nor as active as the Bronze, Narragansett, and Bourbon Red Clubs. Mrs. Walter Hammond, Hastings, Michigan, was the National Black Turkey Club secretary. For reasons unexplained, by September 1940, she had become Mrs. Pauline Rayner. Four excerpts from her notes provide the conclusion to the discussion on Club Notes. Given the grim international situation at the time she was writing, her final comment provides a fitting benediction.

November 1939: Judge P. M. Pierce, Denver, Colorado, was welcomed as a member. An "outstanding turkey judge," Pierce raised "some fine birds and is noted as a good story teller," Mrs. Walter Hammond wrote. He served as the manager of the Denver Poultry Exposition held December 4–7, 1939.

***

July 1940: Secretary Hammond thanked the turkey breeder in Washington who had sent "the box of luscious cherries," but he had neglected to include his name. She reminded club members to submit dues. Shows were coming up, and the club needed to plan for trophies. Mrs. Dorthea Buskirk, Evans Mills, New York, had not raised many turkeys in 1940 because she had not been well. "We assuredly wish you better health, Mrs. Buskirk."

"Please," Mrs. Hammond pleaded, "sharpen that pencil, Black Breeders, and scrawl me a line."

***

September 1940: Mrs. Pauline Rayner (the former Mrs. Hammond), noted that Mr. I. C. Goodson of Ridgehaven Turkey Farm, Chocorua, New Hampshire, had lost his plant in a fire. He had also lost the Black Club ribbon he had won. "So sorry to hear of your loss. The club will be glad to replace the ribbon."

***

March 1941: The March Club Notes began, "Sympathy is extended to Mr. and Mrs. Thomas Feeney of Grand Rapids, Michigan, on the death of their only grandchild."

"We all have our trials and troubles," Mrs. Rayner observed, such as sickness, storms, and tragedies. "But there is one great thing we can all be thankful for. We live in the Good Old United States of America."

From the Club Notes, published monthly in the *American Turkey Journal*, it is apparent that the women raising and exhibiting turkeys derived satisfaction from their clubs and that the clubs served a need, that of building community among turkey breeders with similar interests and concerns. At the fifth annual All-American Turkey Show, the women formed another club. This one, according to the *Grand Forks Herald*, was unique.[3]

[3] *Grand Forks Herald*, February 3, 1928.

CHAPTER 18

The Turkey Hen Club

(with apologies to Gloria Steinem and Betty Friedan)

AT A BANQUET SPONSORED by the Turkey Club Committee of the Grand Forks Commercial Club, held at the Coffee Cup Inn on Thursday evening, February 2, 1928, women exhibitors at the All-American Turkey Show and wives of men who were exhibiting turkeys at the show, formed the "only 'Hen Club' in Existence." A "pioneer club," described by the *Grand Forks Herald* and intended to be "an adjunct to the 'Greatest Turkey Show in the World'": it was the "only 'Hen Club' in the World."[1]

Membership would be limited. Only women exhibiting turkeys at the All-American Turkey Show or wives of men who were exhibiting turkeys at the show were eligible to become members of the Hen Club. Hen Club members would meet only during the week of the All-American Turkey Show, for a banquet (no men allowed), for a business meeting, for luncheons and teas, and for tours of local points of interest. On Friday, February 3, 1928, for example, Hen Club members toured the *Grand Forks Herald* building and the university's Ceramics Department.

The tables at the Coffee Cup Inn for the banquet on February 2, 1928, were decorated with flowers from the Grand Forks Floral Co., "sent for the occasion." The turkey served at the meal, from the largest turkey farm in the state, was a gift from Ray Andrews of Petersburg, North Dakota, "who had it sent in especially for the banquet." Mrs. Ray Andrews, president elect, presided over the meeting until the program, when Mrs. J. C. Simons, Michigan, North Dakota, acted as toastmaster.

---

[1] Mrs. Ray Andrews, Petersburg, North Dakota, was elected president, and Mrs. Melvin Greenwood, Crystal, North Dakota, was elected secretary-treasurer. Miss Grace Baxter, Hazel, South Dakota, was elected chairman of the Entertainment Committee, to be assisted by Mrs. W. J. Janda, St. Hilaire, Minnesota. One member from each of the states represented at the banquet was also named to the Entertainment Committee.

The *Grand Forks Herald* building, Third Street and Kittson, as it appeared during the years of the All-American Turkey Show. Postcard in author's possession.

The evening's program opened with several musical numbers, after which "a great deal of amusement was furnished by the impromptu turkey calling contest." The winner of the contest, Mrs. August Swenson of Gilby, North Dakota, was named the Hen Club's "official caller." Hen Club members then offered a number of "mirth-provoking accounts" and "humorous anecdotes" associated with raising and marketing turkeys. After more musical numbers, Mrs. J. B. Cooley concluded the program "with a short address."[2]

If membership numbers were a measure, the Hen Club was a resounding success. Membership doubled the first year. Forty members had attended the first banquet, held at the Coffee Cup Inn in 1928. In 1929, seventy-five attended the second banquet, held at the Hotel Ryan. According to club members, the banquet was "some affair," and a reporter for the *Herald* was informed that "no mere man would get in on it." Nor would anyone cover the banquet for the paper. The Hen Club banquet was a "social event" and not a feature of the All-American Turkey Show.

The second banquet was held on Thursday evening, January 31, 1929, with President Mrs. Ray Andrews, Petersburg, North Dakota, presiding. During the roll call, each member was asked to give her name, her address, the turkey variety she raised, and "her turkey call"—how she called her flock for feeding or for being penned or housed for the night.

"The methods of summoning the flock varied," as the contestants declared, "from a 'slamming of the screen door,' through the mother tongues of some of the members, to the plaintive 'peep' of the yearling poults." The award for the best call was made on the basis of originality and practicality. Mrs. John Wolf, Merrifield, North Dakota, was judged to have given the best call, and she was awarded the prize cup.

The evening's program included group singing and vocal solos, including "Love's Wondrous Dream" and "I Wonder If Ever the Rose." Mrs. Ida Midgarden, of Grafton, North Dakota, delivered the major address.

[2] Other entertainers included the following: Mrs. Louis Huser of Bantry, North Dakota, related her experiences at her first turkey show. Miss Grace Baxter, Hazel, South Dakota, spoke of what raising Bronze turkeys had taught her, and Mrs. Lars Lovig, Bantry, North Dakota, recounted how she marketed her first turkeys. Mrs. J. B. Cooley gave a reading titled "The Camel's Lament," and Mrs. Frank Bellamy of Drayton, North Dakota, shared humorous experiences from local turkey raisers.

Her keynote address acknowledged "the remarkable spirit of co-operation which was evident at the turkey show between the business men of Grand Forks and the turkey raisers, among the turkey raisers themselves, and that shown by the various educational agencies who through their representatives have done so much in solving problems of club members."

Throughout her remarks, Mrs. Midgarden stressed that it was "through the co-operation between the farmers and the city or townspeople that the farmers' enterprises could be brought to success." She also made the point that it was only with the cooperation of all concerned that "this greatest good could come to those interested in the growth of the turkey industry."

Mrs. Frank Bellamy, Drayton, North Dakota, the second speaker, was among North Dakota's Master Homemakers. Only a short time ago, she observed, those women who wanted to raise turkeys on a large scale by devoting a great deal of time to it and with the intention of doing it scientifically were considered by everyone in their neighborhoods to be "turkey crazy." Fortunately, that time was past.

Before adjourning for the evening, Hen Club members voted to send a petition, signed by every member, to Dr. John Lee Coulter, president of the North Dakota Agricultural College in Fargo, "urging that the services of O. J. Weisner . . . be retained." O. J. Weisner, a poultry expert, was believed to be among the best. Hen Club members had good reason for believing that Dr. Coulter would honor their petition.[3]

At the fourth annual Hen Club dinner, held Thursday evening, January 29, 1931, at the Hotel Ryan, six states were represented: Montana, North Dakota, South Dakota, Minnesota, Iowa, and Missouri. Places at the banquet table were marked with favors and small chocolate turkeys, "significant of the paramount interest of the club members."[4]

---

[3] Dr. Coulter served from 1921 to 1929. According to Elwyn B. Robinson in his *History of North Dakota*, Coulter, a farm boy himself, "spoke the farmer's language and could do the farmer's chores, a fact he liked to demonstrate as he traveled over the state." He cultivated good will when speaking throughout North Dakota and he made "Our State is Our Campus" the school's slogan. Although Weisner had resigned, President Coulter was urged not to accept his resignation. Weisner's services to the state's poultry raisers were invaluable, and an "incalculable loss" would be incurred were he to leave his position.

[4] At the Hen Club's business meeting, held in the city auditorium on Friday, February 1, 1929, Mrs. Frank Bellamy, Drayton, North Dakota, was elected president; Mrs. Ray Andrews, Petersburg, North Dakota, vice president; and Mrs. Mack Burnett of Cummings, North Dakota, secretary/treasurer. Three Hen Club members were elected to the Entertainment Committee: Mrs. William Holmes, Mapes, North Dakota; Mrs. R. N. Hopthen, Michigan, North Dakota; and Mrs. A. G. Nurness, Michigan, North Dakota.

David Dick Slater, "I wonder if ever the Rose" (Published by J. H. Sarway, 1906). Image located at archive.org, call number 83125.

Mrs. Frank Bellamy of Drayton, North Dakota, former Hen Club president, served as toastmaster for the evening. Mrs. W. J. Janda, St. Hilaire, Minnesota, Hen Club president, "in spite of almost total loss of voice," greeted club members, including the twelve new members who were attending the banquet for the first time. Music, including the Hen Club trio, duets, solos, and group singing, was the theme for the evening's program. A state song contest was also inaugurated, and it became a regular feature of future banquets.

Club members from each of the states represented at the banquet were required to choose one or more from their group to sing their state song. Honors went to the three representatives from Iowa. The judges awarded them the trophy "for the spontaneity with which they sang their state song, unaccompanied and without text." Hen Club members met for their annual business meeting in the city auditorium at 10:00 a.m. on Friday, January 30, 1931.

The All-American Turkey Show in 1934 proved to be "a regular 'homecoming,'" and a "jolly good time" was had by all who attended, as reported in the *American Turkey Journal.* Hen Club members were in evidence the entire week, assisting with arrangements and serving as guides for visitors. Mrs. Al C. Johnson, Bath, South Dakota, outgoing Hen Club president, was present all week. An active Hen Club member, she was also looking after a fine exhibit of Bronze turkeys from the Willow Grove Turkey Farm. "Wearing a smile that never rubs off, and always busy with her fine turkeys on exhibit, helping in the social affairs, or following her hobby (boosting the All-American Bronze Club), Mrs. W. J. Janda, St. Hilaire, Minnesota, was present and everywhere. Just say 'I raise Bronze turkeys' and Mrs. Janda will say the rest."[5]

Mrs. Mack Burnett, Cummings, North Dakota, did not speak much, but "how she can work and in any spot or place." As past president of the Hen Club, she had learned the "ropes." She was present the entire week of the show. Another Hen Club past-president, Mrs. Frank Bellamy, Drayton, North Dakota, attended the All-American Turkey Show for a few days, and she attended the Hen Club banquet. Mrs. Ray Anderson, of Petersburg, North Dakota, and the Hen Club's first president, was "always active" in matters connected with the All-American Turkey Show. She attended the show for part of the week, but she did not exhibit any turkeys.

The Hen Club banquet in 1934, the seventh annual, was held in the Hotel Dacotah's Flickertail Room, Mrs. Al C. Johnson of Bath, South Dakota, presiding. Miss Nell Garvich led the assembly in singing.[6] Mem-

---

[5] *American Turkey Journal,* February 1934.

[6] There was also singing by the Hen Club Sextet, consisting of Mrs. William Eddie, Miss Grace Baxter, Mrs. Alvina Bernard, Mrs. Cleve Angen, Mrs. Browell, and Miss Josephine Andrews. Mrs. Bernard also sang a solo, accompanied by Miss Grace Baxter. Betty Zejdlik and Margaret Ann Lukkason of Grand Forks performed a number of dances. Miss Josephine Andrews of Rockford, Illinois, niece of Mr. and Mrs. George W. Hackett, gave two dramatic readings.

Members from four states attended the Hen Club business meeting. Mrs. William Eddie, Northwood, North Dakota, was elected president for 1935, and Mrs. John O. Allen, Radium, Minnesota, was elected secretary/treasurer.

bers voted to hold the annual Hen Club picnic in Grand Forks, as in other years.

The annual Hen Club banquet in 1936 was held at the Hotel Ryan on Wednesday evening, January 22. More than fifty members and guests attended. Special guests included members of the Grand Forks Business and Professional Women's Club. The program consisted of musical numbers, contests, and talks. At the business meeting, Mrs. Godfrey Morris, Neche, North Dakota, was elected president, and Mrs. Roy Vesper, also of Neche, was elected secretary-treasurer, both to serve during the year 1936. Mrs. Mae E. Driscoll of New York was presented with a fine china clock for having come the greatest distance to attend the banquet. Mrs. John Walker was given a heater to warm her car on the long drive back to her home in Williamstown, Missouri.

On Thursday afternoon, the women were entertained by the city's Business and Professional Women's Club at a tea in the Women's League rooms in the university's Davis Hall. "A very dainty and attractive lunch" was served by Domestic Science students of the "practice house," also in Davis Hall. The students lived in and kept house for six weeks as part of their training.

The ladies toured the Home Economics Department and also the weaving rooms, where they saw a number of interesting items. The program prepared for their enjoyment was "greatly enjoyed," and the women visited until time to leave. Everyone was pleased for having been "royally entertained and having again been remembered by the Business and Professional Women's Club."

In January 1939, the *Grand Forks Herald* noted that "Turkey Hens, here for the All-American Turkey Show, flocked to their eleventh annual exclusive dinner Wednesday night at the [Hotel Ryan]."[7] Hen Club members exchanged greetings with members of the Business and Professional Women's Club, and there was a musical program. An exception was made

[7] *Grand Forks Herald,* January 19, 1939.

to a long-standing rule excluding male attendees to allow Everett Mitchell of Chicago, program director of the National Farm and Home Hour, to speak briefly.

Harry Yoder covered the All-American Turkey Show for *Turkey World*. In the February 1940 issue, he noted that "the All-American is famous for its social affairs and this year was no exception." There was "a grand get-together" on Tuesday night, January 16, at the Hotel Dacotah, at which folks met and rubbed shoulders. On Wednesday evening, January 17, "the popular Hen Club held its annual dinner which was all for the ladies—but they did invite the men in later, and we spent many enjoyable minutes together." The big homecoming banquet was held Thursday night, January 18, with over two hundred attendees. It was a "most successful" event, and "as usual, the evening was finished with dancing enjoyed by all."

In the following year, the *American Turkey Journal* noted that the Hen Club, the "unique organization of women turkey growers," had held their annual dinner at the Hotel Ryan during the week of the 1941 All-American Turkey Show.[8] The program included a number of talks, and at the business meeting, officers were elected to serve during 1941.[9]

According to the *Herald*, the "Turkey Hens, the women's fun organization which gets together at the All-American Turkey Show each year," would hold their annual dinner at the Hotel Ryan at 6:30 p.m., Wednesday, January 14. Representatives from the Business and Professional Women's Club were invited. Entertainment would consist of music and stunts.[10]

It could not have been known at the time, but the 1942 banquet would be the Hen Club's last. With the country involved in the war, an era had ended. For the Hen Club, the fun was over, except for the annual Hen Club picnic in July.

---

[8] *American Turkey Journal,* February 1941.

[9] Mrs. George Kirk spoke on "Your Turkey and Mine," and Miss M. Helen Davis spoke on smoked turkeys from the consumer's point of view. Mrs. William Eddie of Northwood, North Dakota, gave a reading titled "The County Agent." A quartette from Central High School, under the direction of Arthur W. Seith, sang several numbers for "the turkey women."

[10] *Grand Forks Herald,* January 13, 1942.

6 AMERICAN TURKEY JOURNAL *June, 1936*

SOME SCENES FROM PREVIOUS ALL-AMERICAN PICNICS

Some prominent "All-Americans" left to right seated: Dyke Page, president; W. W. Blain, secretary; Mack Burnett, floor superintendent. Standing: Alfred Malmberg, Ray Andrews, Harold Herberger, Judge Hackett, Manager of the All-American, H. B. Olin, John Allen, J. C. Sherlock.

Scene at one of the merry picnic tables. And was the food good!

A group from the famous Turkey Hen Club, left to right: Mesdames Bellamy, Janda, Burnett, Hulstrand, Parr, Andrews, Browell, Kelleher, Morris, Allen, McLeod, Kirk and Claney.

Scenes from All-American Picnics. *American Turkey Journal*, June 1936.

Top: C. Dyke Page, All-American Turkey Show President, passes out ice cream to children at a Hen Club picnic. Bottom: All-American Turkey Show Secretary, W. W. Blain, and Vic Cranley carve the watermelons, while President C. Dyke Page watches. Director J. C. Sherlock tests the melon and director Alfred Malmberg looks over Vic Cranley's shoulder at a Hen Club picnic. Image from July 12, 1938; used with permission of Grand Forks Herald.

For twelve years, 1931–1942, the "famous annual picnic of the Turkey Hen Club" was "a sure sign" that summer had arrived. The picnics, held on a Sunday each July in the "deep woods" of Riverside Park in Grand Forks, North Dakota, were of a pattern. Members of "the turkey clan" gathered for an "afternoon of turkey gobbling and talking," some driving five hundred miles or more and others attending from as far away as California and Washington.

Turkeys that would have taken awards at the All-American Turkey Show were served roasted, fried, and smoked and in salads and sandwiches. The picnics were "a paradise for gourmets," with long tables laden with food. The "turkey folks" were treated to music provided by bands from the University of North Dakota, directed by Professor John E. Howard, to "fine, ice cold watermelons," and to ice cream, "enough for an infantry regiment."

These comments on the Hen Club picnics, taken from issues of the *Grand Forks Herald*, the *American Turkey Journal*, and *Turkey World*, cannot alone account for the pleasure the picnics afforded those attending them. Only those who lived through the period and whose experiences were those of the "turkey folks" can appreciate why people drove long distances on dust-covered roads in the heat of summer in the automobiles of the day to attend picnics in a tree-shaded city park.

The years of the Hen Club picnics were years of depression, drought, dust, crop failures, grasshoppers, and record high temperatures. The temperature in Grand Forks on July 12, 1936, the day of the Hen Club picnic, was 106°F.

An account in the *American Turkey Journal* of the Hen Club picnic held on July 8, 1934, put it this way: "Drought, and grasshoppers, and the AAA [Agricultural Adjustment Administration], and all the many other problems that beset the farmer these days meant nothing to the great crowd of turkey folks who journied to Grand Forks, N.Dak., on Sunday July 8th to take part in the 4th annual picnic of the Turkey Hen Club, of the All-American Turkey Show."

Good weather prevailed "for this great annual summer event among the turkey clans." It was good to see the "great spirit of neighborliness existing between these turkey folks," many of whom lived hundreds of miles apart and who saw each other only at the "two great turkey events" of the year—the All-American Turkey Show and the Hen Club picnic.

There was, according to the account, "no such word as [D]epression here. We may talk of our hard times but to join such a feast as the All-American picnic dinner is a fortune in itself. For we are comrades as only they can be who have gone through this same trying and fascinating experience of raising turkeys in a northern climate. Oh! We quite often feel discouraged but the turkey game is too interesting and we come 'smilin' thru."

Mrs. W. J. Janda, St. Hilaire, Minnesota, and secretary of the All-American Bronze club, expressed the same sentiments, but more succinctly, when encouraging people to attend the Hen Club picnic on July 8, 1934. Mrs. Janda assumed that all club members would attend the picnic, because, "having once attended one of the picnics I am sure you wouldn't want to miss another. So pack up your troubles and lunch in your Ford and come to the All-American city."

And, there was this. Many farms of the day lacked trees to cast shade and slow the wind, and water was often in short supply and of poor quality. Before rural electrification, farm families had few means of cooling food, milk, and cream or of keeping fresh meat and butter from turning rancid, to say nothing of having ice with which to cool watermelons or make ice cream.

For good reason, Hen Club picnics were held in Grand Forks's Riverside Park. Located in a gentle bend of the Red River, the park was a natural wood of more than thirty acres, featuring a swimming pool, wading pool, picnic grounds, tennis courts, horseshoe pitches, and ball diamonds. Imagine the pleasure of dressing in good clothes—dresses and hats for the ladies and dress trousers, white shirts, ties, and summer straws for the men—and spending a day with like-minded folks under trees that provided shade and cooled the breeze. An entire day, shaded from the hot sun, protected from the searing wind, free from having to look at wilting crops, and away from the day-after-day drudgery of chores, fence mending, and farm work in work clothes made stiff by sweat and grime. Consider, too, the varieties of good foods, and into the measure, add a band

Riverside Park and swimming pool. From *Grand Forks: Proud People, Proud Heritage* (1988), published by (and used with the permission of) Grand Forks Herald. Original photo submitted by Marien Lien, Grand Forks.

concert, cold watermelon, and ice cream—all were treats seldom available to those living on farms.

Mrs. John O. Allen, Radium, Minnesota, was Hen Club president in 1935. Her comments on the Hen Club picnic held on July 7 that year help to explain why so many looked forward to the picnics and why they derived so much pleasure from them. "The weatherman," Mrs. Allen commented, "smiled on turkey folks as they met in Grand Forks, North Dakota, for the annual All-American picnic. A perfect day for a picnic in Riverside Park after threats of storms and showers for weeks. Everything was grand."

Her concluding remarks and plaintive tone speak volumes. "Yes, it was one good long day but it wasn't long enough. We lingered on as long as possible. Then Good-bye and off for home." What awaited at home for those raising turkeys on family farms were drought, grasshoppers, the AAA, and all the many other problems that beset farmers. But, those who had attended the Hen Club picnic would be sustained by their memories of that perfect day spent among turkey folks under the trees in Riverside Park.

Although the Hen Club picnics were of a pattern, many had features that warrant comment. The picnics and accounts of the day's activities were given prominent place in both the *Grand Forks Herald* and the *American Turkey Journal.*

More than three hundred people, according to the *Herald*, arrived in Grand Forks by train and automobile to attend the second Hen Club picnic held on Sunday, July 10, 1932. "Choice turkeys from three states" were "sacrificed for the occasion, and roast turkey, turkey sandwiches and other farm delicacies" were supplied "in profusion by the picnickers."[11] There were so many cakes, three people were put to work cutting them.

C. Dyke Page, president of the All-American Turkey Show Association and in charge of the picnic, observed that "those turkey hens were not only experts in the turkey yards but also in the kitchen and the lunches they bring to the picnic are enough to set the mouth of a chronic dyspeptic to watering." The All-American Turkey Show Association provided the coffee, cream, and sugar, and Mr. and Mrs. Claude Sipple of the Red River Produce Co. furnished the ice cream. Gathered at the long common table for the main meal at 1:00 p.m., "turkey fanciers" talked "turkey" and made plans for the All-American Turkey Show to be held in January 1933.[12]

A second picnic was held at 5:30 p.m., but, because picnickers had brought "heaping baskets of food," there was still food left to be taken home. The "turkey folks" had eaten until they could eat no more.

Mrs. Al C. Johnson of Bath, South Dakota, was Hen Club president in 1933 and in charge of the third annual Hen Club picnic held on Sunday July 9 that year. More than two hundred people from three states attended, including many from Los Angeles, California. At noon, picnickers "sat down to the long tables that groaned under the weight of delicacies that had been prepared at various farm homes." The delicacies included "many

[11] *Grand Forks Herald*, July 10 and July 11, 1932.

[12] Prominent among those attending the picnic were George W. Hackett, All-American Turkey Show manager; well-known turkey breeders; and representatives of turkey supply companies. Mr. and Mrs. Frank E. Moore and family represented the North Dakota Agricultural College, and O. J. Weisner, Poultry Expert, represented the Agricultural College at Brookings, South Dakota. Iowa's representative was E. J. Palmer, well-known to those in the turkey business.

birds that might have been prize winners at the winter show" had they not been "sacrificed to provide the picnic dinner."

There was no formal program, but exhibitors, officials, and visitors talked "turkey," had "a general good time," and "plenty to eat." Turkey experts attending the picnic and discussing plans for the All-American Turkey Show to be held in January 1934 included show manager George W. Hackett; Frank Moore and O. A. Stevens of the North Dakota Agricultural College; and O. J. Weisner from the Agricultural College at Brookings, South Dakota.

Sunday, July 8, 1934, was the date for "the famous annual picnic of the Turkey Hen Club of the All-American Turkey Show" at Riverside Park in Grand Forks. This summer event, according to the *American Turkey Journal*, never failed "to attract a large and representative group of growers, many traveling as much as 500 miles to attend." No membership was required, and the invitation, "extended cordially" to turkey growers everywhere, "married, unmarried, large growers, small growers and those in between," expected all to attend, "children as well."[13]

Turkey feed and equipment representatives were also invited. Those accepting the cordial invitation brought lunch for themselves and one or two extras. The All-American Turkey Show Association provided the coffee, cream, and sugar. A picnic dinner, "such as the Turkey Hen Club has made famous," was served at 1:00 p.m. "To the great clan of turkey fanciers," the annual picnic was "a real red letter day," and "who could think of a diet."

"The table," according to the description in the *American Turkey Journal*, "seemed a mile long, loaded with good eats that made your mouth water and your heart thankful." The good eats included many roasted turkeys, "all of them fine, toothsome, tasty birds, and further proof" that turkeys could be served "to advantage the year round."[14]

A lunch was served in the late afternoon, and everyone enjoyed the "fine, ice cold watermelons" provided by Harold Herberger, C. Dyke Page, A. I. Hunter, and Vic Cranley. The ice cream "as usual," was furnished by

---

[13] *American Turkey Journal*, June 1934; August–September 1934.

[14] *American Turkey Journal*, July 1934.

the Red River Produce Co. of Grand Forks and it was "a great treat." There was enough of it for a regiment of infantry, but it was all eaten.

The day's entertainment included swimming in the Riverside Park pool and community singing. A new feature added to the program in 1934 was Professor John E. Howard and his German band. "The surprise number was met with great enthusiasm," and it was hoped that the feature would be included in the entertainment at future picnics.

George W. Hackett, All-American Turkey Show manager, could not attend the picnic in 1934 because he was ill. The telegram sent to him expressed best wishes for his health, and it included the comment that "our picnic cannot get along without you and Mrs. Hackett another year. Don't bother with a lunch, just bring yourselves."

The *Grand Forks Herald* of July 7, 1935, included a statement by Mrs. John O. Allen of Radium, Minnesota, Hen Club president in 1935. "If turkeys have ears, they should burn today," Mrs. Allen remarked, "for the Thanksgiving bird will be talked about a good deal this afternoon when the turkey fanciers of the Northwest gather in Riverside Park for the fifth annual picnic of the All-American Turkey Hen Club."

In its August 1935 issue, the *American Turkey Journal* noted that July 7 was a "perfect picnic day and Riverside Park, Grand Forks, was never in finer shape." William Blain, All-American Turkey Show secretary, had selected "the shadiest and most convenient spot in the park and had tables assembled in a long stretch with others handy for the overflow late comers." He also arranged for the coffee, cream, and sugar that was provided by the All-American Turkey Show Association. "The turkey 'clan' began arriving shortly before noon," and Mrs. John O. Allen directed the setting of the tables and served as the "reception committee," along with William Blain, Jack Sherlock, and George W. Hackett.

"After a most sumptuous spread at 1:00 p.m., the afternoon was spent mostly in visiting, the chief purpose of the gathering." There was also a meeting of the directors, exhibitors, and others interested in the All-American Turkey Show. Manager Hackett asked for suggestions for improving the show, but most attendees expressed satisfaction with the show's purpose and with the current management and arrangements.

Reporting on the fifth annual Hen Club picnic in the *American Turkey Journal*, Mrs. Allen commented that she had "never seen so much good food."[15] She wondered why there was "not on the market, a cook book written by our Hen Club. It might be entitled, 'How to Keep a Husband.'" After such a large dinner, she believed, no one could be hungry for weeks, but toward evening people gathered at the tables for more salads, cakes, pies, and coffee. Special treats were the ice cream provided by the Red River Produce Co. and "the fine watermelons," provided by Vic Cranley, salesman for Purina Mills.

The 1935 Hen Club picnic, Mrs. Allen believed, was "a Grand Picnic" and it would "long be remembered as 'the best yet,' and a day well spent." Yes, "it was one good long day but it wasn't long enough. We lingered on as long as possible, then Good-Bye and off for home." Mrs. William Eddie, in her Narragansett Club notes in the *American Turkey Journal*, observed that "large numbers of turkey folks" enjoyed the day "immensely," but "the evening shadows appeared all too soon to remind us it was time to start for home and look after our turkeys."

All were pleased that Manager George W. Hackett could attend the picnic, and everyone enjoyed the poetry he recited, especially two lines from Robert Burns's poem, "To Mary in Heaven." The words were fitting because they expressed what the "turkey folks" were thinking better than they could have expressed it themselves: "Till too, too soon, the glowing West, / Proclaimed the speed of winged day."

The June1936 issue of the *American Turkey Journal* included an announcement that the sixth annual Hen Club picnic would be held on July 12 in "beautiful Riverside Park" in Grand Forks, North Dakota. "Now a traditional annual event in Northwestern turkey annuals" and "a sure sign" that summer had finally arrived, the picnic brought together hundreds of "enthusiastic turkey folks" from several states and Canadian provinces, some "turkey fanciers" driving five hundred miles or more to attend this "summer fiesta."

The picnics, readers of the *American Turkey Journal* were reminded, were sponsored by and held under the auspices of the Hen Club, "a group

---

[15] *American Turkey Journal*, July-August 1935.

of women" who were "prominently active at the All-American Turkey Show held each year." Everyone interested in turkeys was invited.

A prominent *Herald* article, headed by the words "Turkey Hen Picnic," announced on July 12, 1936, that this "traditional event on the turkey breeder's calendar" would "bring together hundreds of them for an afternoon of turkey gobbling and talking." They would "sample each other's wares and subject them to the mature judgement, for, according to custom each of the members brings his own birds, dressed and cooked."

Picnickers brought basket lunches and dishes so that everyone could "help themselves from whatever appeals to individual fancy." The Red River Produce Co. would donate the ice cream and Vic Cranley the iced watermelons. The North Dakota State Mill and Elevator would provide the coffee, cream, and sugar.

Unfortunately, "the terrific heat of Sunday noon [106°F] held down the usual attendance." About seventy-five picnickers braved the heat, some attending from as far away as Beltrami and Minneapolis in Minnesota; Dickinson in North Dakota; and Saskatchewan in Canada. Attendees ranged in age from three months to seventy-seven years.

According to the account in the *American Turkey Journal*, the seventh annual Hen Club picnic was held on July 11, 1937, in perfect weather "amid the beautiful wooded grounds of Riverside Park." The picnic attracted a "large and satisfactory attendance" of "turkey enthusiasts" who spent an enjoyable Sunday "generally making the most of this summer holiday from their turkey tasks."[16]

Picnickers brought basketfuls "of delicious food" that was "laid out on great tables for all to help themselves." The boards of the tables "were groaning with the traditional feast such as only these capable turkey growing ladies know how to prepare. Salads vied with fried chicken and chocolate cakes with prize pickles for honors and at the finish there was no hungry guest who could be located."

About 150 exhibitors, friends, officials, and directors attended the picnic, some from Winnipeg, Manitoba; St. Paul, Minnesota; and Long

---

[16] *American Turkey Journal*, August 1937.

Beach, California. Special visitors included Harry Yoder, representative of *Turkey World*, Mount Morris, Illinois; and Dan Willard, retired agricultural agent of the Great Northern Railway, "a well-known figure at many shows in the past."

"The principal entertainment was the picnic dinner served at one long table," but before the dinner was served, "the ever-welcome Professor John E. Howard" of the University of North Dakota's band and orchestra department and fourteen players in the university's summer school band gave "a tuneful and much appreciated concert." Miss Jeanette Wolf of Grand Forks sang several selections, to the delight of those in the audience.

According to the July 10, 1938, issue of the *Grand Forks Herald*, "from far and near the turkey folks will assemble in Riverside Park this noon bringing baskets of food, many of which will contain roast turkey of which, like pudding, the eating is the test."[17] As was customary, the dinner would be served early in the afternoon on one long table. A hearty lunch would be served late in the afternoon. Grand Forks business concerns would furnish ice cream, watermelons, coffee, cream, and sugar.

Two days later, again according to the *Herald*, about 150 exhibitors and friends attended the picnic, coming from North Dakota, South Dakota, Minnesota, and Saskatchewan, Canada. Because there was no formal program, picnickers devoted their time to reunions of friends, informal discussion of turkey matters, and taking advantage of the facilities provided by the park. Members of the board of directors met to discuss arrangements for the All-American Turkey Show to be held in January 1939. Music was again provided by a band from the university, directed by Professor John E. Howard.[18]

Accounts of national and international events all but crowded mention of the Hen Club picnic from the pages of the *Herald* in 1939. The ninth annual picnic was held on July 9 that year, according to the Grand Forks paper. About 125 "turkey fanciers" from several states attended "the annual outing in Riverside Park."[19]

---

[17] *Grand Forks Herald*, July 10, 1938.

[18] *Grand Forks Herald*, July 12, 1938.

[19] *Grand Forks Herald*, July 11, 1938.

The tenth annual Hen Club picnic was held on July 14, 1940, in Riverside Park in Grand Forks. That more than 175 "turkey folks" attended, the best turnout in years, gave promise of "much interest" in the All-American Turkey Show scheduled for January 1941. The picnic was held in the shadow of World War II that had broken out in Europe the previous autumn. In a three-week campaign that came to be known as a *blitzkrieg*, German ground and air forces had quickly overrun the ill-equipped and outnumbered Polish army.

According to the July 14, 1940, issue of the *Herald*, the picnic was "the only summer meeting of high grade turkey growers in the area," and it would "follow the time honored custom of having no formal program beyond opening with the picnic dinner," after which the "turkey fanciers" would hold reunions, discuss turkey matters, and share comments on the birds they planned to enter in the All-American Turkey Show in January 1941. Also at their disposal were all the park's "amusement facilities."

By 1940, the Hen Club picnic had become "one of the gastronomic events of the summer in Grand Forks." And, according to the *American Turkey Journal*, "as usual the picnic boards groaned with a regular gastronomic blitzkrieg and where the tender young fried spring turkey left off, the light and tasty lemon-filled cakes and finest strawberry pies began. And true to form, after tucking away a goodly portion at the dinner hour, later in the day they hauled the baskets all out again and had "a 4 o'clock coffee hour that was a dandy."[20] Also "as usual," Vic Cranley of Purina Mills provided the cold watermelons; the Red River Produce Co. supplied the ice cream; and the North Dakota State Mill Feed Department donated the coffee, cream, and sugar.

During the afternoon, members of the board of directors held a brief meeting to discuss details of the All-American Turkey Show, scheduled for January 1941. Unable to attend the picnic, Hackett expressed his regrets and extended his greetings and best wishes in a letter that was read to the gathering. Frank Moore, of the North Dakota Agricultural College and now All-American Turkey Show manager, did attend the picnic, as did other officers.

---

[20] *American Turkey Journal*, August 1940.

A feature of the tenth annual Hen Club Club picnic, as noted in the August 1941 issue of *American Turkey Journal*, was the hour-long concert by the National Youth Administration summer school band from the university, directed by Professor John E. Howard. At previous picnics, Professor Howard had come with a "little German band" of eight to twelve members. In 1940, however, "he really spread himself and his concert organization numbered over 40 musicians. They gave a really capable program and were much applauded by the picnickers."[21]

The eleventh annual Hen Club picnic was held on July 13, 1941. An entry in the *American Turkey Journal* noted that "this annual outdoor event constitutes the summer gathering of turkey folks from far and wide who just can't wait for another All-American Turkey Show to roll around to meet together and have a good time."[22] All turkey growers and those engaged in allied industries were invited and welcomed.

More than two hundred "turkey enthusiasts and their families" gathered for this enjoyable summer event, one of the largest crowds ever to attend a Hen Club picnic. Large numbers of picnickers came from Minnesota and North Dakota, and two came from Tacoma, Washington. All brought "great quantities of the most appetizing food," and local firms supplied the "trimmings."[23]

People spent the afternoon enjoying the "bounteous spread and talking turkey." Representatives from concerns dealing with turkey feeds and supplies were on hand, and the All-American Turkey Show board of directors held a brief business meeting. The university's summer school band under the direction of Professor Howard provided the musical entertainment. The picnic disbanded in time for people to attend the ball game. Late in the afternoon it rained, but, because the picnic was already breaking up, "no inconvenience was caused."

On July 12, 1942, an article in the *Grand Forks Herald* announced that "breeders, produce men and just plain turkey fanciers" would gather at noon that day in Riverside Park for the twelfth annual Hen Club picnic. "Usual attendance" ranged from 125 to 150, but, in 1942, more than

---

21. *American Turkey Journal*, August 1941.

22. *American Turkey Journal*, June 1941.

23. Vic Cranley of Purina Mills furnished the "wonderful iced watermelons." The North Dakota State Mill and Elevator provided coffee, cream, sugar, cigarettes, and suckers; and the Red River Produce Co. donated Dixie cups of ice cream.

two hundred people attended, one of the largest crowds in the history of the picnics.[24]

The afternoon session began with the picnic lunch. The Red River Produce Co., Purina Mills, the North Dakota State Mill and Elevator, and the Hubbard Milling Co. provided ice cream, watermelons, coffee, cigars, and "other picnic necessities." At an informal meeting in the afternoon, directors, association members, and breeders discussed problems facing the show and points vital to the turkey industry.

The group decided to proceed with plans for the twentieth All-American Turkey Show, to be held January 19–23, 1943. Formal talks were given by H. L. Schrader of the Marketing Division, United States Department of Agriculture, Washington, D.C., and by Dr. J. H. Longwell, Head of the Animal Husbandry Department at the North Dakota Agricultural College. Both men addressed issues facing the turkey industry. The afternoon's entertainment was provided by the university concert band, directed by Professor John E. Howard.

The 1942 Hen Club picnic was the first since their beginning in 1931 without Page being in charge. He had died on July 6, 1942, from an embolism that developed following an operation for a perforated ulcer. For many years, he had served as the All-American Turkey Show manager and as the All-American Turkey Show Association president. The twelfth annual Hen Club picnic was, in a sense, a dedication to Page because "of his love for and interest in the Association's affairs and his expressed desire during his illness that the picnic go on as scheduled even though he would be unable to attend.

Those attending the twelfth annual Hen Club picnic on July 12, 1942, might have anticipated that it would be the last one. Six months earlier, on December 7, 1941, Japanese forces had attacked Pearl Harbor and the United States had entered World War II. Major events, including the All-American Turkey Shows, were canceled for the duration, many never to be taken up again.

George W. Hackett resigned as editor of the *American Turkey Journal*, and the journal was sold to *Turkey World*. With C. Dyke Page's death,

[24] *Grand Forks Herald*, July 12, 1942.

his printing shop was closed and his publishing business sold. Much to Hackett's dismay, the turkey clubs folded.

The women had enjoyed a good run, a run that lasted nearly twenty years. They and their turkeys had made Grand Forks, North Dakota, the Turkey Capital of the World and the home of the All-American Turkey Show, the largest exclusive turkey show in the world. They had brought the six Standard turkey strains as close to the perfection demanded by the *Standard of Perfection* as was humanly possible.

As businesswomen, they had increased the size of their flocks, and they had convinced consumers to eat more turkey meat and to eat it throughout the year, not just on Thanksgiving and at Christmas. The women were also farm wives, and their turkeys were often the family's sole source of income during the years of the Depression, drought, and crop failure. Their turkey clubs and the Club Notes published monthly in the *American Turkey Journal* appealed to "turkey folks" the world over, and they served to popularize the six Standard turkey strains recognized by the *Standard of Perfection*.

Always the majority of those exhibiting birds at the All-American Turkey Shows, the women formed the exclusive Hen Club, the only club of its kind in the world. They also sponsored the annual Hen Club picnic, one of the "two great turkey events" of the year for "turkey fanciers."

A good run indeed, but it ended with the Hen Club picnic in July 1942. Although the United States had entered World War II the previous December, and many events and activities were canceled for the duration, it was not the war that ended this remarkable run.

The run ended because the women were done in by their own success, as was evident on the large commercial turkey farms and in the meat departments of grocery stores. Consumers were eating more turkey meat, in more ways than as whole birds, and at more times during the year than only on holidays.

Consumers were, however, little concerned with whether the meat was from a Bronze or Narragansett or another of the Standard turkey strains. Nor were they concerned with feather color, pattern, or shading. They did not eat the feathers.

CHAPTER 19

Not Even Coyotes Eat the Feathers

(with apologies to Wile E. Coyote)

THIS ACCOUNT OF THE All-American Turkey Shows is not the one I intended to write, and the conclusion is not the one I expected to reach. I intended to begin the account of the All-American Turkey Shows with the first show, held in 1924, and then discuss each show in turn until concluding with the last show, held in 1942. Such was not to be. The topic asserted itself and dictated how it had to be developed.

Promoters of the first All-American Turkey Show, opening on a wintry day in February 1924, were uncertain about whether this first-ever exclusive turkey show would attract interest and support and whether it would draw exhibitors and their turkeys. They need not have been concerned. Crowds were so large they taxed the capacity of the showroom, and a total of 240 turkeys were entered, 110 more than in any other turkey show ever held in the country. The 164 turkeys in the Bronze class alone outnumbered the turkeys entered in any of the country's major shows.

The success of the first All-American Turkey Show raised questions that needed to be answered, and I realized that I could not address the questions by beginning in 1924 and working towards 1942. I had to begin in 1924 and work backwards, eventually all the way to America's wild turkeys, found nowhere else in the world.

Starting the account of the All-American Turkey Shows with America's wild turkeys, I began working toward 1924. Chapter titles mark my route: "America's Wild Turkeys," "An American Turkey in European Courts," "Carrying Coals to Newcastle: English Colonists Taking Turkeys to America," "The Turkey Drover's Lot is Not a Happy One," and "Turkeys are a lot of darned hard work."

Not until chapter seven, "Fine Feathers Make Fine Birds," had I addressed the questions raised by the success of the first All-American Turkey Show. The first chapters discussed how wild turkeys were domesticated; how they were taken to Europe, where they were further domesticated; why colonists brought turkeys to the New World; why the turkey industry developed in New England and the Atlantic States, before moving to the Midwest and eventually to North Dakota, where there were sufficient numbers of them by 1924 to justify starting what became the largest exclusive turkey show in the world. With subsequent chapters—not chapter one—I gave the account of each of the All-American Turkey Shows.

My original intention, when writing about the All-American Turkey Shows, was also to be objective, just as I had attempted to be when writing about silk trains and quartzite boundary monuments, topics touched upon in the Introduction. Again, the topic dictated that this was not to be.

Romance has always been associated with silk—soft, luxurious, shimmering. Romance has also long been associated with trains, with legendary engineers and their steam locomotives. Silk, for all its sensual properties, is, nevertheless, inanimate. Engineers might have believed that their steam locomotives had personalities, and they may have cursed them for being perverse and temperamental, but steam locomotives for all of that were inanimate. Silk trains combined the romance of silk with the romance of trains to produce a romance all their own. Silk trains also, however, were inanimate.

Each quartzite monument on the boundary between North Dakota and South Dakota is distinctive, with markings and features not shared with any of the others. I cannot explain the hold that the monuments have on me, a hold that was forged more than forty years ago that continues to draw me to them. For all that the monuments mean to me, they are, nevertheless, inanimate.

Not so the All-American Turkey Shows. Becoming caught up in the shows, attending them as it were by doing research on them, I could not remain objective. Over the course of nearly twenty years, I developed a familiarity with George W. Hackett, Walter Burton, Harry Yoder, C. Dyke Page, John Howard, and Frank E. Moore. I learned to know Mrs. W. J. Janda and her infant daughter Marion Joan, the John O. Allens, the Roy

Utnes, the Ray Andrews, Mrs. William Eddie, Sadie B. Caldwell, Grace Baxter, and George E. Lamm.

Along with others, I read the newsy (sometimes gossipy) Club Notes in the *American Turkey Journal*. Not allowed to enter the Hen Nest or attend the Hen Club banquets, I did, however, attend the Hen Club picnics. In short, I could not mingle with turkey folk for nearly twenty years and remain entirely objective.

No sooner had the All-American Turkey Shows begun than scoffers joined skeptics in predicting that people would soon tire of them and that the shows would not last. At best, they gave the All-American Turkey Shows five years. Indignant at the very thought, George W. Hackett, the heart and soul of the All-American Turkey Shows, believed that the shows would never end and that breeders would never want to discontinue raising Standard variety turkeys. Neither the All-American Turkey Shows nor the breeders of the Standard varieties lived up to Hackett's expectations.

By the eve of World War II, Standard variety turkeys were losing their pride of place, and in 1943 the All-American Turkey Shows were cancelled for the duration, never to be started up again. World War II was, however, the occasion, not the cause, for ending the All-American Turkey Shows and for displacing the Standard varieties of turkeys.

Once the shows had fulfilled the purposes for which they were intended, there would be little reason for continuing to hold them. Those purposes had largely been fulfilled by the time the country became involved in World War II. The All-American Turkey Shows had been done in by their own success.

King Wheat still reigned in North Dakota in 1942, but farmers were diversifying their operations, and more farm families were raising turkeys to supplement income from crops and other livestock. Because turkeys had proved their worth and raising them had become profitable enterprises, more people were raising them in ever-larger numbers.

Having brought the Standard varieties as close to perfection as was humanly possible, breeders were causing headaches and sleepless nights for All-American Turkey Show judges. More people were eating more turkey, and they were eating it throughout the year, not only during the holidays.

That people were eating more turkey and that producing turkeys was profitable was not lost on those whom George W. Hackett derisively termed "the productionists," those raising turkeys by the thousands in large-scale operations. The productionists, however, were raising meat-type birds, not those of the *Standard* varieties. They were producing for the market, not for the purpose of competing in turkey shows with the hope that their birds would be cooped in the Court of Honor. These operations were in business for the purpose of making money, not for the purpose of improving turkeys to meet the unforgiving standards specified in the *Standard of Perfection*. In short, productionists produced turkeys for their meat, not for their feathers. People did not eat feathers.

Bowing to pressure, for example, brought by breeders of meat-type birds who had long derided the All-American Turkey Shows for being "feather shows," officials provided for a Dressed Division. Admitting dressed turkeys for competition and judging did not, however, bode well for the future of the All-American Turkey Show nor for the future of Standard varieties.

Standard varieties are distinguished by the color, pattern, and shading of their feathers. Dressed birds of all varieties look very much alike and their flesh tastes very much the same. Even experienced turkey judges might have found it difficult to determine whether a dressed bird was a Bronze, a Narragansett, or a Bourbon Red.

Once admitted for competition and judging at the All-American Turkey Shows, dressed birds soon outnumbered live birds and their numbers increased with the introduction of Single Packs and Special Packs. Soon, the highlight of each All-American Turkey Show was not the crowning of the Grand Champion live bird, but the auction on Friday afternoon of the Grand Champion of the Dressed Division.

The All-American Turkey Shows scored publicity coups that could not be matched by any other turkey or livestock shows in the country. It did not speak well for the future of the shows and for the future of Standard varieties, however, that the most newsworthy coups were scored, not by live birds, but by dressed birds.

The Grand Champion in the All-American Turkey Show's Dressed Division in 1936 was a twenty-two-pound Bronze young tom, described in *Turkey World* as a "superb bird, perfect, or as nearly perfect as a turkey could be." Purchased for $25.00, this "nearly perfect" bird was displayed in the famous Dempsey Restaurant in New York City.

The Grand Champion dressed turkey in the 1937 show, a 34½-pound heavy Bronze tom, was shipped by Railway Express to President Franklin D. Roosevelt. The Reserve Champion, a Narragansett young hen, was sent to Secretary of Agriculture Henry A. Wallace. The entire dressed bird exhibit, weighing 8,500 pounds, was purchased by the Great Atlantic and Pacific Tea Co. and placed on exhibit in Washington, D.C.

At the All-American Turkey Show in 1942, the Dressed Division's Grand Champion, a Broad Breast old hen weighing twenty-two pounds, was auctioned off for $7.40 per pound or $162.80, a record price for Champion dressed meat.

By the time the All-American Turkey Shows were cancelled for the duration of World War II, people were eating more turkey to be sure, and they were eating it throughout the year. But, according to an article in the October 1940 issue of the *American Turkey Journal*, consumers cared little for *Standard of Perfection* specifications, feather color and shading and conformity, and they were the "court of last resort."

Consumers wanted turkeys that appealed to them for their size, shape, and finish. When they served their families whole birds, rolled turkey breasts, or drumsticks, they did not care whether the meat was that of a Bronze, Narragansett, or Bourbon Red.

Coyotes did not care either.

Not even they ate the feathers.

BIBLIOGRAPHICAL ESSAY

I introduce this bibliographic essay with parallels to the bibliographical essay in my book *The Quartzite Boundary* because the problems encountered when researching the two topics were similar. Researching *The All-American Turkey Show*, much like researching *The Quartzite Boundary*, was a pioneering effort, as almost nothing has been written about the project.

In ten years of research on the quartzite boundary markers, I found only three references to it in print. In about the same number of years of research on the All-American Turkey Show, I found fewer than a half-dozen references to them in print. On both topics, therefore, my research would have to be almost entirely in primary materials, if there were any, if I could locate them, and if I could gain access to them.

Unfortunately, when researching the All-American Turkey Show, I was denied what should have been my most valuable resource—the records of the All-American Turkey Show Association. The Association must have kept records. The Association had a slate of officers, including a secretary, and its offices were in the Grand Forks City Hall. The All-American Turkey Show was held for one week each year, but the business of the show continued throughout the year.

An event attracting hundreds of turkeys from all over the United States and Canada and thousands of exhibitors and visitors from all over the country required extensive planning and careful management. The volume of correspondence must have been considerable, and vendors, prizes, awards, finances, notices, reservations, employees—all had to be arranged for and in place when the show opened.

If the All-American Turkey Show Association kept records, I have been unable to locate them, and because few people have even heard of the All-American Turkey Show, it is unlikely that I will find someone who would know whether records were kept and where they might be. It is more than speculation that, if records of the All-American Turkey Show were kept, they have been lost.

When Grand Forks was almost entirely inundated by the flood of 1997, forcing the evacuation of almost the entire population of the city, every building in the heart of the city had flooded basements and several feet of water on their main floors, including the County Courthouse, the City Hall, the offices of the *Grand Forks Herald*, and the Chamber of Commerce. For convenience and safekeeping, court records, tax records, city files, newspaper and photograph collections—and possibly All-American Show records?—were often stored in basements. All were lost in the flood.

In the absence of the All-American Turkey Show records, I relied on the sources available to me—the *Grand Forks Herald*, the *American Turkey Journal*, *Turkey World*, and the *Annual Reports* of William R. Page, County Agent, Grand Forks County. These sources were used extensively and almost exclusively for research on the shows.

Because I had issues of the *Grand Forks Herald* for all the years that the All-American Turkey Shows were held, 1924–1942, this source was not only the most extensive, it was also the most helpful. The paper usually began its coverage of the All-American Turkey Show with announcements and other notices in November, in keeping with the observance of Thanksgiving.

February issues sometimes included articles or letters written by those reminiscing about the show just concluded, recounting events occurring during the show, or sharing the harrowing experiences they had when returning to their homes after the show. A word of caution when relying on the *Grand Forks Herald* when researching the All-American Turkey Shows: reporters and typesetters did not always accurately identify individuals, spell names, or record events. They (and other sources) often inserted a hyphen in the show's title, making the first two words, "All-American," into an adjectival phrase, a phrase adopted herein for consistency.

The *American Turkey Journal* was an especially rich source of research for the last ten years of the All-American Turkey Show because, by design, the journal was devoted to covering the shows. George W. Hackett's editorials were as informative as they were interesting; and his was the final word on all things turkey—from the need to preserve the Standard varieties; to feather color, pattern, and shading; care and feeding; and breeding and exhibiting. He was also the recognized authority on how to stage and manage turkey shows and on what judges should look for when judging turkeys and awarding points.

Even those not breeding and exhibiting turkeys found the *American Turkey Journal*'s Club Notes, ads, and other features to be interesting. The material in the *American Turkey Journal* was not only interesting, it was well written and painstakingly edited.

Material in *Turkey World* on the All-American Turkey Shows provide a different perspective from that gained by reading the *Grand Forks Herald* or the *American Turkey Journal,* both of which were published in Grand Forks, North Dakota. Harry Yoder of *Turkey World* became a regular at the All-American Turkey Shows, and it was refreshing to read his often humorous accounts of the shows.

For copies of *Turkey World*, I am indebted to Mary Ella Jerome of Barron, Wisconsin. Daughter of Wallace H. Jerome, among the biggest names in the turkey industry, Mary Ella Jerome maintains her father's large collection of journals and other materials relating to turkeys. Interested in the All-American Turkey Shows through her father, Mary Ella Jerome graciously allowed me to use her father's collection of *Turkey World* for my research.

William R. Page, county agent for Grand Forks County, arranged the Education Sessions for the All-American Turkey Shows. His annual reports, papers, and biographical materials are located in the Institute for Regional Studies, North Dakota State University, Fargo, North Dakota. His comments on the All-American Turkey Shows add a dimension not found in either the *Grand Forks Herald* or in the *American Turkey Journal.* Always informative, his accounts of the shows are usually candid, often humorous, and sometimes cynical. Some reports are comparable to diary entries, seemingly intended for his eyes alone.

Besides the *Grand Forks Herald*, the *American Turkey Journal*, *Turkey World*, and William R. Page's *Annual Reports*, only the Grand Forks *City Directory* and *North Dakota: A Guide to the Northern Prairie State* (New York: Oxford University Press, 1938) contained references to the All-American Turkey Show. The other material, referred to in the text or cited in this bibliographical essay, was included in the research to provide information on turkeys and other poultry and to provide the historical context into which I could place accounts of the All-American Turkey Shows.

To refresh my appreciation for the importance of poultry on family farms, I steeped myself in the farm journals I read when growing up on a South Dakota farm, including *The Farmer, The Dakota Farmer, Farm Journal,* and *Successful Farming.* All the journals contained sections devoted to poultry, often addressed to farm women.

The cover story in the March 1940 issue of *Successful Farming* was on Mrs. Gale Craig of Ovid, Michigan, and her picture was on the magazine's cover. Mrs. Craig's poultry supplied 20 percent of the farm's annual income. Clara Sutton was a nationally known writer on poultry, and her columns were published in *The Farmer*. She often attended the All-American Turkey Shows, and so welcome was she that officials and exhibitors commented that the shows were not the same when she was not in attendance.

For books specifically on turkeys, I used Andrew F. Smith, *The Turkey: An American Story* (Chicago and Urbana: University of Illinois Press, 2006); Karen Davis, *More Than a Meal: The Turkey in History, Myth, Ritual, and Reality* (New York: Lantern Books, 2001); and D. Philipp Sponenberg, Jeanette Berander, and Alison Martin, of the Livestock Conservancy, *An Introduction to Heritage Breeds: Saving and Raising Rare-Breed Livestock and Poultry* (North Adams, MA: Storey Publishing, 2014). In addition to providing information on turkeys, Andrew F. Smith and Karen Davis demolished many of the myths associated with turkeys, for example, that they figured prominently in "the first Thanksgiving."

The Livestock Conservancy's discussions on turkeys in *An Introduction to Heritage Breeds* were especially helpful when describing the six Standard turkey varieties that were admitted for judging at the All-American Turkey Shows and when explaining why the feather color, pattern, and shading were so difficult to perfect.

Andrew Smith, in his book *The Turkey*, gave the subtitle, "How the Turkey Lost Its Flavor," to one of his chapter titles. Turkeys lost their flavor when meat-type birds replaced the Standard varieties shown at the

All-American Turkey Shows. The book *Heritage Breeds* attested to this in a delightful passage that linked Thanksgiving to saving heritage breeds. "Thanksgiving turkeys can save heritage breeds," according to the writer, and "the production of pasture-raised Thanksgiving Day birds by small-scale farmers for local sale has been essential in increasing demand for heritage turkeys. Robust and physically sound heritage breed turkeys thrive in pasture systems, growing into attractive delicious birds that delight consumers."

Two books were used to describe the role that women played in raising the turkeys that provided farm income when drought, disease, and grasshoppers decimated crops. The best of these turkeys must have been among those displayed at the All-American Turkey Shows. Barbara Handy-Marchello's *Women of the Northern Plains* (St. Paul, MN: Minnesota Historical Society Press, 2005), received the Caroline Bancroft History Prize. The book is an account of "the realities of women's lives on the harsh northern plains," according to one reviewer, and it emphasizes that poultry and dairy were the domains of the farm wife. Carrie Young's *Nothing to Do but Stay* (New York: Bantam Doubleday Dell Publishing Group, Inc., 1993) contains an (often humorous) account of the author's mother raising turkeys in western North Dakota. In his review, author and humorist Garrison Keillor described the book as "truthful, funny, modest of manner and full of feeling" and "a testament to our ancestors on the prairie."

In my research, for many reasons, I included Paula Nelson's *The Prairie Winnows Out Its Own: The West River Country of South Dakota in the Years of Depression and Dust* (Iowa City: University of Iowa Press, 1996). Nelson's book recounts conditions in western South Dakota during the years of drought and the Depression, largely through the words of those living in the area. The settlers' stories reveal that a strong bond existed between them and the land, a bond so strong that no adversity could break it. Nelson's descriptions of people and conditions in South Dakota's west river country also describe people and conditions in North Dakota's west river country during the years of drought and the Depression, the years that paralleled the All-American Turkey Shows.

It is instructive, and perhaps not coincidental, that both Paula Nelson and Elwyn B. Robinson used the verb "winnow" in their histories. To winnow is to separate the chaff from the grain by tossing both into the air and allowing the wind to blow the chaff away, leaving the clean grain. For Nelson, the wind doing the winnowing was the prairie. In South Dakota's west river country, she wrote in her Preface, "the Depression was a blast furnace from which a diminished population emerged, hardened yet still hopeful, scathed but undefeated."

For Robinson, in his *History of North Dakota*, the wind doing the winnowing was "the conditions of existence" encountered by those who settled the state, the conditions that contributed to producing what he termed "the North Dakota character." The image produced in readers' minds by both authors is that of settlers being tossed into the air by the unseen hands of threshers. Those lacking in fortitude (my mother would use the word "grit"), the chaff, were blown away by the wind of drought, the Depression, and the conditions of existence. Those with courage and determination, the clean grain, could not be blown away. They, unlike all the John Steinbeck Joads who made for California, fell back to the threshing floor to remain on the land.

I also used Paula Nelson's book because among the settlers' stories she recounts were those of the farm women who raised turkeys. As was recounted by many farm women during the years of drought and depression, it was the income from their turkeys that sustained the family, paid the bills, and forestalled mortgage foreclosure when drought, disease, and grasshoppers destroyed crops.

I turned to Paula Nelson's book as well because the town of Philip is located in South Dakota's west river country. George E. Lamm was from Philip and he and his mode of transporting his prize-winning turkeys became legendary. Lamm used first his turkey "hauler," a trailer towed behind his coupe, and then his "All-American Express," a truck, to transport his prizewinning Bronze turkeys on the six-hundred mile trip from Philip to the All-American Turkey Show in Grand Forks, sometimes driving through the night in order to be on time when the show opened on Monday morning. Temperatures during the week of the shows were often -35 F. to -45 F. Those who respected Dakota winters knew what it meant for Lamm to drive six-hundred miles on dirt and gravel roads in the dead of winter in vehicles likely lacking heaters and defrosters.

Paula Nelson's treatment of South Dakota's west river country and its residents during the years of drought, dust, and the Great Depression is that of a historian. For that of a novelist, one should read Feike Feikema's (Frederick Manfred) *The Golden Bowl* (St. Paul, MN: Webb Publishing Co., 1944). Feikema wrote of the Thor family attempting to eke out a living on a farm in southwestern South Dakota. Members of the Thor family had also forged a bond with the land so strong they could look with optimism beyond the arid, devastated landscape to glimpse the promise of "a golden bowl of plenty." They were not, as the Psalmist said, "like the chaff which the wind driveth away." The Thors would remain on their farm in what Paula Nelson refers to as "next year country."

A number of works were used for the research on the development of the turkey industry in England, what Andrew Smith refers to as "The English Turkey; or, How the Turkey Cooked the Christmas Goose." Naomi Riches in *The Agricultural Revolution in Norfolk* (Chapel Hill: University of North Carolina Press, 1937) describes the turkey strains developed in Norfolk and explains why the area was so well suited for the production of turkeys.

Also helpful for the many facets of the turkey industry that the work covered was Herbert Myrick, ed., *Turkeys and How to Grow Them: A Treatise on the Natural History and Origin of the Name of Turkeys, the Various Breeds, and the Best Methods to Insure Success in the Business of Turkey Growing* (New York: Orange Judd, 1897). John Gay, *Fables* (London: J. F. and C. Rivington, 1792) was useful for literary references to turkeys.

Best known for his novels, *Robinson Crusoe* (1719) and *Moll Flanders* (1722), Daniel Defoe also wrote accounts of the turkey industry in England that are as delightful to read as they are informative. Defoe's eye for detail is evident in his *Tour Thro' the Whole Island of Great Britain, Divided Into Circuits or Journies* (1724) and *A Tour Through England and Wales* (London: Everyman Edition, 1928). His vivid descriptions read like documentaries and readers can imagine themselves sitting beside him on the seat of his carriage and listening to his running commentaries or standing with him on the roadside, hailing the turkey drovers and counting the turkeys as they crossed the River Stour on the Stratford Bridge on their way to London markets.

A number of writers (including Daniel Defoe) described turkey drives, and Karen Davis in *More Than a Meal* compared them to cattle drives. For details on the range cattle industry and for descriptions of the cattle drives that Karen Davis mentioned but did not describe, I relied on the classics: Ernest Staples Osgood, *The Day of the Cattleman* (Minneapolis: The University of Minnesota Press, 1929) and Walter Prescott Webb, *The Great Plains* (New York: Grosset & Dunlap, 1931).

Three works on North Dakota history provided the context in which to place the All-American Turkey Shows held between 1924 and 1942. They were Elwyn B. Robinson, *History of North Dakota* (Lincoln: University of Nebraska Press, 1966); D. Jerome Tweton and Daniel F. Rylance, *The Years of Despair: North Dakota in the Depression* (Grand Forks, ND: The Oxcart Press, 1973); and Robert P. Wilkins and Wynona H. Wilkins, *North Dakota: A Bicentennial History* (New York: W. W. Norton & Company, Inc., 1977).

Two books, one based on a diary and one an autobiographical novel, helped to set the tone for the manuscript. Both books were set in Depression North Dakota, both dealt with the struggles of farm families during years of crop failure, and both described daily life under extremes of weather. The first words in Ann Marie Low's *Dust Bowl Diary* (Lincoln: University of Nebraska Press, 1984) are "Talk about wind! Most of the scenery is in the air." In her diary, Ann Marie Low confides her thoughts on the loss of her beloved Stony Brook country to the Arrowwood National Wildlife Refuge and the effects of drought, the Depression, and government programs on her family and neighbors.[1]

Lois Phillips Hudson's *The Bones of Plenty* (Boston: Little, Brown and Company, 1962), "a novel to touch the heart and sear the mind," according to the *New York Times Book Review*, was set on George Armstrong Custer's farm near Jamestown, North Dakota. Custer battled not only drought, grasshoppers, rust, and runs of bad luck, but also a flinty, penny-pinching landlord. In recounting Custer's struggles, according to a

---

[1] *The New York Times Book Review* described Low's book as a "moving and informative account of a decade (1927–37) of drought and depression in North Dakota." Other reviewers referred to Ann Marie Low's "irrepressible spirit," and the *Los Angeles Times Book Review* described Low's book as a "lovingly detailed, sometimes humorous and often painful account of a ravaged land."

reviewer, Hudson may have written "the farm novel of the Great Drought of the 1920s and 1930s and the Great Depression." Although set in North Dakota, *The Bones of Plenty*, according to the *New York Times Book Review*, presents "the frightful disaster that closed thousands of rural banks and drove farmers off their farms, the hopes and savings of a lifetime in ruins about them."

BIBLIOGRAPHY

Books

American Poultry Association. *Standard of Perfection.* Rochester, New York: American Poultry Association, 1905. https://archive.org/details/cu31924003039314/page/n5/mode/2up. First edition published 1874; current edition published 2015.

Davis, Karen. *More Than a Meal: The Turkey in History, Myth, Ritual, and Reality.* New York: Lantern Books, 2001.

Defoe, Daniel, *A Tour Through England and Wales.* London: Everyman Edition, 1928.

__________. *Tour Thro' the Whole Island of Great Britain, Divided into Circuits or Journies* (1724). London: Everyman Edition, 1928.

Federal Writers' Project. *North Dakota: A Guide to the Northern Prairie State.* Fargo: Knight Printing Co., 1938.

Feikema, Feike (Fredrick Manfred). *The Golden Bowl.* St. Paul, MN: Webb Publishing Co., 1944.

Gay, John, *Fables.* London: J. F. and C. Rivington, 1792.

Grand Forks City Directory, 1924–1942.

Handy-Marchello, Barbara. *Women of the Northern Plains.* St. Paul, MN: Minnesota Historical Society Press, 2005.

Hudson, Lois Phillips. *The Bones of Plenty.* Boston: Little, Brown and Company, 1962.

Low, Ann Marie. *Dust Bowl Diary.* Lincoln: University of Nebraska Press, 1984.

Myrick, Herbert. *Turkeys and How to Grow Them: A Treatise on the Natural History and Origen of the Name of Turkeys, the various Breeds, and the best Methods to ensure success in the business of Turkey Growing.* New York: Orange Judd, 1897.

Nelson, Paula. *The Prairie Winnows Out its Own: The West River Country of South Dakota in the Years of Depression and Dust.* Iowa City: University of Iowa press, 1996.

Osgood, Ernest Staples. *The Day of the Cattleman.* Minneapolis: University of Minnesota Press, 1929.

Riches, Naomi. *The Agricultural Revolution in Norfolk.* Chapel Hill: University of North Carolina Press, 1937.

Robinson, Elwyn B. *History of North Dakota.* Lincoln: University of Nebraska Press, 1966.

Smith, Andrew F. *The Turkey: An American Story.* Chicago and Urbana: University of Illinois Press, 2006.

Sponenberg, D. Phillip, Jeanette Berander, and Alison Martin. *An Introduction to Heritage Breeds: Saving and Raising Rare-Breed Livestock and Poultry.* North Adams, MA: Storey Publishing, 2014.

Tweton, D. Jerome and Daniel F. Rylance. *The Years of Despair: North Dakota in the Depression.* Grand Forks: The Oxcart Press, 1973.

Webb, Walter Prescott. *The Great Plains.* New York: Grosset & Dunlap, 1931.

Wilkins, Robert P. and Wynona H. Wilkins. *North Dakota: A Bicentennial History.* New York: W. W. Norton & Company, Inc., 1977.

Young, Carrie. *Nothing to Do but Stay.* New York: Bantam Doubleday Dell Publishing Group, Inc., 1993.

Journals and Magazines

American Turkey Journal
The Dakota Farmer
Farm Journal
The Farmer
Successful Farming
Turkey World

Manuscript Collections

Special Collections, Chester Fritz Library, University of North Dakota.

William R. Page Papers. 1937–1991, MSS 241, Institute for Regional Studies, North Dakota State University, Fargo, ND.

Yoder, Harry. Mary Ella Jerome Private Collection, Barron, WI.

Newspapers

Grand Forks (ND) *Herald*

Web

Carradice, Phil. "The drovers of Wales." BBC Blog, Wales History. https://www.bbc.co.uk/blogs/waleshistory/2012/03/drovers_of_wales.html

INDEX

Page numbers in **bold** type indicate photographs or illustrations.

B

C

H

K

L

M

S

T

U

V

W

Y

Z

APPENDIX A

Common Terms and Definitions Used in Judging Turkey Features

To inform the uninitiated and to remind turkey breeders of how judges assessed and awarded points, the *Standard of Perfection*, the *American Turkey Journal*, and other publications provided helpful descriptions of a turkey's features that figured in judging and definitions of the terms used in evaluations. The following is a sampling of those descriptions and definitions.

barring: bars or stripes extending across a feather

beard: a tuft of coarse, bristly hair-like feathers projecting from the upper portion of the breast

breed: a race of domestic fowls that maintains distinctive characteristics of shape, growth, and temperament. A broader term than "variety," the breed includes varieties.

caruncles: naked, fleshy protuberances across the head of a turkey

carunculated: having caruncles

conformation: the form, shape, or outline of a bird or its dressed carcass

fluff: the soft, downy plumage on the abdomen, between the thighs, and on the posterior parts of fowls; also the soft, downy part of a feather

foreign color: any color that differs from the color prescribed as a part of the plumage of a Standard bred fowl

penciling: the narrow transverse or concentric lines or stripes on a feather; may run straight across or may follow the outlines of the feather, taking a concentric form

primaries: the longest feathers of the wing, growing between the pinions and secondaries, hidden when the wing is folded; otherwise known as flight feathers

saddle: the rear part of the back of a male bird, extending to the tail and covered with the saddle feathers

secondaries: long, large quill feathers, between the first and second joints of the wing, nearest the body, visible when the wing is folded

shank: the lower part of the leg

sweepstake prize: the prize awarded to a specimen or group of specimens winning the highest honor in either a variety, breed, or class, or a combination of classes

symmetry: perfection of proportion, harmony of all parts or sections of a fowl, viewed as a whole with regard to the Standard type of the breed it represents

tail coverts: curved feathers in front of and at the sides of the tail

variety or strain: a subdivision of a breed that reproduces with marked regularity; a term used to distinguish fowls having the Standard shape and other characteristics of the breed to which they belong, but differing in color of plumage and other features from other groups of the same breed

wattle: fleshy, brightly colored growths at the sides and base of the beak

APPENDIX B

Male and Female Turkey Shape Specifications

Under the heading Shape of Male and Female, toms and hens had to conform to the specifications stated in the *Standard of Perfection*. The following are some examples of those specifications.

back: broad, sloping from neck in a convex curve to tail

beak: strong, curved, well set in head

beard: long, bristly, prominent in adult males; not required in females

body and fluff: long, deep through middle, broad its entire length, well-rounded; fluff, moderately short

breast: broad, deep, full, well-rounded, carried well forward; most prominent in adult males

eyes: full, oval, prominent

head: long, deep, broad, carunculated, with tubular leader at base of beak, its size subject to extension or contraction, according to mood of specimen.

legs and toes: lower thighs, long, stout; shanks large, moderately long, strong; toes, straight, strong, well-spread

neck: long, erect, gracefully curved, blending into back

tail: rather long, feathers broad, carried low in a continued graceful curve in line with back; coverts, broad and abundant, extending well onto tail

throat-wattle: heavily carunculated

wings: large, powerful, smoothly folded, carried well up on sides

Note: deformed wings disqualify a bird

APPENDIX C

Summary of Standards for Bronze Turkeys

To acquaint the uninitiated with the standards to which breeders were held and those for which judges awarded points, it would be helpful to see a summary of the standards as they were described in the *Standard of Perfection*. Presented here is a summary of weight standards for Bronze turkeys, which will serve as an example of the specificity with which each variety was described.

The Bronze, "the ideal turkey in all respects" and "the finest type turkey, barring none" was, according to Andrew F. Smith, "the most popular in the United States during the nineteenth and early twentieth centuries." The *Countryside Daily* noted that the Bronze turkey was what most people described when asked, "what does a turkey look like." In the first All-American Turkey Show held in Grand Forks, North Dakota, in 1924, more than twice as many Bronze turkeys competed for awards and prizes as all the other varieties combined. By the time of the All-American Turkey Show in 1936 and discriminating breeders had continued to breed for "perfection," forty-three of them entered a total of 196 turkeys for judging in the Bronze division, including 28 yearling toms, 67 young toms, and 58 young hens. The Bronze was the largest and heaviest of the varieties, and individual birds had to conform to the Standard Weights given in the *Standard of Perfection:*

Adult tom (more than two years old)	36 pounds
Yearling tom	33 pounds
Young tom (December 1 to January 1)	25 pounds
Adult hen (more than two years old)	20 pounds
Yearling hen	20 pounds
Young hen (December 1 to January 1)	16 pounds

Note: When two specimens were both over Standard weight and equal in all other points, the one nearest Standard weight won.

Color of Male

wings: Fronts and bows, rich, brilliant copperish bronze, ending in a narrow band of black; coverts, bright, rich copperish bronze, forming a beautiful broad, bronze band across wings when folded, feathers terminating in a distinct black band, forming a glossy, ribbon-like mark, which separates them from secondaries; primaries, each feather, throughout its entire length, alternately crossed with distinct, parallel black and white bars of equal width, running straight across the feathers; flight coverts, barred similar to primaries; secondaries, dull black, alternately crossed with distinct parallel black and white bars, the black bar taking on a rich bronze cast on the shorter top secondaries, the white bar becoming less distinct; an edging of brown in secondaries being very objectionable.

back: From neck to middle of back, a rich, brilliant copperish bronze, each feather terminating in a narrow black band, extending across end; from middle of back to tail-coverts, black, each feather having a broad, brilliant copperish bronze band extending across it near the end, the more bronze the better, the feathers ending in a distinct black band, gradually narrowing as the tail coverts are approached.

tail: Main-tail and greater coverts, dull black, each feather evenly and distinctly marked transversely with parallel lines of brown; each feather having a wide bronze band extending across it near the end (the more bronze on this band the better), a narrow but distinct band of black bordering the bronze, and terminating in a wide edging of pure white; coverts, dull black, each feather evenly and distinctly marked transversely with parallel lines of brown, each feather having a very wide bronze band extending across it near the end, a narrow but distinct band of black bordering the bronze, terminating in a wide edging of pure white. The more distinct the colors throughout the whole plumage, the better.

breast: Rich, brilliant, copperish bronze; feathers on lower part of breast, approaching the body, terminate in a black band extending across the end.

body and fluff: Body, black, each feather with a wide, brilliant copperish bronze band extending across it near the end, a narrow but distinct band of black bordering the bronze and terminating in a narrow edging of pure white; fluff, dull black.

legs and toes: Lower thighs, similar to breast, but less brilliant in shade; shanks and toes, in mature specimens, pinkish; in young specimens, dark, approaching black.

Color of Female

beak, eyes, throat-wattle, legs and toes: Same as male.

plumage: Similar to that of male, except an edging of white on feathers of back, wing-bows, wing-coverts, breast, and body, which edging should be narrow in front, gradually widening as it approaches the rear of the specimen.

Disqualifications

White feathers in any part of the plumage; wings showing one or more primary or secondary feathers clear black or brown, or absence of white or gray bars more than one-half of the length of primaries; color of back, clear black; feather or feathers in tail or main-tail coverts, clear black, brown or gray; white or gray bars showing on main-tail feathers beyond greater main-tail coverts, except the terminating wide edging of white; and complete absence of white edging on breast of female.

Note: The following defects should be cut severely: Absence of one or more primary or secondary wing feathers; absence of one or more center main-tail feathers; white or gray bars, other than the terminating wide edging of white, showing on base of main-tail feathers; absence of black bands on one or more of the large main-tail coverts; decidedly crooked breastbone.

On the male the surface should be of an iridescent bronze on neck, back, wing bows and on breast and edged with a narrow black band, except on back where the band should be prominent, approximately three sixteenths of an inch in width, high on the back, and gradually becoming narrower as the tail section is approached, and finishing with clear-cut white edging over lower saddle; this edging growing wider on lesser and greater coverts. (Tail coverts are the secondary feathers extending farthest up on main tail.) There should also be a wide edging of white at the end of the main tail effecting a circular trimming of white on the tail when spread.

Other important tail markings are: the transverse lines of rich, mahogany brown bars on both main tail and greater coverts. This is called penciling and should be as regular in form and extend as far toward the base of the tail as possible. The proper shade of mahogany, brown penciling harmonizes with the rich bronze and produces a beautiful effect. Black shafts and centers of main tail feathers which project out toward the end of the tail are very objectionable.

The "rainbow band," a wide band of bronze which has the appearance of being laid over a wider band of black, with a narrow edge of the black showing on either side of the bronze, should be of rich, copperish bronze and equal in width throughout the entire circle of the tail spread. The greater coverts should be crossed near the end with wide black bands, through which a band of bronze effects what is known as a "double rainbow." In many cases, and even on some of our very best specimens, the covert black bands are marked with large bronze "spots" instead of full bands. While bands of bronze are what the Standard calls for, good strong spots over good black bands are preferred over complete bands over indistinct black bands. The side fluff of the male should be edged with white.

The shape and color of the female is similar to that of the male, size considered, and of course, less masculine in general appearance. Her color markings are the same, except that she should have white edging in every section, but very narrow, high on the breast, upper back and wing bows, but growing wider as the tail and rear fluff is approached.

Defects in both sexes which are hard to overcome are: brown or brick color in end of tail, tan color in coverts, "overflow" on back and in coverts, and lack of black bands on females. But none of these are disqualifications and can be penalized as defects only. Bronze which is too dark to be brilliant, or having a green sheen, is undesirable and should be discounted according to degree of defect.

Color disqualifications are: lack of any bronze on back, either sex, lack of white edging on breast of female; gray barring at base of tail extending beyond greater coverts; primaries without barring for more than half the length of secondaries, wholly black or brown or white feather or feathers in any part of plumage.

Bronze breeders of long experience have found it difficult, if not impossible, to breed from a single mating, both male and female that come nearest to present Standard requirements. With the required white edging on the female comes the undesirable "fringe" on the back of males from the same matings, and with the males with the desired width of black band on back, comes the female with little or no black bands. This problem is one for each breeder to solve and furnishes a source for unlimited experimenting to obtain these desired color characteristics. But all this is sauce for the true fancier and spurs him on to greater achievements. That is how Bronze turkeys were made.

APPENDIX D

Special Prizes, 1924

Special Prizes included $5 for the largest turkey; $5 for the second-best display of turkeys from each of the four states of North Dakota, South Dakota, Montana, and Minnesota; $5 for the largest number of turkeys displayed by one exhibitor from Grand Forks County; and $5 for the largest number of Bronze and White Holland turkeys by one exhibitor from outside Grand Forks County.

Other cash Special Prizes in the Bronze category were $5 for the best display consisting of not fewer than four turkeys, $2 for the best yearling tom, and $2 for the best yearling hen. In the White Holland category, $5 for the best display consisting of not fewer than four turkeys, $2 for the best yearling tom, and $2 for the best yearling hen. For the Bourbon Reds, $5 for the best display consisting of not fewer than four turkeys, $2 for the best yearling tom, and $2 for the best yearling hen.

Special Sweepstakes prizes included ribbons for the best Bronze, White Holland, and Bourbon Red toms and hens, and the Grand Sweepstakes Ribbons and $2 went to the show's best turkey and the show's best pen. A ribbon and $5 were awarded to the exhibitor with the largest and best entry of turkeys.

Examples of Unique Prizes

The M. M. Johnson Incubator Company, maker of the Old Trusty Incubator, offered one of its units for the best four Bronze turkeys displayed by one exhibitor; and the Kiepper Kuper Company of Milwaukee, Wisconsin, offered one dozen shipping boxes (fifteen-egg size) for the best White Holland tom under a year old. The George W. Hackett and Company Poultry Supply Dealers of Minneapolis offered a four-gallon "drop bottom" water fountain for the best Bourbon Red younger than one year. The Northwest Turkey Breeders' Association offered prizes only to those who had paid membership dues of $1 per year: copies of the 1923 edition of the *Standard of Perfection*, $2.50 value, for the best Bronze, Bourbon Red, and White Holland hens over a year old. The Association also offered copies of the book *Turkey Raising*, valued at $1.85, for the best White Holland, Bronze, and Bourbon Red yearling toms.

The W. E. Stanfield Company of St. Paul, Minnesota, provided $1 tubes of Stanfield's Famous Lice Kill for the Bronze and White Holland young toms and young hens that had placed second in their classes. The *Northwestern Poultry Journal*, Minneapolis, Minnesota, Ed L. Hayes, editor, provided a three-year subscription to the journal, a $1 value, to every exhibitor at the All-American Turkey Show who had paid the Entry Fee. The journal also provided a prize of $1.50 for each pen of one tom and four hens of the same variety. The Northwest Turkey Breeders' Association offered a silver loving cup for the show's best turkey. The winner would be known as the Grand Champion of the All-American Turkey Show.

APPENDIX E

Brief Biographies of the Directors and Trustees Listed on the All-American Turkey Show Articles of Incorporation

Alfred Malmberg of Crookston, Minnesota, was among the premier Bronze turkey breeders in the Red River Valley. He exhibited as many as twenty-five Bronze at each of the All-American Turkey Shows, where his birds were consistent winners, and he held the record for the breeder exhibiting the largest number of turkeys at the show. Feed companies and suppliers vied for Malmberg's endorsement, and in 1928 he was elected president of the Northwest Turkey Breeders' Association.

Ray Andrews of Petersburg, North Dakota, another Bronze breeder, was often featured with his turkeys in the *Grand Forks Herald*. Andrews was in charge of the sales department of the All-American Turkey Show, and in 1928 he was elected second vice president of the Northwest Turkey Breeders' Association. Andrews's wife was president of the All-American Turkey Hen Club.

Victor Hartl, New Rockford, North Dakota, was elected secretary-treasurer of the Northwest Turkey Breeders' Association in 1928. His Bronze tom was the Grand Champion of the 1926 All-American Turkey Show, and the caption of the Grand Champion's photograph in the *Grand Forks Herald* noted that its coloring "was extremely beautiful, his bronze feathers gleaming with all the colors of the rainbow."

Manville A. Johnson, Michigan, North Dakota, is another deserving of comment. In 1928, he was elected first vice president of the Northwest Turkey Breeders' Association.

G. S. Mairs, of Lisbon, North Dakota, dubbed himself "Chief Turkey Wrangler" of the All-American Turkey Shows. In charge of tending the turkeys in their coops, he saw to their feed and water. Mairs kept a large flock of Bourbon Reds on his Lisbon farm.

John C. Sherlock was manager of Hamm Co., located in East Grand Forks, Minnesota. In the 1928 City Directory, Hamm Co. was listed as Distributors of Hamm's Products, Bowey's Inc. Crushed Fruits and Syrups, Wholesale Candies, Cigars and Smokers' Sundries. Sherlock was known as a deal maker and as "an esteemed and enterprising" individual whose "tireless promotional efforts' had helped to develop the city and the northern Red River Valley. He is credited with getting the American Beet Sugar Company to locate Minnesota's largest and most modern beet sugar factory in East Grand Forks. The plant, built in 1926, employed five hundred workers during harvest and processed beets from twenty thousand acres in 1928.

Roy F. Bridgeman, vice president of the North Dakota Dairymen's Association, was manager of Bridgeman–Russell Creamery, the city's largest dairy plant. The plant produced Primus Butter, Delicious Velvet Ice Cream, and Pasturized Purity Milk. Cream from the large trade area around Grand Forks arrived at the plant by rail, as did eggs. Butter and eggs were shipped from the plant by the carload. In 1928, Bridgeman was president of the Commercial Club of Grand Forks, the civic organization from which the All-American Turkey Show received much of its support, and William W. Blain was the secretary. The headquarters of the All-American Turkey Show were in the Commercial Club's office.

Isadore Papermaster in 1928 was manager of the Pantorium, listed in the City Directory as North Dakota's Leading Cleaners. The plant, "Master Cleaners since 1900," specialized in "Cleaning, Dyeing, Repairing of Men's, Women's and Children's Garments, Household Goods, and Pleating." Isadore Papermaster was the son of Rabbi Benjamin Papermaster, among the city's most public-spirited and civic-minded citizens. Rabbi Papermaster arrived in Grand Forks in 1890 from his native Lithuania with only a Torah scroll and a few belongings. Resisting entreaties to serve elsewhere, Rabbi Papermaster remained in Grand Forks and served the Congregation of the Children of Israel until his death in 1934. Isadore Papermaster did not become a rabbi, but his civic mindedness and public service emulated those of his father.

Daniel F. McGowan was listed under Mortgage Loans and Insurance in the 1928 City Directory, with an office in the First National Bank building. McGowan was manager of the North Dakota state fair in Grand Forks.

APPENDIX F

The *American Turkey Journal* in Brief

The February 1934 issue, Vol. II, No. 12, of the *American Turkey Journal*, establishes the tone and form for the more than one hundred issues that would follow. Therefore, it stands here as representative of features and content for the series.

The *American Turkey Journal* averaged thirty-four to forty pages in length. Color was reserved for the cover, which always featured something associated with the All-American Turkey Show.

Several Bronze breeders had full-page ads in the issue, including Reiman Turkey Farms, Inc., Planada, California ("You Can Always Tell a Reiman Bird"); Mr. And Mrs. O. J. Shelton, Pomona, California (Selling high quality Toms weighing 27–34 lbs. at 8 months as low as $8); and George E. Lamm, Philip, South Dakota (Write for Free Mating and Our Low 1934 Prices!).

Some breeders preferred smaller ads. Mr. And Mrs. Martin Ellingson, Evansville, Minnesota, raised Ellingson's Narragansetts (BREEDING STOCK FOR SALE AT 1934 PRICES); Sadie B. Caldwell, Broughton, Kansas, bred Sadie's Bourbons (POULTS AND EGGS FROM PEDIGREED BREEDING STOCK); and Mrs. Homer Price, Newark, Ohio, prided herself on her White Hollands (SPECIAL MATINGS OF FIRST PRIZE WINNERS furnish highest quality stock for Exhibitors and Breeders).

Mrs. Homer Price, Newark, Ohio, raised White Holland turkeys and offered EGG AND DAY OLD POULTS (Specialty Breeder, Snow White Strain). J. J. Quam, Beltrami, Minnesota, bred Narragansetts (WE RAISE OUR OWN WINNERS). The Quam ad included the notation, "Our breeders are A. P. A. Certified and mated for best results." The notation "A. P. A. Certified" carried weight, and fortunate was the breeder who could add "Inspected by" or "Certified by" George W. Hackett to the notation. Hackett's name and reputation added even more weight.

A. P. A. stood for the American Poultry Association, the oldest poultry organization in the United States, having been founded in 1873. Shortly after its founding, the Association published the *American Standard of Perfection*, the most widely used and respected handbook on poultry breed and variety standards.

To earn A. P. A. Certification, a turkey flock had to be inspected, on the premises, by a licensed flock inspector. The inspector had to inspect all turkeys in the flock visually, and he had to inspect a minimum percentage physically. Every turkey in the flock had to appear healthy, with no visibly ill specimens. A. P. A. Certified breeding stock commanded premium prices, because the purchaser was assured that the turkeys met the specifications for the variety in the *Standard of Perfection.*

APPENDIX G
"Turkeys . . . Show Many Characteristics"

In the February 2, 1927, edition of the *Grand Forks Herald*, a perceptive individual who had attended the All-American Turkey Show shared his waggishly worded observations with readers in a column titled, "Turkeys at All-American Show Many Characteristics." The individual observed that "[p]robably every characteristic of the human race is exemplified by the birds at the All-American Turkey Show now being held in Grand Forks." The following characteristics are excerpts from the column.

No Sex Appeal

Three flighty young hens in the next booth to him haven't stirred his interest one bit, though they have all the earmarks of the most thoroughly unscrupulous vamps that the turkey world could offer. They are blond, a very delicately auburn red, very young, very sentimental and talkative and most certainly beautiful. Voluble conversation on their part failed to arouse the interest of the gentleman next door and they put him down on their black lists as most certainly not a gentleman.

Blabber Mouth

The loudest mouthed turkey in the world got himself entangled on a judge's table Tuesday afternoon and raised the very devil because he was subjected to such an ordeal. "He's the dumbest bird in the show," the judge said, "and he doesn't understand the honor of being here at all."

Because he's dumb hasn't stopped his linguistic qualities at all nor hampered them in the way of vocabulary. He is fluent, perverse, and all but sacrilegious.

The Young Sheiks

The feathered protypes of Beau Brummell are present in force at the All-American Show. They are the most noticeable, the most beautiful young toms that the world of gobblers has to offer, and the trouble with them all is that they realize it all too acutely and allow no one to forget it for a moment. All the way down the line are the finest examples of sullenness, cheerfulness, laziness, amiability, stupidity, asininity that even a college classroom would have difficulty in matching.

Turkeys are a lot better in exhibiting human nature than chickens. They are more closely related, one breed to the next, and they don't exhibit the traits of Latin, Scotch, or British ancestry along with their normal disposition. Only one thing can be said of them all. They're all good Americans.

ABOUT THE AUTHOR

Photo by Jacque Lawrence

Until his recent retirement, Dr. Gordon L. Iseminger was the longest-serving—fifty-seven years—faculty member at the University of North Dakota. Iseminger is a graduate of Augustana College, the University of South Dakota, and the University of Oklahoma. His work has appeared in *North Dakota Quarterly*, *Minnesota History*, *Agricultural History*, *Pennsylvania History*, *The Journal of American History*, the *Middle East Journal*, and the *Encyclopedia of the Great Plains*. Iseminger has served on the Grand Forks Historic Preservation Commission, and he was named Chester Fritz Distinguished Professor in 2003. Iseminger is the 2015 recipient of the Distinguished Contribution to Preservation of the Cultural Heritage of the Northern Plains Award, presented by the Center for Western Studies. In 2020, his years-long contributions were recognized by the State Historical Society of North Dakota with the Heritage Profile Honor Award.

ABOUT THE PRESS

North Dakota State University Press (NDSU Press) exists to stimulate and coordinate interdisciplinary regional scholarship. These regions include the Red River Valley, the state of North Dakota, the plains of North America (comprising both the Great Plains of the United States and the prairies of Canada), and comparable regions of other continents. We publish peer reviewed regional scholarship shaped by national and international events and comparative studies.

Neither topic nor discipline limits the scope of NDSU Press publications. We consider manuscripts in any field of learning. We define our scope, however, by a regional focus in accord with the press's mission. Generally, works published by NDSU Press address regional life directly, as the subject of study. Such works contribute to scholarly knowledge of region (that is, discovery of new knowledge) or to public consciousness of region (that is, dissemination of information or interpretation of regional experience). Where regions abroad are treated, either for comparison or because of ties to those North American regions of primary concern to the press, the linkages are made plain. For nearly three-quarters of a century, NDSU Press has published substantial trade books, but the line of publications is not limited to that genre. We also publish textbooks (at any level), reference books, anthologies, reprints, papers, proceedings, and monographs. The press also considers works of poetry or fiction, provided they are established regional classics or they promise to assume landmark or reference status for the region. We select biographical or autobiographical works carefully for their prospective contribution to regional knowledge and culture. All publications, in whatever genre, are of such quality and substance as to embellish the imprint of NDSU Press.

Our name changed to North Dakota State University Press in January 2016. Prior to that, and since 1950, we published as the North Dakota Institute for Regional Studies Press. We continue to operate under the umbrella of the North Dakota Institute for Regional Studies, located at North Dakota State University.